AF580879

ADVANCES IN THE HOMOTOPY ANALYSIS METHOD

ADVANCES IN THE HOMOTOPY ANALYSIS METHOD

Editor

Shijun Liao

Shanghai Jiao Tong University, China

NEW JERSEY • LONDON • SINGAPORE • BEIJING • SHANGHAI • HONG KONG • TAIPEI • CHENNAI

Published by

World Scientific Publishing Co. Pte. Ltd.
5 Toh Tuck Link, Singapore 596224
USA office: 27 Warren Street, Suite 401-402, Hackensack, NJ 07601
UK office: 57 Shelton Street, Covent Garden, London WC2H 9HE

Library of Congress Cataloging-in-Publication Data
Advances in the homotopy analysis method / edited by Shijun Liao, professor, deputy director of the State Key Lab of Ocean Engineering, Shanghai Jiao Tong University, China.
pages cm
Includes bibliographical references.
ISBN 978-9814551243 (hard cover : alk. paper)
1. Homotopy theory. I. Liao, Shijun, 1963– editor of compilation.
QA612.7.A375 2014
514'.24--dc23
2013028624

British Library Cataloguing-in-Publication Data
A catalogue record for this book is available from the British Library.

In-house Editor: Angeline Fong

Printed in Singapore by World Scientific Printers.

Preface

The homotopy is a fundamental concept in topology, which can be traced back to Jules Henri Poincaré (1854–1912), a French mathematician. Based on the homotopy, two methods have been developed. One is the homotopy continuation method dating back to 1930s, which is a global convergent numerical method mainly for nonlinear algebraic equations. The other is the homotopy analysis method (HAM) proposed in 1990s by Shijun Liao, the editor of this book, which is an analytic approximation method with guarantee of convergence, mainly for nonlinear differential equations.

Different from perturbation techniques which are strongly dependent upon small/large physical parameters (i.e. perturbation quantities), the HAM has nothing to do with any small/large physical parameters at all. Besides, many analytic approximation methods, such as "Lyapunov artificial small parameter method", "Adomian decomposition method" and so on, are only special cases of the HAM. Unlike other analytic approximation techniques, the HAM provides us great freedom and flexibility to choose equation-type and solution expression of high-order approximation equations. Notice that "the essence of mathematics lies entirely in its freedom", as pointed out by Georg Cantor (1845–1918). Most importantly, different from all of other analytic approximation methods, the HAM provides us a convenient way to guarantee the convergence of approximation series by means of introducing the so-called "convergence-control parameter". In fact, it is the convergence-control parameter that differs the HAM from all other analytic approximation methods. As a result, the HAM is generally valid for various types of equations with high nonlinearity, especially for those without small/large physical parameters.

Since 1992 when the early HAM was first proposed by Liao, the HAM has been developing greatly in theory and applied successfully to numerous types of nonlinear equations in lots of different fields by scientists, researchers, engineers and graduated students in dozens of countries. All of these verify the originality, novelty, validity and generality of the HAM.

So, it is necessary to describe, although briefly, the current advances of the HAM in both theory and applications. This is the first motivation of the book, whose chapters are contributed by the leading researchers in the HAM coming from seven countries.

Any truly new method should give something novel and/or better. In the past 20 years, hundreds of articles related to the HAM were published in various fields, and some new solutions were indeed found by means of the HAM. Thus, it is now the time to suggest some valuable but challenging nonlinear problems to the HAM community. This is the second motivation of the book. Some of these problems are very famous, with a long history. Hopefully, the above-mentioned freedom and flexibility of the HAM might create some novel ideas and inspire brave, enterprising, young researchers with stimulated imagination to attack them with satisfactory results. I personally believe that the applications of the HAM on these famous, challenging problems might not only indicate the great potential of the HAM, but also lead to great modifications of the HAM in theory.

A brief review of the HAM is given in Chapter 1, with some suggested challenging problems. The fascinating "Predictor HAM" and "Spectral HAM" are described in Chapters 2 and 3, respectively. Some interesting theoretical works on the auxiliary linear operator, convergence-control parameter and convergence of approximation series are described in Chapters 4 and 5. An attractive application of the HAM about flows of nanofluid is given in Chapter 6. A charming application of the HAM for time-fractional boundary-value problem is illustrated in Chapter 7. The HAM-based Maple package `NOPH` 1.0.2 (`http://numericaltank.sjtu.edu.cn/NOPH.htm`) for periodic oscillations and limit cycles of nonlinear dynamic systems with various applications is described in Chapter 8. The HAM-based Mathematica package `BVPh` 2.0 (`http://numericaltank.sjtu.edu.cn/BVPh.htm`) for coupled nonlinear ordinary differential equations and its applications are given in Chapter 9. Both of them are easy-to-use, user-friendly, and free available online with user's guide. They can greatly simplify some applications of the HAM.

It is a great pity that it is impossible to describe, even briefly, the whole advances of the HAM in theory and applications in such a book. Here, I would like to express my sincere and truthful acknowledgements to all of the HAM community for their great contributions to the HAM.

Shijun Liao
June 2013, Shanghai

Contents

Preface v

1. Chance and Challenge: A Brief Review of Homotopy Analysis Method 1
S.-J. Liao

2. Predictor Homotopy Analysis Method (PHAM) 35
S. Abbasbandy and E. Shivanian

3. Spectral Homotopy Analysis Method for Nonlinear Boundary Value Problems 85
S. Motsa and P. Sibanda

4. Stability of Auxiliary Linear Operator and Convergence-Control Parameter 123
R. A. Van Gorder

5. A Convergence Condition of the Homotopy Analysis Method 181
M. Turkyilmazoglu

6. Homotopy Analysis Method for Some Boundary Layer Flows of Nanofluids 259
T. Hayat and M. Mustafa

7. Homotopy Analysis Method for Fractional Swift–Hohenberg Equation 291

S. Das and K. Vishal

8. HAM-Based Package NOPH for Periodic Oscillations of Nonlinear Dynamic Systems 309

Y.-P. Liu

9. HAM-Based Mathematica Package BVPh 2.0 for Nonlinear Boundary Value Problems 361

Y.-L. Zhao and S.-J. Liao

Chapter 1

Chance and Challenge: A Brief Review of Homotopy Analysis Method

Shijun Liao

Shanghai Jiao Tong University, Shanghai 200240, China

sjliao@sjtu.edu.cn

A brief review of the homotopy analysis method (HAM) and some of its current advances are described. We emphasize that the introduction of the homotopy, a basic concept in topology, is a milestone of the analytic approximation methods, since it is the homotopy which provides us great freedom and flexibility to choose equation type and solution expression of high-order approximation equations. Besides, the so-called "convergence-control parameter" is a milestone of the HAM, too, since it is the convergence-control parameter that provides us a convenient way to guarantee the convergence of solution series and that differs the HAM from all other analytic approximation methods. Relations of the HAM to the homotopy continuation method and other analytic approximation techniques are briefly described. Some interesting but challenging nonlinear problems are suggested to the HAM community. As pointed out by Georg Cantor (1845–1918), "the essence of mathematics lies entirely in its freedom". Hopefully, the above-mentioned freedom and great flexibility of the HAM might create some novel ideas and inspire brave, enterprising, young researchers with stimulated imagination to attack them with satisfactory, better results.

Contents

1.1. Background . . . 2
1.2. A brief history of the HAM . . . 5
1.3. Some advances of the HAM . . . 10
1.3.1. Generalized zeroth-order deformation equation . . . 10
1.3.2. Spectral HAM and complicated auxiliary operator . . . 13
1.3.3. Predictor HAM and multiple solutions . . . 15
1.3.4. Convergence condition and HAM-based software . . . 16
1.4. Relationships to other methods . . . 18
1.5. Chance and challenge: some suggested problems . . . 20
1.5.1. Periodic solutions of chaotic dynamic systems . . . 21
1.5.2. Periodic orbits of Newtonian three-body problem . . . 22
1.5.3. Viscous flow past a sphere . . . 24
1.5.4. Viscous flow past a cylinder . . . 25
1.5.5. Nonlinear water waves . . . 26
References . . . 29

1.1. Background

Physical experiment, numerical simulation and analytic (approximation) method are three mainstream tools to investigate nonlinear problems. Without doubt, physical experiment is always the basic approach. However, physical experiments are often expensive and time-consuming. Besides, models for physical experiments are often much smaller than the original ones, but mostly it is very hard to satisfy all similarity criterions. By means of numerical methods, nonlinear equations defined in rather complicated domain can be solved. However, it is difficult to gain numerical solutions of nonlinear problems with singularity and multiple solutions or defined in an infinity domain. By means of analytic (approximation) methods, one can investigate nonlinear problems with singularity and multiple solutions in an infinity interval, but equations should be defined in a simple enough domain. So, physical experiments, numerical simulations and analytic (approximation) methods have their inherent advantages and disadvantages. Therefore, each of them is important and useful for us to better understand nonlinear problems in science and engineering.

In general, exact, closed-form solutions of nonlinear equations are hardly obtained. Perturbation techniques [1–4] are widely used to gain analytic approximations of nonlinear equations. Using perturbation methods, many nonlinear equations are successfully solved, and lots of nonlinear phenomena are understood better. Without doubt, perturbation methods make great contribution to the development of nonlinear science. Perturbation methods are mostly based on small (or large) physical parameters, called perturbation quantity. Using small/large physical parameters, perturbation methods transfer a nonlinear equation into an infinite number of sub-problems that are mostly linear. Unfortunately, many nonlinear equations do not contain such kind of perturbation quantities at all. More importantly, perturbation approximations often quickly become invalid when the so-called perturbation quantities enlarge. In addition, perturbation techniques are so strongly dependent upon physical small parameters that we have nearly no freedom to choose equation type and solution expression of high-order approximation equations, which are often complicated and thus difficult to solve. Due to these restrictions, perturbation methods are valid mostly for weakly nonlinear problems in general.

On the other side, some non-perturbation methods were proposed long ago. The so-called "Lyapunov's artificial small-parameter method" [5] can trace back to the famous Russian mathematician Lyapunov (1857–1918), who first rewrote a nonlinear equation

$$\mathcal{N}[u(\mathbf{r},t)] = \mathcal{L}_0[u(\mathbf{r},t)] + \mathcal{N}_0[u(\mathbf{r},t)] = f(\mathbf{r},t), \tag{1.1}$$

where $\mathbf{r}$ and t denote the spatial and temporal variables, $u(\mathbf{r},t)$ a unknown function, $f(\mathbf{r},t)$ a known function, $\mathcal{L}_0$ and $\mathcal{N}_0$ are linear and nonlinear operator, respectively, to such a new equation

$$\mathcal{L}_0[u(\mathbf{r},t)] + q\,\mathcal{N}_0[u(\mathbf{r},t)] = f(\mathbf{r},t), \tag{1.2}$$

where q has *no* physical meaning. Then, Lyapunov regarded q as a small parameter to gain perturbation approximations

$$u \approx u_0 + u_1\,q + u_2\,q^2 + u_3\,q^3 + \cdots = u_0 + \sum_{m=1}^{+\infty} u_m\,q^m, \tag{1.3}$$

and finally gained approximation

$$u \approx u_0 + \sum_{m=1}^{+\infty} u_m, \tag{1.4}$$

by setting $q = 1$, where

$$\mathcal{L}_0[u_0(\mathbf{r},t)] = f(\mathbf{r},t), \quad \mathcal{L}_0[u_1(\mathbf{r},t)] = -\mathcal{N}_0[u_0(\mathbf{r},t)], \cdots \tag{1.5}$$

and so on. It should be emphasized that one has *no* freedom to choose the linear operator $\mathcal{L}_0$ in Lyapunov's artificial small-parameter method: it is exactly the linear part of the whole left-hand side of the original equation $\mathcal{N}[u] = f$, where $\mathcal{N} = \mathcal{L}_0 + \mathcal{N}_0$. Thus, when $\mathcal{L}_0$ is complicated or "singular" (for example, it does not contain the highest derivative), it is difficult (or even impossible) to solve the high-order approximation equation (1.5). Besides, the convergence of the approximation series (1.4) is *not* guaranteed in general. Even so, Lyapunov's excellent work is a milestone of analytic approximation methods, because it is independent of the existence of physical small parameter, even though it first regards q as a "small parameter" but finally enforces it to be 1 that is however *not* "small" strictly from mathematical viewpoints.

The so-called "Adomian decomposition method" (ADM) [6–8] was developed from the 1970s to the 1990s by George Adomian, the chair of the Center for Applied Mathematics at the University of Georgia, USA. Adomian rewrote (1.1) in the form

$$\mathcal{N}[u(\mathbf{r},t)] = \mathcal{L}_A[u(\mathbf{r},t)] + \mathcal{N}_A[u(\mathbf{r},t)] = f(\mathbf{r},t), \tag{1.6}$$

where $\mathcal{L}_A$ often corresponds to the highest derivative of the equation under consideration, $\mathcal{N}_A[u(\mathbf{r},t)]$ gives the left part, respectively. Approximations of the ADM are also given by (1.4), too, where

$$\mathcal{L}_A[u_0(\mathbf{r},t)] = f(\mathbf{r},t), \;\; \mathcal{L}_A[u_m(\mathbf{r},t)] = -A_{m-1}(\mathbf{r},t), \;\; m \geq 1, \tag{1.7}$$

with the so-called Adomial polynomial

$$A_k(\mathbf{r},t) = \frac{1}{k!}\left\{\frac{\partial^k}{\partial q^k}\mathcal{N}_A\left[\sum_{n=0}^{+\infty} u_n(\mathbf{r},t)\; q^n\right]\right\}\Bigg|_{q=0}. \tag{1.8}$$

Since the linear operator $\mathcal{L}_A$ is simply the highest derivative of the considered equation, it is convenient to solve the high-order approximation equations (1.7). This is an advantage of the ADM, compared to "Lyapunov's artificial small-parameter method" [5]. However, the ADM does *not* provides us freedom to choose the linear operator $\mathcal{L}_A$, which is restricted to be related only to the highest derivative. Besides, like "Lyapunov's artificial small-parameter method" [5], the convergence of the approximation series (1.4) given by the ADM is still *not* guaranteed.

Essentially, both of the "Lyapunov's artificial small parameter method" and the "Adomian decomposition method" transfer a nonlinear problem into an infinite number of linear sub-problems, *without* small *physical* parameter. However, they have two fundamental restrictions. First, one has *no* freedom and flexibility to choose the linear operators $\mathcal{L}_0$ or $\mathcal{L}_A$, since $\mathcal{L}_0$ is exactly the linear part of $\mathcal{N}$ and $\mathcal{L}_A$ corresponds to the highest derivative, respectively. Second, there is *no* way to guarantee the convergence of the approximation series (1.4). The second ones is more serious, since divergent approximations are mostly useless. Thus, like perturbation methods, the traditional non-perturbation methods (such as Lyapunov's artificial small parameter method and the ADM) are often valid for weakly nonlinear problems in most cases.

In theory, it is very valuable to develop a new kind of analytic approximation method which should have the following characteristics:

(1) it is *independent* of small physical parameter;
(2) it provides us great *freedom* and *flexibility* to choose the equation-type and solution expression of high-order approximation equations;
(3) it provides us a convenient way to *guarantee* the convergence of approximation series.

One of such kind of analytic approximation methods, namely the "homotopy analysis method" (HAM) [9–17], was developed by Shijun Liao from 1990s to 2010s, together with contributions of many other researchers in theory and applications. The basic ideas of the HAM with its brief history are described below.

1.2. A brief history of the HAM

The basic ideas of "Lyapunov's artificial small-parameter method" can be generalized in the frame of the homotopy, a fundamental concept of topology. For a nonlinear equation

$$\mathcal{N}[u(\mathbf{r},t)] = f(\mathbf{r},t), \tag{1.9}$$

Liao [9] propose the so-called "homotopy analysis method" (HAM) by using the homotopy, a basic concept in topology:

$$(1-q)\mathcal{L}[\varphi(\mathbf{r},t;q)-u_0(\mathbf{r},t)] = c_0\, q\, H(\mathbf{r},t)\ \{\mathcal{N}[\varphi(\mathbf{r},t;q)] - f(\mathbf{r},t)\}\,, \tag{1.10}$$

where $\mathcal{L}$ is an auxiliary linear operator with the property $\mathcal{L}[0] = 0$, $\mathcal{N}$ is the nonlinear operator related to the original equation (1.9), $q \in [0,1]$ is the embedding parameter in topology (called the homotopy parameter), $\varphi(\mathbf{r},t;q)$ is the solution of (1.10) for $q \in [0,1]$, $u_0(\mathbf{r},t)$ is an initial guess, $c_0 \neq 0$ is the so-called "convergence-control parameter", and $H(\mathbf{r},t)$ is an auxiliary function that is non-zero almost everywhere, respectively. Note that, in the frame of the homotopy, we have great *freedom* to choose the auxiliary linear operator $\mathcal{L}$, the initial guess $u_0(\mathbf{r},t)$, the auxiliary function $H(\mathbf{r},t)$, and the value of the convergence-control parameter c_0.

When $q = 0$, due to the property $\mathcal{L}[0] = 0$, we have from (1.10) the solution

$$\varphi(\mathbf{r}, t; 0) = u_0(\mathbf{r}, t). \tag{1.11}$$

When $q = 1$, since $c_0 \neq 0$ and $H(\mathbf{r}, t) \neq 0$ almost everywhere, Eq. (1.10) is equivalent to the original nonlinear equation (1.9) so that we have

$$\varphi(\mathbf{r}, t; 1) = u(\mathbf{r}, t), \tag{1.12}$$

where $u(\mathbf{r}, t)$ is the solution of the original equation (1.9). Thus, as the homotopy parameter q increases from 0 to 1, the solution $\varphi(\mathbf{r}, t; q)$ of Eq. (1.10) varies (or deforms) *continuously* from the initial guess $u_0(\mathbf{r}, t)$ to the solution $u(\mathbf{r}, t)$ of the original equation (1.9). For this sake, Eq. (1.10) is called *the zeroth-order deformation equation.*

Here, it must be emphasized once again that we have great freedom and flexibility to choose the auxiliary linear operator $\mathcal{L}$, the auxiliary function $H(\mathbf{r}, t)$, and especially the value of the convergence control parameter c_0 in the zeroth-order deformation equation (1.10). In other words, the solution $\varphi(\mathbf{r}, t; q)$ of the zeroth-order deformation equation (1.10) is also dependent upon all[a] of the auxiliary linear operator $\mathcal{L}$, the auxiliary function $H(\mathbf{r}, t)$ and the convergence-control parameter c_0 as a whole, even though they have *no* physical meanings. This is a key point of the HAM, which we will discuss in details later. Assume that $\mathcal{L}$, $H(\mathbf{r}, t)$ and c_0 are properly chosen so that the solution $\varphi(\mathbf{r}, t; q)$ of the zeroth-order deformation equation (1.10) always exists for $q \in (0, 1)$ and besides it is analytic at $q = 0$, and that the Maclaurin series of $\varphi(\mathbf{r}, t; q)$ with respect to q, i.e.

$$\varphi(\mathbf{r}, t; q) = u_0(\mathbf{r}, t) + \sum_{m=1}^{+\infty} u_m(\mathbf{r}, t)\; q^m \tag{1.13}$$

converges at $q = 1$. Then, due to (1.12), we have the approximation series

$$u(\mathbf{r}, t) = u_0(\mathbf{r}, t) + \sum_{m=1}^{+\infty} u_m(\mathbf{r}, t). \tag{1.14}$$

Substituting the series (1.13) into the zeroth-order deformation equation (1.10) and equating the like-power of q, we have the high-order approximation equations for $u_m(\mathbf{r}, t)$, called the mth-order deformation equation

$$\mathcal{L}[u_m(\mathbf{r}, t) - \chi_m u_{m-1}(\mathbf{r}, t)] = c_0\; H(\mathbf{r}, t)\; R_{m-1}(\mathbf{r}, t), \tag{1.15}$$

[a]More strictly, $\varphi(\mathbf{r}, t; q)$ should be replaced by $\varphi(\mathbf{r}, t; q, \mathcal{L}, \mathrm{H}(\mathbf{r},\mathrm{t}), c_0)$. Only for the sake of simplicity, we use here $\varphi(\mathbf{r}, t; q)$, but should always keep this point in mind.

where

$$R_k(\mathbf{r},t) = \frac{1}{k!}\left\{\frac{\partial^k}{\partial q^k}\left(\mathcal{N}\left[\sum_{n=0}^{+\infty}(\mathbf{r},t)\,q^n\right] - f(\mathbf{r},t)\right)\right\}\Bigg|_{q=0}, \tag{1.16}$$

with the definition

$$\chi_k = \begin{cases} 0, \text{ when } k \leq 1, \\ 1, \text{ when } k \geq 2. \end{cases} \tag{1.17}$$

For various types of nonlinear equations, it is easy and straightforward to use the theorems proved in Chapter 4 of Liao's book [11] to calculate the term $R_k(\mathbf{r},t)$ of the high-order deformation equation (1.15).

It should be emphasized that the HAM provides us great freedom and flexibility to choose the auxiliary linear operator $\mathcal{L}$ and the initial guess u_0. Thus, different from all other analytic methods, the HAM provides us great freedom and flexibility to choose the equation type and solution expression of the high-order deformation equation (1.15) so that its solution can be often gained without great difficulty. Notice that "the essence of mathematics lies entirely in its freedom", as pointed out by Georg Cantor (1845–1918). More importantly, the high-order deformation equation (1.15) contains the convergence-control parameter c_0, and the HAM provides great freedom to choose the value of c_0. Mathematically, it has been proved that the convergence-control parameter c_0 can adjust and control the convergence region and ratio of the approximation series (1.14). For details, please refer to Liao [10, 12, 13] and especially § 5.2 to § 5.4 of his book [11]. So, unlike all other analytic approximation methods, the convergence-control parameter c_0 of the HAM provides us a convenient way to guarantee the convergence of the approximation series (1.14). In fact, it is the convergence-control parameter c_0 that differs the HAM from all other analytic methods.

At the mth-order of approximation, the optimal value of the convergence-control parameter c_0 can be determined by the minimum of residual square of the original governing equation, i.e.

$$\frac{d\mathcal{E}_m}{dc_0} = 0, \tag{1.18}$$

where

$$\mathcal{E}_m = \int_\Omega \left\{\mathcal{N}\left[\sum_{n=0}^{m} u_n(\mathbf{r},t)\right] - f(\mathbf{r},t)\right\}^2 d\Omega. \tag{1.19}$$

Besides, it has been proved by Liao [16] that a homotopy series solution (1.14) must be one of solutions of considered equation, as long as it is convergent. In other words, for an arbitrary convergence-control parameter $c_0 \in \mathbf{R}_c$, where

$$\mathbf{R}_c = \left\{ c_0 : \lim_{m \to +\infty} \mathcal{E}_m(c_0) \to 0 \right\} \tag{1.20}$$

is an interval, the solution series (1.14) is convergent to the true solution of the original equation (1.9). For details, please refer to Liao [16] and Chapter 3 of his book [11].

In summary, the HAM has the following advantages:

(a) it is *independent* of any small/large physical parameters;
(b) it provides us great *freedom* and large *flexibility* to choose equation type and solution expression of linear high-order approximation equations;
(c) it provides us a convenient way to *guarantee* the convergence of approximation series.

In this way, nearly *all* restrictions and limitations of the traditional non-perturbation methods (such as Lyapunov's artificial small parameter method [5], the Adomian decomposition method [6–8], the δ-expansion method [18] and so on) can be overcome by means of the HAM.

Besides, it has been generally proved [10, 12, 13] that the Lyapunov's artificial small parameter method [5], the Adomian decomposition method [6–8] and the δ-expansion method [18] are only special cases of the HAM for some specially chosen auxiliary linear operator $\mathcal{L}$ and convergence-control parameter c_0. Especially, the so-called "homotopy perturbation method" (HPM) [19] proposed by Jihuan He in 1998 (six years later after Liao [9] proposed the early HAM in 1992) was only a special case of the HAM when $c_0 = -1$, and thus has "nothing new except its name" [20]. Some results given by the HPM are divergent even in the whole interval except the given initial/boundary conditions, and thus "it is very important to investigate the convergence of approximation series, otherwise one might get useless results", as pointed out by Liang and Jeffrey [21]. For details, see § 6.2 of Liao's book [11]. Thus, the HAM is more general in theory and widely valid in practice for more of nonlinear problems than other analytic approximation techniques.

In calculus, the famous Euler transform is often used to accelerate convergence of a series or to make a divergent series convergent. It is interesting that one can derive the Euler transform in the frame of the HAM, and give a similar but more general transform (called the generalized Euler transform), as shown in Chapter 5 of Liao's book [11]. This provides us a theoretical cornerstone for the validity and generality of the HAM.

The introduction of the so-called "convergence-control parameter" c_0 in the zeroth-order deformation equation (1.10) is a *milestone* for the HAM. From physical viewpoint, the "convergence-control parameter" c_0 has no physical meanings so that convergent series of solution given by the HAM must be *independent* of c_0. This is indeed true: there exists such a region $\mathbf{R}_c$ that, for arbitrary $c_0 \in \mathbf{R}_c$, the HAM series converges to the true solution of the original equation (1.9), as illustrated by Liao [10, 11]. However, if $c_0 \notin \mathbf{R}_c$, the solution series diverges! So, from a mathematical viewpoint, the "convergence-control parameter" is a key point of the HAM, which provides us a convenient way to guarantee the convergence of the solution series. In fact, it is the so-called "convergence-control parameter" that differs the HAM from all other analytic approximation methods.

The introduction of the basic concept homotopy in topology is also a *milestone* of the analytic approximation methods for nonlinear problems. It is the homotopy that provides us great freedom and large flexibility to choose the auxiliary linear operator $\mathcal{L}$ and initial guess u_0 in the zeroth-order deformation equation (1.10), which determine the equation type and solution expression of the high-order deformation equations (1.15). Besides, it is the homotopy that provides us the freedom to introduce the so-called "convergence-control parameter" c_0 in (1.10), which becomes now a cornerstone of the HAM. Note that it is impossible to introduce such kind of "convergence-control parameter" in the frame of perturbation techniques and the traditional non-perturbation methods (such as Lyapunov's artificial small parameter, Adomian decomposition method and so on).

The freedom on the choice of the auxiliary linear operator $\mathcal{L}$ is so large that the *second-order* nonlinear Gelfand equation can be solved conveniently (with good agreement with numerical results) in the frame of the HAM even by means of a *forth-order* auxiliary linear operator (for two dimensional Gelfand equation) or a *sixth-order* auxiliary linear operator (for three dimensional Gelfand equation), respectively, as illustrated by Liao [14]. Although it is true that the auxiliary linear operator (with the same highest order of derivative as that of considered problem) can be chosen straightforwardly in most cases, such kind of freedom of the HAM should be taken into account sufficiently by the HAM community when necessary, especially for some valuable but challenging problems (some of them are suggested below in § 1.5).

In addition, by means of the above-mentioned freedom of the HAM, the convergence of approximation solution can be greatly accelerated in the frame of the HAM by means of the iteration, the so-called homotopy-Padé technique and so on. For details, please refer to § 2.3.5 to § 2.3.7 of Liao's book [11].

Indeed, "the essence of mathematics lies entirely in its freedom", as pointed out by Georg Cantor (1845–1918).

Such kind of great freedom of the HAM should provide us great possibility to solve some open questions. One of them is described below. The solution of the high-order deformation equation (1.15) can be expressed in the form

$$u_m(\mathbf{r},t) = -\chi_m u_{m-1}(\mathbf{r},t) + \mathcal{L}^{-1}\left[c_0 \; H(\mathbf{r},t) \; R_{m-1}(\mathbf{r},t)\right], \tag{1.21}$$

where $\mathcal{L}^{-1}$ is the inverse operator of $\mathcal{L}$. For a few auxiliary linear operator $\mathcal{L}$, its inverse operator is simple. However, in most cases, it is not straightforward to solve the above linear differential equation. Can we directly choose (or *define*) the inverse auxiliary linear operator $\mathcal{L}^{-1}$ so as to solve (1.15) conveniently? This is possible in the frame of the HAM, since in theory the HAM provides us great freedom and large flexibility to choose the auxiliary linear operator $\mathcal{L}$. If successful, it would be rather efficient and convenient to solve the high-order deformation equation (1.15). This is an interesting but open question for the HAM community, which deserves to be studied in details. Note that some interesting problems are suggested in § 1.5.

1.3. Some advances of the HAM

Since 1992 when Liao [9] proposed the early HAM, the HAM has been developing greatly in theory and applications, due to the contributions of many researchers in dozens of countries. Unfortunately, it is impossible to describe all of these advances in details in this brief review, and even in this book. In fact, the HAM has been successfully applied to numerous, various types of nonlinear problems in science, engineering and finance. So, we had to focus on a rather small port of these advances here.

1.3.1. *Generalized zeroth-order deformation equation*

The starting point of the use of the HAM is to construct the so-called zeroth-order deformation equation, which builds a connection (i.e. a continuous mapping/deformation) between a given nonlinear problem and a relatively much simpler linear ones. So, the zeroth-order deformation equation is a base of the HAM.

Given a nonlinear equation, we have great freedom and large flexibility in the frame of the HAM to construct the so-called zeroth-order deformation equation using the concept homotopy in topology. Especially, the convergence-control parameter c_0 plays an important role in the frame of the HAM. So, it is natural to enhance the ability of the so-called "convergence control" by means of introducing more such kind of auxiliary parameters. Due to the above-mentioned freedom and flexibility of the HAM, there are numerous approaches to do so. For example, we can construct such a kind of zeroth-order deformation equation with $K+1$ convergence-control parameters:

$$\begin{aligned}&(1-q)\mathcal{L}\left[\varphi(\mathbf{r},t;q)-u_0(\mathbf{r},t)\right]\\&=\left(\sum_{n=0}^{K} c_n\, q^{n+1}\right) H(\mathbf{r},t)\left\{\mathcal{N}[\varphi(\mathbf{r},t;q)]-f(\mathbf{r},t)\right\},\end{aligned}\tag{1.22}$$

where $\varphi(\mathbf{r},t;q)$ is the solution, $\mathcal{N}$ is a nonlinear operator related to an original problem $N[u(\mathbf{r},t)]=f(\mathbf{r},t)$, $q\in[0,1]$ is the homotopy parameter, u_0 is an initial guess, $\mathcal{L}$ is an auxiliary linear operator, $H(\mathbf{r},t)$ is an auxiliary function which is nonzero almost everywhere, and

$$\mathbf{c}=\{c_0,c_1,\cdots,c_K\}$$

is a vector of $(K+1)$ non-zero convergence-control parameters, respectively. Note that, when $K=0$, it gives exactly the zeroth-order deformation equation (1.10).

The corresponding high-order deformation equation reads

$$\mathcal{L}[u_m(\mathbf{r},t)-\chi_m u_{m-1}(\mathbf{r},t)]=H(\mathbf{r},t)\sum_{n=0}^{\min\{m-1,K\}} c_n\, R_{m-1-n}(\mathbf{r},t),\tag{1.23}$$

where $R_n(\mathbf{r},t)$ and χ_n are defined by the same formulas (1.16) and (1.17), respectively. When $K=0$, the above high-order deformation equation (1.23) is exactly the same as (1.15). At the mth-order of approximation, the optimal convergence-control parameters are determined by the minimum of the residual square of the original equation, i.e.

$$\frac{d\mathcal{E}_m}{dc_n}=0,\quad 0\le n\le\min\{m-1,K\},\tag{1.24}$$

where $\mathcal{E}_m$ is defined by (1.19). For details, please refer to Chapter 4 of Liao's book [11]. When $K\to+\infty$, it is exactly the so-called "optimal homotopy asymptotic method" [22]. So, the "optimal homotopy asymptotic method" [22] is also a special case of the HAM, as shown in § 3.2.2 and § 6.3 of Liao's book [11].

In theory, the more the convergence-control parameters, the larger the ability to control the convergence of the HAM series. However, it is found [16] that much more CPU times is needed in practice when more convergence-control parameters are used. In most cases, one optimal convergence-control parameter is good enough to gain convergent results by means of the HAM. Considering the computational efficiency, one up to three convergence-control parameters are generally suggested in the frame of the HAM. For details, please refer to § 2.3.3, § 2.3.4, § 4.6.1 and Chapter 3 of Liao's book [16].

It should be emphasized once again that, in the frame of the homotopy in topology, we have rather great freedom and large flexibility to construct the so-called zeroth-order deformation equation. In theory, given a nonlinear equation $\mathcal{N}[u(\mathbf{r},t)] = f(\mathbf{r},t)$, we can always properly choose an initial guess $u_0(\mathbf{r},t)$ and an auxiliary linear operator $\mathcal{L}$ to construct such a zeroth-order deformation equation in a rather general form

$$\mathcal{A}\left[u_0(\mathbf{r},t), \mathcal{L}, \varphi(\mathbf{r},t;q), \mathbf{c}; q\right] = 0 \tag{1.25}$$

that it holds

$$\varphi(\mathbf{r},t;0) = u_0(\mathbf{r},t), \quad \text{when } q = 0, \tag{1.26}$$

and

$$\varphi(\mathbf{r},t;1) = u(\mathbf{r},t), \quad \text{when } q = 1, \tag{1.27}$$

i.e., when $q = 1$ the zeroth-order deformation equation (1.25) is equivalent to the original nonlinear equation $\mathcal{N}[u(\mathbf{r},t)] = f(\mathbf{r},t)$. Using the theorems given in Chapter 4 of Liao's book [11], it is easy to gain the corresponding high-order deformation equations. Here,

$$\mathbf{c} = \{c_0, c_1, \cdots, c_K\}$$

is a vector of convergence-control parameters, whose optimal values are determined by the minimum of residual square of the original equation. Note that (1.25) is rather general: the zeroth-order deformation equations (1.10) and (1.22), and even Eq. (1.2) for Lyapunov's artificial small parameter method, are only special cases of (1.25). Some commonly used zeroth-order deformation equations are described in § 4.3 of Liao's book [11] as special cases of the generalized zeroth-order deformation equation (1.25).

In theory, there are an infinite number of different ways to construct a zeroth-order deformation equation (1.25). Therefore, in the frame of the

HAM, we indeed have huge freedom and flexibility. Such kind of freedom and flexibility comes from the homotopy, a basic concept in topology. In theory, this kind of freedom and flexibility provides us great ability to solve some interesting but challenging nonlinear problems (some of them are suggested below in § 1.5 of this chapter), if we can clearly know how to use them in a proper way with stimulated imagination!

In practice, it is suggested to firstly use the zeroth-order deformation equation (1.10), since it works for most of nonlinear problems, as illustrated by Liao [10, 11]. If unsuccessful, one can further attempt a little more complicated zeroth-order deformation equations, such as (1.22). Finally, we emphasize once again that, in theory, one has huge freedom to construct a zeroth-order deformation equation (1.25) satisfying both of (1.26) and (1.27), as long as one clearly knows how to use such kind of freedom.

1.3.2. *Spectral HAM and complicated auxiliary operator*

Although the HAM provides us great freedom to choose the auxiliary linear operator $\mathcal{L}$, it might be difficult to solve the linear high-order deformation equation (1.15) or (1.23) *exactly*, if $\mathcal{L}$ is complicated. This is mainly because most of linear differential equations have no closed-form solutions, i.e. their solutions are mostly expressed by an infinite series. So, in order to exactly solve high-order deformation equations in the frame of the HAM, we often should choose a reasonable but simple enough auxiliary linear operator $\mathcal{L}$. This, however, restricts the applications of the HAM.

This is the main reason why only a few simple auxiliary linear operators, such as

$$\mathcal{L}u = u', \quad \mathcal{L}u = xu' + u, \quad \mathcal{L}u = u' + u, \quad \mathcal{L}u = u'' + u$$

and so on, have been mostly used in the frame of the HAM, where the prime denotes the differentiation with respect to x. These auxiliary linear operators correspond to some fundamental functions such as polynomial, exponential, trigonometric functions and their combination.

There are many special functions governed by linear differential equations. Although many solutions can expressed by these special functions, they are hardly used in the frame of the HAM up to now, because the corresponding high-order deformation equations often become more and more difficult to solve. This is a pity, since in theory the HAM indeed provides us freedom to use special functions to express solutions of many nonlinear differential equations. Currently, Van Gorder [23] made an interesting attempt in this direction. In the frame of the HAM, Van Gorder [23] expressed analytic approximations of the Fitzhugh–Nagumo equation by means of error function, Gaussian function and so on. The key is that Van Gorder [23] chose such an auxiliary linear operator

$$\mathcal{L}u = u'' + \left(\frac{2z^2 - 1}{z}\right) u',$$

where the prime denotes the differentiation with respect to z, and especially such a proper auxiliary function $H(z) = z|z|$, that the corresponding high-order deformation equations can be solved easily. For details, please refer to Van Gorder [23], Vajravelu and Van Gorder [24] and § 4.6 of this book. This example illustrates once again that the HAM indeed provides us great freedom, i.e. lots of possibilities. The key is how to use such kind of freedom!

Generally speaking, solution of a complicated linear ODE/PDE should be expressed in a series with an infinite number of terms. Mathematically, such a series leads to the larger and larger difficulty to gain higher-order analytic approximations of a nonlinear problem. Fortunately, from physical viewpoint, it is often accurate enough to have analytic approximations with many enough terms. Currently, using the Schmidt-Gram process, Zhao, Lin and Liao [25] suggested an effective truncation technique in the frame of the HAM, which can be used to greatly simplify the right-hand side of the high-order deformation equations, such as (1.15) and (1.23), prior to solving them. In this way, much CPU time can be saved, even without loss of accuracy.

In 2010, Motsa *et al.* [26, 27] suggested the so-called "spectral homotopy analysis method" (SHAM) using the Chebyshev pseudospectral method to solve the linear high-order deformation equations and choosing the auxiliary linear operator $\mathcal{L}$ in terms of the Chebyshev spectral collocation differentiation matrix [28]. In theory, any a continuous function in a bounded interval can be best approximated by Chebyshev polynomial. So, the SHAM provides us larger freedom to choose the auxiliary linear operator $\mathcal{L}$ and initial guess in the frame of the HAM. It is valuable to expand the SHAM for nonlinear partial differential equations. Besides, it is easy to employ the optimal convergence-control parameter in the frame of the SHAM. Thus, the SHAM has great potential to solve more complicated nonlinear problems, although further modifications in theory and more applications are needed. For the details about the SHAM, please refer to [26, 27] and Chapter 3 of this book.

Chebyshev polynomial is just one of special functions. There are many other special functions such as Hermite polynomial, Legendre polynomial, Airy function, Bessel function, Riemann zeta function, hypergeometric functions, error function, Gaussian function and so on. Since the HAM provides us extremely large freedom to choose auxiliary linear operator $\mathcal{L}$ and initial guess, it should be possible to develop a "generalized spectral HAM" which can use proper special functions for some nonlinear problems. Especially, combined the SHAM [26, 27] with the above-mentioned truncation technique suggested by Zhao, Lin and Liao [25], it would be possible to use, when necessary, more complicated auxiliary linear operators in the frame of the HAM so that some difficult nonlinear problems can be solved.

1.3.3. *Predictor HAM and multiple solutions*

Many nonlinear boundary value problems have multiple solutions. In general, it is difficult to gain these dual solutions by means of numerical techniques, mainly because dual solutions are often strongly dependent upon initial conditions but we do not know how to choose them exactly. Comparatively speaking, it is a little more convenient to use analytic approximation methods to search for multiple solutions of nonlinear problems, since analytic methods admit unknown variables in initial guess.

For example, let us consider a second-order nonlinear differential equation of a two-point boundary value problem:

$$\mathcal{N}[u(x)] = 0, \quad u(0) = a, \quad u(1) = b, \tag{1.28}$$

where $\mathcal{N}$ is a 2nd-order nonlinear differential operator, a and b are known constants, respectively. Assume that there exist multiple solutions $u(x)$. These multiple solutions must have something different. Without loss of generality, assume that they have different first-order derivative $u'(0) = \sigma$, where σ is unknown.

Obviously, different initial guess $u_0(x)$ might lead to multiple solutions. Fortunately, the HAM provides us great freedom to choose initial guess $u_0(x)$. As mentioned before, such kind of freedom is one cornerstone of the HAM. So, in the frame of the HAM, it is convenient for us to choose such an initial guess $u_0(x)$ that it satisfies not only the two boundary conditions $u(0) = a$, $u(1) = b$ but also the additional condition $u'(0) = \sigma$. In this way, the initial guess $u_0(x)$ contains an unknown parameter σ, called by Liao (see Chapter 8 of [11]) the *multiple-solution-control parameter*. Then, the analytic approximations gained by the HAM contain at least *two* unknown auxiliary parameters: the convergence-control parameter c_0 and the multiple-solution-control parameter σ. As suggested by Liao (see Chapter 8 of [11]), the optimal values of c_0 and σ can be determined by the minimum of the residual square of governing equations. In this way, multiple solutions of some nonlinear differential equations can be gained, as illustrated by Liao (see Chapter 8 of [11]).

In the frame of the HAM, Abbasbandy and Shivanian [29, 30] developed a differential but rather interesting approach to gain dual solutions, namely the *Predictor HAM* (PHAM). For simplicity, let us use the same equation (1.28) as an example to describe its basic ideas. First of all, an additional condition such as $u'(0) = \sigma$ is introduced with the unknown parameter σ. Then, in the frame of the HAM, one solves the nonlinear differential equation $\mathcal{N}[u(x)] = 0$, but with the two boundary conditions $u'(0) = \sigma$ and $u(1) = b$. Then, $u(0)$, the HAM approximation at $x = 0$, contains at least *two* unknown parameters: one is the so-called *convergence-control parameter* c_0, the other is $\sigma = u'(0)$, called the *multiple-solution-control parameter* by Liao (see Chapter 8 of [11]) in the above-mentioned approach. Substituting the expression of $u(0)$ into the boundary condition $u(0) = a$ gives a nonlinear algebraic equation about c_0 and σ. From the physical viewpoint, $\sigma = u'(0)$ has physical meanings, but the convergence-control parameter c_0 does not. If the order of approximation is high enough, one can gain convergent, accurate enough multiple values of σ for properly chosen values of c_0 in a finite interval, as illustrated in [29, 30]. In this way, one can find multiple solutions of a given nonlinear problem. For details, please refer to Chapter 2 of this book.

In the frame of the HAM, some new branches of solutions for viscous boundary-layer flows were found [31, 32], and the multiple equilibrium-states of resonant waves in deep water [33] and in finite water depth [34] were discovered for the first time, to the best of author's knowledge. All of these illustrate the potential, novelty and validity of the HAM to give something new and different. This is a superiority of the HAM to numerical methods and some other analytic approximation techniques. Certainly, it is valuable to apply the HAM to discover some new solutions of other nonlinear problems!

1.3.4. *Convergence condition and HAM-based software*

Theoretically speaking, the HAM indeed provides us great freedom to choose initial guess, auxiliary linear operator, convergence-control parameter, equation-type and solution-expression of high-order deformation equation, and so on. However, it is still not very clear how to use these freedom in the frame of the HAM, mainly because little mathematical theorems have been proved in an abstract way.

Some studies on the stability of auxiliary linear operator and convergence-control parameter of the HAM are described in Chapter 4 of this book. Some current works about convergence condition of the HAM series are described in Chapter 5.

It should be especially emphasized that Park and Kim [35, 36] successfully applied the HAM to solve a few classic problems in finance, and gave convergence conditions for their analytic approximations. It is rather interesting that they even gave an error estimation for their analytic approximations in [36]. Currently, Park and Kim used the HAM to solve an abstract linear problem with respect to bounded linear operators from a Banach space to a Banach space, and rigorously proved that the homotopy solution exists in the sense that a series of the problem converges in a Banach norm sense if the linear operator satisfies some mild conditions. Their fantastic works are very important, and might pioneer a new research direction and style (i.e. abstract proof) in the frame of the HAM. Such kind of abstract mathematical theorems in the frame of the HAM are more valuable and useful, if nonlinear governing equations and especially the influence of the convergence-control parameter on the convergence could be considered.

On the other side, the HAM has been successfully applied to numerous nonlinear problems in various fields of science and engineering. These applications show the general validity and novelty of the HAM. Unfortunately, it is impossible to mention all of them here in details. As examples among these numerous applications, a HAM-based approach about boundary-layer flows of nanofluid is given in Chapter 6 of this book. In addition, an application of the HAM for time-fractional boundary-value problem is illustrated in Chapter 7.

To simplify some applications of the HAM, two HAM-based software were developed. The HAM-based Maple package NOPH (version 1.0.2) for periodic oscillations and limit cycles of nonlinear dynamic systems is described in Chapter 8 of this book with various applications. It is free available online at

`http://numericaltank.sjtu.edu.cn/NOPH.htm`

with a simple user's guide. Besides, the HAM-based Mathematica package BVPh (version 2.0) for coupled nonlinear ordinary differential equations with boundary conditions at multiple points are given in Chapter 9 of this book. It is free available online at

http://numericaltank.sjtu.edu.cn/BVPh.htm

with a simple user's guide and some examples of application. Both of these two HAM-based software are easy-to-use and user-friendly. They greatly simplify some applications of the HAM, and are especially helpful for the beginners of the HAM.

1.4. Relationships to other methods

In pure mathematics, the homotopy is a fundamental concept in topology and differential geometry. The concept of homotopy can be traced back to Jules Henri Poincaré (1854–1912), a French mathematician. A homotopy describes a kind of continuous variation (or deformation) in mathematics. For example, a circle can be continuously deformed into a square or an ellipse, the shape of a coffee cup can deform continuously into the shape of a doughnut but cannot be distorted continuously into the shape of a football. Essentially, a homotopy defines a connection between different things in mathematics, which contain same characteristics in some aspects. In pure mathematics, the homotopy is widely used to investigate existence and uniqueness of solutions of some equations, and so on.

In applied mathematics, the concept of homotopy was used long ago to develop some numerical techniques for nonlinear algebraic equations. The so-called "differential arc length homotopy continuation method" were proposed in 1970s by Keller [37, 38]. However, the global homotopy methods can be traced as far back as the work of Lahaye [39] in 1934. To solve a nonlinear algebraic equation $f(x) = 0$ by means of the homotopy continuation method, one first constructs such a homotopy

$$H(x, q) = q\, f(x) + (1 - q)\, g(x), \tag{1.29}$$

where $q \in [0, 1]$ is the homotopy parameter, $g(x)$ is a function for which a zero is known or readily obtained. As discussed by Wayburn and Seader [40], the choice of $g(x)$ is arbitrary, but the two most widely used functions are the Newton homotopy

$$H(x, q) = q\, f(x) + (1 - q) \left[f(x) - f(x_0) \right],$$

and the fixed-point homotopy

$$H(x, q) = q\, f(x) + (1 - q) \left(x - x_0 \right),$$

where x_0 is an arbitrary starting point. The locus of solutions defines the homotopy path that is tracked with some continuation method. Consequently, homotopy continuation methods consist not only of the homotopy equation itself, but also the homotopy path tracking method, i.e. of some continuation strategy.

Homotopy continuation methods are usually based upon differentiating the homotopy equation (1.29) with respect to the arc length s, which gives the equation

$$\frac{\partial H}{\partial x}\frac{dx}{ds}+\frac{\partial H}{\partial q}\frac{dq}{ds}=0. \tag{1.30}$$

Taking into account the arc-length relation

$$\left(\frac{dx}{ds}\right)^2+\left(\frac{dq}{ds}\right)^2=1$$

and the initial condition $H(x_0,0)=0$, we obtain an initial value problem. Then, path tracking based on the initial value problem is *numerically* carried out with a predictor-corrector algorithm to gain a solution of the original equation $f(x)=0$. Some elegant theorems of convergence are proved in the frame of the homotopy continuation method. For details of the homotopy continuation method, please refer to [41–49].

Unlike the homotopy continuation method that is a global convergent *numerical* method mainly for nonlinear *algebraic* equations, the HAM is a kind of *analytic* approximation method mainly for nonlinear *differential* equations. So, the HAM is essentially different from the homotopy continuation method, although both of them are based on the homotopy, the basic concept of the topology. Note that the HAM uses much more complicated homotopy equation (1.10) or (1.22) than (1.29) for the homotopy continuation method. Furthermore, the HAM provides larger freedom to choose the auxiliary linear operator $\mathcal{L}$. Most importantly, the so-called convergence-control parameter c_0 is introduced *for the first time*, to the best of our knowledge, in the homotopy equation (1.10) or (1.22) so that the HAM provides us a convenient way to guarantee the convergence of series series. Note that the homotopy equation (1.29) of the homotopy continuation method does not contain such kind of convergence-control parameter at all. So, the convergence-control parameter c_0 is indeed a novel. In fact, it is the convergence-control parameter c_0 which differs the HAM from all other analytic approximation methods.

In addition, the HAM logically contains many other analytic approximation methods and thus is rather general. For example, it has been generally proved [10–13] that the Lyapunov's artificial small parameter method [5], the Adomian decomposition method [6–8], the δ-expansion method [18] are only special cases of the HAM for some specially chosen auxiliary linear operator $\mathcal{L}$ and convergence-control parameter c_0. Furthermore, the so-called "optimal homotopy asymptotic method" [22] developed in 2008 is also a special case of the homotopy equation (1.22) when $K \to +\infty$, too, as pointed out by Liao (see § 6.3 of Liao's book [11]).

Especially, the so-called "homotopy perturbation method" (HPM) [19] proposed by Jihuan He in 1998 (six years later after Liao [9] proposed the early HAM in 1992) was only a special case of the HAM when $c_0 = -1$, as proved in [20], and thus it has "nothing new except its name" [20]. Some results given by the HPM are divergent even in the whole interval except the given initial/boundary conditions, and thus "it is very important to investigate the convergence of approximation series, otherwise one might get useless results", as pointed out by Liang and Jeffrey [21]. For more details, see § 6.2 of Liao's book [11].

In addition, even the famous Euler transform in calculus can be derived in the frame of the HAM (see Chapter 5 of Liao's book [11]). This provides us a theoretical cornerstone for the validity and generality of the HAM.

In summary, based on the concept of homotopy topology, the HAM is a novel analytic approximation method for highly nonlinear problems, with great freedom and flexibility to choose equation-type and solution expression of high-order approximation equations and also with a convenient way to guarantee the convergence, so that it might overcome restrictions of perturbation techniques and other non-perturbation methods.

1.5. Chance and challenge: some suggested problems

Any truly new methods should give something novel and/or different, or solve some difficult problems that can not be solved with satisfaction by other methods.

Unlike other analytic approximation methods, the HAM provides us great freedom and flexibility to choose equation-type and solution expression of high-order approximation equations, and especially a simple way to guarantee the convergence of solution series. Thus, the HAM provides us a large possibility and chance to give something novel or different, and to attack some difficult nonlinear problems. For example, some new solutions [31, 32] of boundary-layer flows have been found by means of the HAM, which had been neglected even by numerical techniques and had been never reported. Some analytic approximations for the optimal exercise boundary of American put option were given, which are valid from a couple of years (see [50, 51]) up to even 20 years (see Chapter 13 of Liao's book [11]) prior to expiry, and thus much better than the asymptotic/perturbation approximations that are often valid only for a couple of days or weeks. Besides, the HAM has been successfully employed to solve some complicated nonlinear PDEs: the multiple equilibrium-states of resonant waves in deep water [33] and in finite water depth [34] were discovered by means of the HAM for the first time, to the best of our knowledge, which greatly deepen and enrich our understandings about resonant waves.

All of these successful applications show the originality, validity and generality of the HAM for nonlinear problems, and encourage us to apply the HAM to attack some famous, challenging nonlinear problems. Some of these problems are suggested below for the HAM community, especially for brave, enterprising, young researchers.

1.5.1. *Periodic solutions of chaotic dynamic systems*

It is well known that chaotic dynamic systems have the so-called "butterfly effect" [52, 53], say, the computer-generated numerical simulations have sensitive dependence to initial conditions (SDIC). For example, the nonlinear dynamic system of Lorenz equations [52]

$$\dot{x} = \sigma\,(y - x)\,, \tag{1.31}$$

$$\dot{y} = r\,x - y - x\,z, \tag{1.32}$$

$$\dot{z} = x\,y - b\,z, \tag{1.33}$$

has chaotic solution in case of $r = 28, b = 8/3$ and $\sigma = 10$ for most of given initial conditions x_0, y_0, z_0 of x, y, z at $t = 0$. However, for some special initial conditions such as

$$x_0 = -13.7636106821,\ \ y_0 = -19.5787519424,\ \ z_0 = 27;$$
$$x_0 = -9.1667531454,\ \ y_0 = -9.9743951128,\ \ z_0 = 27;$$
$$x_0 = -13.5683173175,\ \ y_0 = -19.1345751139,\ \ z_0 = 27,$$

the above dynamic system of Lorenz equation has unstable periodic solutions, as reported by Viswanath [54].

A periodic solution $u(t)$ with the period T has the property

$$u(t) = u(t + n\,T)$$

for *arbitrary* time $t \geq 0$ and *arbitrary* integers n, even if $t \to +\infty$ and $n \to \infty$. This property *cannot* be checked strictly by means of numerical approaches, since all numerical integration simulations are gained in a finite interval of time. Naturally, a periodic solution should be expressed analytically by periodic base functions such as trigonometric functions. So, theoretically speaking, it is inherently better to use analytic approximation methods to search for periodic solutions of chaotic dynamic systems than numerical ones.

In fact, as illustrated by Liao in Chapter 13 of his book [10], the HAM can be employed to gain periodic solution of nonlinear dynamic systems. Can we employ the HAM to gain the above-mentioned unstable periodic solutions of Lorenz equation found by Viswanath [54]? More importantly, it would be very interesting if the HAM could be employed to find some *new* periodic solutions of Lorenz equation with physical parameters leading to chaos! This is mainly because Lorenz equation is one of the most famous ones in nonlinear dynamics and nonlinear science.

1.5.2. *Periodic orbits of Newtonian three-body problem*

Let us consider one of the most famous problem in mechanics and applied mathematics: the Newtonian three-body problem, say, the motion of three celestial bodies under their mutual gravitational attraction. Let x_1, x_2, x_3 denote the three orthogonal axes. The position vector of the ith body is expressed by $\mathbf{r}_i = (x_{1,i}, x_{2,i}, x_{3,i})$, where $i = 1, 2, 3$. Let T and L denote the characteristic time and length scales, and m_i the mass of the ith body, respectively. Using Newtonian gravitation law, the motion of the three bodies are governed by the corresponding non-dimensional equations

$$\ddot{x}_{k,i} = \sum_{j=1, j\neq i}^{3} \rho_j \frac{(x_{k,j} - x_{k,i})}{R_{i,j}^3}, \quad k = 1, 2, 3, \tag{1.34}$$

where

$$R_{i,j} = \left[\sum_{k=1}^{3} (x_{k,j} - x_{k,i})^2\right]^{1/2} \tag{1.35}$$

and

$$\rho_i = \frac{m_i}{m_1}, \quad i = 1, 2, 3 \tag{1.36}$$

denotes the ratio of the mass.

According to H. Poincaré, orbits of three-body problem are unintegrable in general. Although chaotic orbits of three-body problems widely exist, three families of periodic orbits were found:

(1) the Lagrange–Euler family, dating back to the analytical solutions in the 18th century (one recent orbit was given by Moore [55]);
(2) the Broucke–Hadjidemetriou–Hénon family, dating back to the mid-1970s [56–61];
(3) the Figure-8 family, discovered in 1993 by Moore [55] and extended to the rotating cases [62–65].

Note that nearly all of these reported periodic orbits are planar. In 2013, Šuvakov and Dmitrašinović [66] found that there exist four classes of planar periodic orbits of three-body problem, with the above three families belonging to one class. Besides, they reported three new classes of planar periodic orbits and gave the corresponding initial conditions for each class. For the details of their 15 planar periodic orbits, please refer to the gallery [67].

Šuvakov and Dmitrašinović [66] found these new classes of planar periodic orbits by means of an iterative numerical integration approach without using multiple precision. So, it is unknown whether or not the numerical simulations depart the corresponding periodic orbits for rather large time, i.e. $t \to \infty$. As mentioned before, it is better and more natural to express a periodic solution $u(t)$ with the period T in series of periodic base functions (with the same period T) so that $u(t) = u(t + nT)$ can hold for arbitrary integer n and arbitrary time t even if $t \to \infty$. Thus, it is valuable to apply the HAM to *double check* all of the reported periodic orbits in [66], and more importantly, to find some completely *new* periodic orbits!

Note that nearly all of the periodic orbits of three-body problem reported up to now are planar. Therefore, it is valuable and interesting if the HAM can be applied to find some periodic orbits of Newtonian three-body problems, which are *not* planar, i.e. *three* dimensional. Mathematically speaking, we should determine such unknown initial positions $\mathbf{r}_1, \mathbf{r}_2, \mathbf{r}_3$, unknown initial velocities $\dot{\mathbf{r}}_1, \dot{\mathbf{r}}_2, \dot{\mathbf{r}}_3$ and unknown corresponding mass-ratios ρ_1, ρ_2, ρ_3 of three bodies in the frame of the HAM that Eqs. (1.34) have periodic solution $x_{k,i}(t) = x_{k,i}(t + nT)$ for arbitrary time t and integer n, where T is the unknown corresponding period to be determined, and $i, k = 1, 2, 3$. This is a valuable, interesting but challenging problem for the HAM community.

1.5.3. *Viscous flow past a sphere*

One of the most famous, classical problem in fluid mechanics is the steady-state viscous flow past a sphere [68–75], governed by the Navier-Stokes equation, i.e. a system of nonlinear partial differential equations. Consider the steady-state viscous flow past a sphere in a uniform stream. How large is the drag of the sphere due to the viscosity of fluid?

To study the steady-state viscous flow past a sphere, the spherical coordinates $\vec{r} = (r, \theta, \phi)$ is often used. Since the problem has axial symmetry, one can use the Stokes stream function $\psi(r, \theta)$ defined through the following relations:

$$v_r = \frac{1}{r^2 \sin(\theta)} \psi_\theta, \; v_\theta = -\frac{1}{r \sin(\theta)} \psi_r, \; v_\phi = 0. \tag{1.37}$$

The stream function $\psi(r, \theta)$ is governed by the dimensionless equation

$$D^4 \psi = \frac{R}{r^2} \left[\frac{\partial(\psi, D^2\psi)}{\partial(r, \mu)} + 2D^2 \psi L \psi \right], \tag{1.38}$$

subject to the boundary conditions

$$\psi(1, \mu) = 0, \tag{1.38a}$$

$$\left. \frac{\partial \psi(r, \mu)}{\partial r} \right|_{r=1} = 0, \tag{1.38b}$$

$$\lim_{r \to \infty} \frac{\psi(r, \mu)}{r^2} = \frac{1}{2}(1 - \mu^2), \tag{1.38c}$$

where $R = aU_\infty / \nu$ is the Reynolds number and

$$\mu \equiv \cos(\theta), \tag{1.39}$$

$$D^2 \equiv \frac{\partial^2}{\partial r^2} + \frac{1 - \mu^2}{r^2} \frac{\partial^2}{\partial \mu^2}, \tag{1.40}$$

$$L \equiv \frac{\mu}{1 - \mu^2} \frac{\partial}{\partial r} + \frac{1}{r} \frac{\partial}{\partial \mu}. \tag{1.41}$$

Here, a denotes the radius of the sphere and U_∞ the uniform stream velocity at infinity, respectively, according to the notation of Proudman and Pearson [71]. As mentioned by Liao [74], the drag coefficient reads

$$C_D = \frac{4}{R} \int_{-1}^{1} \left. \left(-p \, \mu + \frac{\partial^2 \psi}{\partial r^2} \right) \right|_{r=1} d\mu, \tag{1.42}$$

where the pressure p is given by

$$p = -\int_{\mu}^{1} \frac{1}{(1 - \mu^2)} \left. \frac{\partial^3 \psi}{\partial r^3} \right|_{r=1} d\mu. \tag{1.43}$$

Unfortunately, neither the linearization method [68–70] nor the perturbation techniques [71, 72] can provide an analytic approximation of drag coefficient C_D valid for $R_d > 3$, where $R_d = d\,U_\infty/\nu = 2R$ for the diameter d of the sphere. Especially, the 3rd-order multiple-scale perturbation approximation of C_D given by Chester and Breach [72] was valid even in a smaller interval of Reynolds number than the 2nd-order multiple-scale perturbation result of Proudman and Pearson [71]. This implies the invalidity of perturbation methods for this famous problem. So, "the idea of using creeping flow to expand into the high Reynolds number region has not been successful", as pointed out by White in his textbook [76]. Besides, the method of renormalization group can not essentially modify these analytic results [75], either.

In 2002, Liao [74] employed the HAM to solve the steady-state viscous flow past a sphere and gained a analytic approximation of drag coefficient C_D, which agree well with experimental data in a considerably larger interval $R_d \leq 30$. However, the corresponding experiments indicate that the steady-state viscous flow past a sphere exists until $R_d \approx 100$. So, strictly speaking, this HAM result given in [74] is not satisfactory.

Theoretically speaking, it is very interesting and valuable if one can give an accurate enough analytic result of the drag coefficient C_D valid for the steady-state viscous flow past a sphere up to $R_d \approx 100$, mainly because it is one of the most famous, classical problems in fluid mechanics with a history of more than 150 year!

Can we solve this famous, classical problem by means of the HAM?

1.5.4. *Viscous flow past a cylinder*

The steady-state viscous flow past an infinite cylinder is also one of the most famous, classical problems in fluid mechanics with a long history. For the steady-state viscous flow past an infinite cylinder, it is natural to use cylindrical coordinates $\vec{r} = (r, \theta, z)$. Since the problem is two dimensional, it is convenient to use the Lagrangian stream function $\psi(r, \theta)$ defined by Proudman and Pearson [71]:

$$u_r = \frac{1}{r}\frac{\partial \psi}{\partial \theta}, \; u_\theta = -\frac{\partial \psi}{\partial r}, \; u_z = 0. \tag{1.44}$$

The stream function $\psi(r, \theta)$ is governed by

$$\nabla_r^4 \psi(r, \theta) = -\frac{R}{r}\frac{\partial(\psi, \nabla_r^2)}{\partial(r, \theta)}, \tag{1.45}$$

subject to the boundary conditions

$$\psi(r=1,\theta) = 0, \tag{1.45a}$$

$$\left.\frac{\partial\psi(r,\theta)}{\partial r}\right|_{r=1} = 0, \tag{1.45b}$$

$$\lim_{r\to\infty}\frac{\psi(r,\theta)}{r} = \sin(\theta), \tag{1.45c}$$

where

$$\nabla_r^2 \equiv \frac{\partial^2}{\partial r^2} + \frac{1}{r}\frac{\partial}{\partial r} + \frac{1}{r^2}\frac{\partial^2}{\partial\theta^2}, \quad \nabla_r^4 \equiv \nabla_r^2\nabla_r^2.$$

Here, $R = aU_\infty/\nu$ is the Reynolds number, a and U_∞ denote the radius of cylinder and the uniform stream velocity at infinity, respectively.

As reviewed in [75], neither the linearization method nor perturbation technique can give good analytic approximation of the drag coefficient C_D of a cylinder for $R \geq 3$. In fact, even the method of renormalization group cannot modify these results greatly [75]. So, it is still an open question.

Theoretically speaking, it is valuable to gain an accurate analytic expression of drag coefficient C_D valid for large Reynolds number up to $R \approx 40$, beyond which the periodic Von Karmán vortex occurs. This is mainly because it is one of the most famous, historical problem in fluid mechanics.

Can this famous, classical problem be solved by means of the HAM?

1.5.5. *Nonlinear water waves*

The HAM has been successfully applied to solve some nonlinear wave equations. Especially, in the frame of the HAM, the multiple equilibrium-states of resonant waves in deep water [33] and in finite water depth [34] were discovered for the first time, to the best of the author's knowledge. Thus, the HAM provides us a convenient tool to investigate some complicated wave problems.

Strictly speaking, water waves are governed by Euler equation with two nonlinear boundary conditions satisfied on an unknown free surface, which however are rather difficult to solve in general. Based on the exact Euler equation, some simplified wave models for shallow water waves, such as the KdV equation [77], Boussinesq equation [78], Camassa–Holm (CH) equation [79], and so on, are derived by assuming the existence of some small physical parameters in shallow water. Although these shallow water wave equations are much simpler than the exact Euler equation, they can well explain many physical phenomena, such as soliton waves, wave propagations and interactions in shallow water, wave breaking, and so on.

For example, the celebrated Camassa–Holm (CH) equation [79]

$$u_t + 2\omega u_x - u_{xxt} + 3uu_x = 2u_x u_{xx} + uu_{xxx}, \tag{1.46}$$

subject to the boundary condition

$$u = 0, \; u_x = 0, \; u_{xx} = 0, \quad \text{as } x \to \pm\infty, \tag{1.47}$$

can model both phenomena of soliton interaction and wave breaking (see [80]), where $u(x,t)$ denotes the wave elevation, x, t are the temporal and spatial variables, ω is a constant related to the critical shallow water wave speed, the subscript denotes the partial differentiation, respectively. Mathematically, the CH equation is integrable and bi-Hamiltonian, thus possesses an infinite number of conservation laws in involution [79]. In addition, it is associated with the geodesic flow on the infinite dimensional Hilbert manifold of diffeomorphisms of line (see [80]). Thus, the CH equation (1.46) has many intriguing physical and mathematical properties. As pointed out by Fushssteiner [81], the CH equation (1.46) even "has the potential to become the new master equation for shallow water wave theory".

Especially, when $\omega = 0$, the CH equation (1.46) has the peaked solitary wave

$$u(x,t) = c \exp(-|x - c\,t|),$$

which was found first by Camassa and Holm [79]. The first derivative of the peaked solitary wave is discontinuous at the crest $x = c\ t$. Like the CH equation, many shallow water equations admit peaked and/or cusped solitary waves. These equations with peaked and/or cusped solitary waves have been widely investigated mathematically, and thousands of related articles have been published. However, to the best of the author's knowledge, peaked and cusped solitary waves have *never* been gained directly from the exact Euler equation! This is very strange. Logically speaking, since these simplified equations (like the CH equation) are good enough approximations of the Euler equation in shallow water, the exact Euler equation should also admit the peaked and/or cusped solitary waves as well.

Can we gain such kind of peaked and/or cusped solitary waves of the exact wave equation by means of the HAM, if they indeed exist? Either positive or negative answers to this question have important scientific meanings. If such kind of peaked solutions of the exact wave equation indeed exist, it can greatly enrich and deepen our understandings about peaked solitary waves. If the peaked solitary waves given by the exact wave equation exists mathematically but is impossible in physics, we had to check the physical validity of the peaked solitary waves. So, this is an interesting and valuable work, although with great challenge. For some attempts in this direction, please refer to Liao [82], who proposed a generalized wave model based on the symmetry and the fully nonlinear wave equations, which admits not only the traditional waves with smooth crest but also peaked solitary waves. It is found that the peaked solitary waves satisfy Kelvin's theorem everywhere. Besides, these peaked solitary waves include the famous peaked solitary waves of the Camassa–Holm equation. So, the generalized wave model [82] is consistent with the traditional wave theories. It is found [82] that the peaked solitary waves have some unusual characteristics quite different from the traditional ones, although it is still an open question whether or not they are reasonable in physics if the viscosity of fluid and the surface tension are considered.

In addition, the so-called "rogue wave" [83, 84] is a hot topic of nonlinear waves. Certainly, it is valuable to apply the HAM to do some investigations in this field.

In summary, it is true that the problems suggested above are indeed difficult, but very valuable and interesting in theory. In fact, there are many such kind of interesting but difficult problems in science, engineering and finance. It should be emphasized that, unlike all other analytic approximation methods, the HAM provides us great freedom and flexibility to choose equation-type and solution expression of high-order approximation equations, and besides a convenient way to guarantee the convergence of solution series. As pointed out by Georg Cantor (1845–1918), "the essence of mathematics lies entirely in its freedom". Hopefully, the great freedom and flexibility of the HAM might create some novel ideas and inspire some brave, enterprising, young researchers with stimulated imagination to attack them with satisfactory, much better results.

Chance always stays with challenges!

Acknowledgment

This work is partly supported by National Natural Science Foundation of China (Approval No. 11272209), the Foundation for Shanghai Leading Scientists, and State Key Laboratory of Ocean Engineering (Approval No. GKZD010056).

References

[1] A.H. Nayfeh, *Perturbation Methods.* John Wiley & Sons, New York, 1973.

[2] M. Van Dyke, *Perturbation Methods in Fluid Mechanics.* Parabolic, Stanford, 1975.

[3] E.J. Hinch, *Perturbation Methods.* Cambridge University Press, Cambridge, 1991.

[4] J.A. Murdock, *Perturbations: Theory and Methods.* John Wiley & Sons, New York, 1991.

[5] A.M. Lyapunov, *General Problem on Stability of Motion* (English translation). Taylor & Francis, London, 1992 (the original one was published in 1892 in Russian).

[6] G. Adomian, Nonlinear stochastic differential equations, *J. Math. Anal. Appl.* **55**: 441–452 (1976).

[7] G. Adomian, A review of the decomposition method and some recent results for nonlinear equations, *Comput. Math. Appl.* **21**: 101–127 (1991).

[8] G. Adomian, *Solving Frontier Problems of Physics: The Decomposition Method.* Kluwer Academic Publishers, Boston, 1994.

[9] S.J. Liao, *On the proposed homotopy analysis techniques for nonlinear problems and its application.* Ph.D. dissertation, Shanghai Jiao Tong University, 1992.

[10] S.J. Liao, *Beyond Perturbation: Introduction to the Homotopy Analysis Method.* Chapman & Hall/CRC Press, Boca Raton, 2003.

[11] S.J. Liao, *Homotopy Analysis Method in Nonlinear Differential Equations.* Springer & Higher Education Press, Heidelberg, 2012.

[12] S.J. Liao, An explicit, totally analytic approximation of Blasius' viscous flow problems, *Int. J. Non-Linear Mech.* **34**: 759–778 (1999).

[13] S.J. Liao, On the homotopy analysis method for nonlinear problems, *Appl. Math. Comput.* **147**: 499–513 (2004).

[14] S.J. Liao and Y. Tan, A general approach to obtain series solutions of nonlinear differential equations, *Stud. Appl. Math.* **119**: 297–354 (2007).

[15] S.J. Liao, Notes on the homotopy analysis method: some definitions and theorems, *Commun. Nonlinear Sci. Numer. Simulat.* **14**: 983–997 (2009).

[16] S.J. Liao, An optimal homotopy-analysis approach for strongly nonlinear differential equations, *Commun. Nonlinear Sci. Numer. Simulat.* **15**: 2315–2332 (2010).

[17] S.J. Liao, On the relationship between the homotopy analysis method and Euler transform, *Commun. Nonlinear Sci. Numer. Simulat.* **15**: 1421–1431, 2010.

[18] A.V. Karmishin, A.T. Zhukov and V.G. Kolosov, *Methods of Dynamics Calculation and Testing for Thin-walled Structures* (in Russian). Mashinostroyenie, Moscow, 1990.

[19] J.H. He, An approximate solution technique depending upon an artificial parameter, *Commun. Nonlinear Sci. Numer. Simulat.* **3**: 92–97 (1998).

[20] M. Sajid and T. Hayat, Comparison of HAM and HPM methods for nonlinear heat conduction and convection equations. *Nonlin. Anal. B.* **9**: 2296–

2301 (2008).

[21] S.X. Liang and D.J. Jeffrey, Comparison of homotopy analysis method and homotopy perturbation method through an evaluation equation, *Commun. Nonlinear Sci. Numer. Simulat.* **14**: 4057–4064 (2009).

[22] V. Marinca and N. Herişanu: Application of optimal homotopy asymptotic method for solving nonlinear equations arising in heat transfer, *Int. Commun. Heat Mass.* **35**: 710–715 (2008).

[23] R.A. Van Gorder: Gaussian waves in the Fitzhugh-Nagumo equation demonstrate one role of the auxiliary function $H(x)$ in the homotopy analysis method, *Commun. Nonlinear Sci. Numer. Simulat.* **17**: 1233–1240 (2012).

[24] K. Vajravelu and R.A. Van Gorder, *Nonlinear Flow Phenomena and Homotopy Analysis: Fluid Flow and Heat Transfer.* Springer, Heidelberg, 2012.

[25] Y.L. Zhao, Z.L. Lin and S.J. Liao, An iterative HAM approach for nonlinear boundary value problems in a semi-infinite domain, *Comput. Phys. Commun.* **184**: 2136–2144 (2013).

[26] S.S. Motsa, P. Sibanda and S. Shateyi, A new spectral homotopy analysis method for solving a nonlinear second order BVP, *Commun. Nonlinear Sci. Numer. Simulat.* **15**: 2293–2302 (2010).

[27] S.S. Motsa, P. Sibanda, F.G. Auad and S. Shateyi, A new spectral homotopy analysis method for the MHD Jeffery-Hamel problem, *Computer & Fluids.* **39**: 1219–1225 (2010).

[28] W.S. Don and A. Solomonoff, Accuracy and speed in computing the Chebyshev collocation derivative. *SIAM J. Sci. Comput.* **16**: 1253–1268 (1995).

[29] S. Abbasbandy and E. Shivanian, Prediction of multiplicity of solutions of nonlinear boundary value problems: Novel application of homotopy analysis method, *Commun. Nonlinear Sci. Numer. Simulat.* **15**: 3830–3846 (2010).

[30] S. Abbasbandy and E. Shivanian, Predictor homotopy analysis method and its application to some nonlinear problems, *Commun. Nonlinear Sci. Numer. Simulat.* **16**: 2456–2468 (2011).

[31] S.J. Liao, A new branch of solutions of boundary-layer flows over an impermeable stretched plate, *Int. J. Heat Mass Tran.* **48**: 2529–2539 (2005).

[32] S.J. Liao and E. Magyari, Exponentially decaying boundary layers as limiting cases of families of algebraically decaying ones, *Z. angew. Math. Phys.* **57**, 777–792 (2006).

[33] S.J. Liao, On the homotopy multiple-variable method and its applications in the interactions of nonlinear gravity waves, *Commun. Nonlinear Sci. Numer. Simulat.* **16**: 1274 – 1303 (2011).

[34] D.L. Xu, Z.L. Lin, S.J. Liao and M. Stiassnie, On the steady-state fully resonant progressive waves in water of finite depth, *J. Fluid Mech.* **710**: 379–418 (2012).

[35] S.H. Park and J.H. Kim, Homotopy analysis method for option pricing under stochastic volatility, *Appl. Math. Lett.* **24**: 1740–1744 (2011).

[36] S.H. Park and J.H. Kim, A semi-analytic pricing formula for lookback options under a general stochastic volatility model, *Statistics and Probability Letters*, **83**: 2537 – 2543 (2013).

[37] H.B. Keller, Numerical solution of bifurcation and nonlinear eigenvalue prob-

lems. In P. Rabinowitz, *Application of Bifurcation Theory* (p. 359), Academic Press, New York, 1977.

[38] H.B. Keller, Global homotopies and Newton methods. In C. de Boor, *et al.*, *Recent Advances in Numerical Analysis* (p. 73), Academic Press, New York, 1978.

[39] E. Lahaye, Une méthode de résolution d'une catégorie déquations transcendantes. *Comptes Rendus de l'Académie des Sciences Paris*, **198**: 1840–1842 (1934).

[40] T.L. Wayburn and J.D. Seader, Homotopy continuation methods for computer-aided process design, *Computers in Chemical Engineering.* **11**: 7 (1987).

[41] B.C. Eaves, Homotopies for computation of fixed points, *Math. Programming.* **3**: 1–22 (1972).

[42] S.N. Chow, J. Mallet-Paret and J.A. Yorke, Finding zeros of maps: Homotopy methods that are constructive with probability one, *Math. Comp.* **32**: 887–899 (1978).

[43] L.T. Watson, Solving the nonlinear complementarity problem by a homotopy method, *SIAM J. Control Optim.* **17**: 36–46 (1979).

[44] J.C. Alexander, T.Y. Li and J.A. Yorke, Piecewise smooth homotopies, in *Homotopy Methods and Global Convergence*, pp. 1–14, B.C. Eaves, F.J. Gould, H.O. Peitgen and M.J. Todd, eds., Plenum, New York, 1983.

[45] L.T. Watson, M. Sosonkina, R.C. Melville, A.P. Morgan and H.F. Walker, Algorithm 777: HOMPACK90: A suite of Fortran 90 codes for globally convergent homotopy algorithms, *ACM Trans. Math. Software.* **23**: 514–549 (1997).

[46] T.Y. Li, Numerical solution of multivariate polynomial systems by homotopy continuation methods, *Acta Numerica.* **6**: 399–436 (1997).

[47] T.Y. Li, Solving polynomial systems by the homotopy continuation method, *Handbook of Numerical Analysis*, Vol. XI, pp. 209–304, edited by P. G. Ciarlet, North-Holland, Amsterdam, 2003.

[48] T. Lee, C. Tsai and T.Y. Li, HOM4PS-2.0: A software package for solving polynomial systems by the polyhedral homotopy continuation method, *Computing.* **83**: 109–133 (2008).

[49] T.Y. Li and C.H. Tsai, HOM4PS-2.0para: Parallelization of HOM4PS-2.0 for solving polynomial systems, *Parallel Computing.* **35**: 226–238 (2009).

[50] S.P. Zhu, An exact and explicit solution for the valuation of American put options, *Quant. Financ.* **6**: 229–242 (2006).

[51] J. Cheng, S.P. Zhu and S.J. Liao, An explicit series approximation to the optimal exercise boundary of American put options, *Commun. Nonlinear Sci. Numer. Simulat.* **15**: 1148–1158 (2010).

[52] E.N. Lorenz, Deterministic non-periodic flows, *J. Atmos. Sci.* **20**: 130–141 (1963).

[53] E.N. Lorenz, *The Essence of Chaos.* University of Washington Press, Seattle, 1995.

[54] D. Viswanath, The fractal property of the Lorenz attractor, *Physica D.* **190**: 115–128 (2004).

[55] C. Moore, Braids in classical dynamics, *Phys. Rev. Lett.* **70**: 3675–3679 (1993).
[56] R. Broucke and D. Boggs, Periodic orbits in the Planar General Three-Body Problem, *Celest. Mech.* **11**: 13–38 (1975).
[57] J.D. Hadjidemetriou and T. Christides, Families of periodic orbits in the planar three-body problem, *Celest. Mech.* **12**: 175–187 (1975).
[58] J.D. Hadjidemetriou, The stability of periodic orbits in the three-body problem, *Celest. Mech.* **12**: 255–276 (1975).
[59] R. Broucke, On relative periodic solutions of the planar general three-body problem, *Celest. Mech.* **12**: 439–462 (1975).
[60] M. Hénon, A family of periodic solutions of the planar three-body problem, and their stability, *Celest. Mech.* **13**: 267–285 (1976).
[61] M. Hénon, Stability of interplay motions, *Celest. Mech.* **15**: 243–261 (1977).
[62] M. Nauenberg, Periodic orbits for three particles with finite angular momentum, *Phys. Lett. A.* **292**: 93–99 (2001).
[63] A. Chenciner, J. Fejoz and R. Montgomery, Rotating Eights: I. the three Γ_i families, *Nonlinearity.* **18**: 1407 (2005).
[64] R. Broucke, A. Elipe and A. Riaguas, On the figure-8 periodic solutions in the three-body problem, *Chaos, Solitons Fractals.* **30**: 513–520 (2006).
[65] M. Nauenberg, Continuity and stability of families of figure eight orbits with finite angular momentum, *Celest. Mech.* **97**: 1–15 (2007).
[66] M. Šuvakov and V. Dmitrašinović, Three classes of newtonian three-body planar periodic orbits, *Phys. Rev. Lett.* **110**: 114301 (2013).
[67] http://suki.ipb.ac.rs/3body/.
[68] G.G. Stokes, On the effect of the internal friction of fluids on the motion of pendulums, *Trans. Cambridge Philos. Soc.* **9**: 8 (1851).
[69] C.W. Oseen, Über die Stokessche Formel und die verwandte Aufgabe in der Hydrodynamik, *Arkiv. Mat. Astron. Phsik.* **6**: 29 (1910).
[70] S. Goldstein, The steady flow of viscous fluid past a fixed spherical obstacle at samll Reynolds numbers, *Proc. R. Soc. London, Ser. A.* **123**: 225 (1929).
[71] I. Proudman and J.R.A. Pearson, Expansions at small Reynolds numbers for the flow past a sphere and a circular cylinder, *J. Fluid Mech.* **2**: 237 (1957).
[72] W. Chester and D.R. Breach, On the flow past a sphere at low Reynolds number, *J. Fluid Mech.* **37**: 751–760 (1969).
[73] E.R. Lindgren, The motion of a sphere in an incompressible viscous fluid at Reynolds numbers considerably less than one, *Phys. Scr.* **60**: 97 (1999).
[74] S.J. Liao, An analytic approximation of the drag coefficient for the viscous flow past a sphere, *Int. J. Non-Linear Mech.* **37**: 1–18 (2002).
[75] J. Veysey and N. Goldenfeld, Simple viscous flow: From boundary layers to the renormalization group, *Rev. Mod. Phys.* **79**: 883 (2007).
[76] F.M. White, *Viscous Fluid Flow.* McGraw-Hill, New York, 1991.
[77] D.J. Korteweg and G. de Vries, On the change of form of long waves advancing in a rectangular canal, and on a new type of long stationary waves, *Phil. Mag.* **39**: 422–443 (1895)
[78] J. Boussinesq, Théorie des ondes et des remous qui se propagent le long d'un

canal rectangulaire horizontal, en communiquant au liquide contenu dans ce canal des vitesses sensiblement pareilles de la surface au fond, *Journal de Math'ematiques Pures et Appliquées.* **17**: 55–108 (1872).

[79] R. Camassa and D.D. Holm, An integrable shallow water equation with peaked solitons, *Phys. Rev. Lett.* **71**: 1661–1664 (1993).

[80] A. Constantin, Existence of permanent and breaking waves for a shallow water equation: a geometric approach, *Ann. Inst. Fourier, Grenoble.* **50**: 321–362 (2000).

[81] B. Fuchssteiner, Some tricks from the symmetry-toolbox for nonlinear equations: generalizations of the Camassa-Holm equation, *Physica D.* **95**: 296–343 (1996).

[82] S.J. Liao, Do peaked solitary water waves indeed exist? *ArXiv:1204.3354*, 2012.

[83] I. Didenkulova and E. Pelinovsky, Rogue waves in nonlinear hyperbolic systems (shallow-water framework), *Nonlinearity.* **24**: R1–R18 (2011).

[84] A. Calini and C. M. Schober, Dynamical criteria for rogue waves in nonlinear Schrödinger models, *Nonlinearity.* **25**: R99–R116 (2012).

Chapter 2

Predictor Homotopy Analysis Method (PHAM)

Saeid Abbasbandy* and Elyas Shivanian†

Department of Mathematics
Imam Khomeini International University
Ghazvin, Iran
**abbasbandy@yahoo.com*, †*shivanian@sci.ikiu.ac.ir*

This chapter introduces a method to predict the multiplicity of the solutions of nonlinear boundary value problems. This procedure can be easily applied on non-linear ordinary differential equations with boundary conditions so that it calculates effectively the all branches of the solutions (on the condition that, there exist such solutions for the problem) analytically at the same time. In this manner, for practical use in science and engineering, this method might give new unfamiliar class of solutions which is of fundamental interest.

Contents

2.1. Preliminaries . . . 36
2.2. Description of the method . . . 37
  2.2.1. Zeroth-order deformation equation . . . 38
  2.2.2. High-order deformation equation . . . 39
  2.2.3. Prediction of the multiple solutions . . . 40
2.3. Convergence analysis . . . 48
2.4. Some illustrative models . . . 54
  2.4.1. Nonlinear problem arising in heat transfer . . . 55
  2.4.2. Strongly nonlinear Bratu's equation . . . 60
  2.4.3. Nonlinear reaction-diffusion model . . . 66
  2.4.4. Mixed convection flows in a vertical channel . . . 71
2.5. Concluding remarks . . . 79
References . . . 81

2.1. Preliminaries

Many of the mathematical modeling of the physical phenomena in science and engineering often lead to nonlinear differential equations. There are a lot of methods, from the past up to now, to give numerically approximate solutions of nonlinear differential equations such as Euler method, Runge-Kutta method, multistep method, Taylor series method, Hybrid methods, family of finite difference methods [1, 2], family of finite element methods [3], meshless methods, differential quadrature, spectral methods [4–6] etc. There are also many of methods which give analytically approximate solutions like for example perturbation methods [7, 8], the artificial small parameter method [9], the δ-expansion method [10], and recently noteworthy techniques consists of the Adomian decomposition method [11], the variational iteration method [12] and so on. In this regard, one may ask the question: Do the approximate methods enable to predict multiplicity of solutions of the nonlinear differential equations? In the other words, can we forecast existence of multiple solutions of nonlinear equations by approximate methods and at the same time obtain all branches of solutions? Answer to this question is somehow difficult by concerning that the approximate methods usually converge to one solution by one initial guess that is exactly meaning of "convergence". Nevertheless, the present chapter is going to infract this convention by intellectual using homotopy analysis method.

The purpose of the present chapter is to introduce a method, probably for the first time, to predict the multiplicity of the solutions of nonlinear boundary value problems. This procedure can be easily applied on nonlinear ordinary differential equations with boundary conditions. This method, as will be seen, besides anticipating of multiplicity of the solutions of the nonlinear differential equations, calculates effectively all branches of the solutions (on the condition that, there exist such solutions for the problem) analytically at the same time. In this manner, for practical use in science and engineering, this method might give new unfamiliar class of solutions which is of fundamental interest and furthermore, the proposed approach convinces to apply it on nonlinear equations by today's powerful symbolic software programs so that it does not need tedious stages of evaluation and can be used without studying the whole theory. In fact, this technique has new point of view to homotopy analysis method. As it is well known in the frame of HAM, the convergence-controller parameter plays important role to guarantee the convergence of the solutions of nonlinear differential equations. It is shown that the convergence-controller parameter plays a fundamental role in the prediction of multiplicity of solutions and all branches of solutions are obtained simultaneously by one initial approximation guess, one auxiliary linear operator and one auxiliary function.

Notwithstanding both the perturbation and non-perturbation methods such as the artificial small parameter method, the δ-expansion method, the Adomian decomposition method and variational iteration method cannot provide us with a simple way to adjust and control the convergence region and rate of given approximate series, the HAM can guarantee the convergence of the series solutions by convergence-controller parameter $\hbar$. In fact, this trait makes HAM to be different from the others analytical techniques which are used to approach to just one solution and so that possibly to lose the other solutions. Therefore in this way, present procedure maybe generates new class of solutions for further physical interpretations in engineering and sciences.

The legitimacy and reliability of the method is checked by its application to four important nonlinear equations namely nonlinear heat transfer equation [13, 14], strongly nonlinear Bratu's equation [15, 16], nonlinear reaction-diffusion equation [17] and the model of mixed convection flows in a vertical channel [18, 19]. All these nonlinear problems admit multiple solutions for some values of parameters of the equations.

2.2. Description of the method

To illustrate the procedure consider the following nonlinear differential equation:

$$\mathcal{N}\left[u\left(r\right)\right]=0, \qquad r\in\Omega, \tag{2.1}$$

with boundary conditions

$$\mathcal{B}\left(u,\frac{\partial u}{\partial n}\right)=0, \quad r\in\Gamma, \tag{2.2}$$

where $\mathcal{N}$ is general nonlinear operator, $\mathcal{B}$ is a boundary operator, and Γ is the boundary of the domain Ω. The crucial step of the technique is that the boundary value problem (2.1) and (2.2) should be replaced by equivalent problem so that the conditions (2.2) involve an unknown parameter like δ (*prescribed parameter*) and are split to

$$\mathcal{B}'\left(u,\delta,\frac{\partial u}{\partial n}\right)=0, \quad r\in\Gamma \quad \text{and} \quad u\left(\alpha\right)=\beta, \tag{2.3}$$

where $u(\alpha) = \beta$ is the forcing condition that comes from original conditions (2.2). Now, homotopy analysis method is applied on the problem (2.1) with the conditions (2.3) except forcing condition as follows:

$$\mathcal{N}[u(r)] = 0, \ r \in \Omega, \tag{2.4}$$

$$\mathcal{B}'\left(u, \delta, \frac{\partial u}{\partial n}\right) = 0, r \in \Gamma. \tag{2.5}$$

2.2.1. *Zeroth-order deformation equation*

We suppose that all the solutions $u = u(r)$ of problem (2.4) can be expressed by the set of base functions $\{\omega_i(r), \ i = 0, 1, 2, \ldots\}$ in the form

$$u = u(r) \ = \sum_{n=0}^{+\infty} a_n \omega_n(r), \tag{2.6}$$

where a_n are coefficients to be determined.

Let $u_0(r, \delta)$ denote an initial approximation guess of the exact solution $u(r)$ which satisfies boundary conditions (2.5) automatically. Also, as that is well known in the frame of HAM, assume $\hbar \neq 0$ denote convergence-controller parameter, $H(r) \neq 0$ an auxiliary function, and $\mathcal{L}$ an auxiliary linear operator. Now using $p \in [0, 1]$ as an embedding parameter, we construct the general zeroth-order deformation equation and the corresponding boundary conditions as follow:

$$(1 - p)\mathcal{L}[\varphi(r, \delta; p) - u_0(r, \delta)] = p\hbar H(r)\mathcal{N}[\varphi(r, \delta; p)], \tag{2.7}$$

$$\mathcal{B}'\left(\varphi(r, \delta; p), \delta, \frac{\varphi(r, \delta; p)}{\partial n}\right) = 0, r \in \Gamma, \tag{2.8}$$

where $\varphi(r, \delta; p)$ is an unknown function to be determined. When $p = 0$, the zeroth-order deformation equation (2.7) becomes

$$\mathcal{L}[\varphi(r, \delta; 0) - u_0(r, \delta)] = 0, \tag{2.9}$$

which gives $\varphi(r, \delta; 0) = u_0(r, \delta)$. When $p = 1$, Eq. (2.7) leads to

$$\mathcal{N}[\varphi(r, \delta; 1)] = 0, \tag{2.10}$$

which is exactly the same as the original Eq. (2.1) provided that $\varphi(r, \delta; 1) = u(r, \delta)$.

We now expand the function $\varphi(r,\delta;p)$ in a Taylor series to the embedding parameter p. This Taylor expansion can be written in the form

$$\varphi(r,\delta;p) = u_0(r,\delta) + \sum_{m=1}^{+\infty} u_m(r,\delta)p^m, \tag{2.11}$$

where

$$u_m(r,\delta) = \frac{1}{m!}\frac{\partial^m \varphi(r,\delta;p)}{\partial p^m}, \qquad m = 0,1,2,\ldots,+\infty. \tag{2.12}$$

As it is well known in during the frame of HAM [20, 21], when the linear operator $\mathcal{L}$, the initial approximation $u_0(r,\delta)$, the auxiliary parameter $\hbar \neq 0$, and the auxiliary function $H(r) \neq 0$ are chosen properly, the series (2.11) converges for $p=1$, and thus

$$u(r,\delta) = u_0(r,\delta) + \sum_{m=1}^{+\infty} u_m(r,\delta) = \sum_{n=0}^{+\infty} a_n\omega_n(r), \tag{2.13}$$

will be the solution of the nonlinear problem (2.4) and (2.5) as will be proved later.

2.2.2. *High-order deformation equation*

Assume that the linear operator $\mathcal{L}$, the initial approximation $u_0(r,\delta)$, and the auxiliary function $H(r) \neq 0$ are chosen properly (it is worth mentioning here that $\hbar \neq 0$ so-called convergence-controller parameter will be determined later), the unknown functions $u_m(r,\delta)$ in Eq. (2.13) can be determined with the aid of the high-order deformation equations as follows. At first we define the vector $\vec{u}_n = \{u_0(r), u_1(r), \ldots, u_n(r)\}$ then, differentiating the zeroth-order deformation equation (2.7) m times with respect to the embedding parameter p, dividing it by $m!$, setting subsequently $p=0$ and taking into account the boundary conditions (2.8), one obtains the mth-order deformation equation

$$\mathcal{L}\left[u_m(r,\delta) - \chi_m u_{m-1}(r,\delta)\right] = \hbar H(r) R_m(\vec{u}_{m-1}, r, \delta), \tag{2.14}$$

subject to the boundary conditions

$$\frac{\partial^m}{\partial p^m}\mathcal{B}'\left(\varphi(r,\delta;p), \delta, \frac{\varphi(r,\delta;p)}{\partial n}\right)\Bigg|_{p=0} = 0,\ r \in \Gamma, \tag{2.15}$$

where

$$\chi_m = \begin{cases} 0, & m \leq 1, \\ 1, & m > 1, \end{cases} \tag{2.16}$$

and

$$R_m\left(\vec{u}_{m-1},r,\delta\right)=\frac{1}{(m-1)!}\left.\frac{\partial^{m-1}\mathcal{N}\left[\varphi\left(r,\delta;p\right)\right]}{\partial p^{m-1}}\right|_{p=0}$$
$$=\frac{1}{(m-1)!}\left.\frac{\partial^{m-1}\mathcal{N}\left[\sum_{n=0}^{n=+\infty}u_n(r,\delta)p^n\right]}{\partial p^{m-1}}\right|_{p=0}. \tag{2.17}$$

The high-order deformation equation (2.14) obviously, is just the ordinary differential equation with boundary condition (2.15) and, can be easily solved by using some symbolic software programs such as Mathematica or Maple. In this way, starting by $u_0(r,\delta)$, we obtain the functions $u_m(r,\delta)$ for $m=1,2,3,\ldots$ from Eqs. (2.14) and (2.15) successively. Accordingly, the Mth-order approximate solution of the problems (2.4) and (2.5) is given by

$$u\left(r,\delta\right)\approx U_M(r,\delta,\hbar)=u_0\left(r,\delta\right)+\sum_{m=1}^{M}u_m(r,\delta)=\sum_{n=0}^{M'}a_n\omega_n\left(r\right). \tag{2.18}$$

2.2.3. *Prediction of the multiple solutions*

It is noteworthy to indicate that up to this stage, the linear operator $\mathcal{L}$, the initial approximation $u_0(r,\delta)$, and the auxiliary function $H(r)\neq 0$ have been chosen properly so that the series solutions (2.18) would be convergence. However, there are still two unknown parameters in series (2.18) namely δ (prescribed parameter) and $\hbar$ (convergence-controller parameter) which should be determined. It is essential that existence of unique or multiple solutions in terms of the basic functions (2.6) for the original boundary value problem (2.1) depends on the fact whether the forcing condition (2.3) $(u\left(\alpha\right)=\beta)$, admits unique or multiple values for the formally introduced parameter δ in the boundary conditions (2.3). This stage is called *rule of multiplicity of solutions* that is a criterion in order to know how many solutions the boundary value problem (2.1) admits. The so-called *rule of multiplicity of solutions* is applied as follows:
Consider the Mth-order approximate solution (2.18) and set in the forcing condition (2.3) $(u\left(\alpha\right)=\beta)$, so the following equation is derived

$$u(\alpha)\approx U_m\left(\alpha,\delta,\hbar\right)=\beta. \tag{2.19}$$

The above equation has two unknown parameters namely δ, and $\hbar$ which controls the convergence of the HAM series (2.18). It is a basic feature of HAM that the series solution (2.18) converges at $r=\alpha$ only in that range of $\hbar$, where the parameter δ does not change with the variation of $\hbar$. This means that in the plot of δ as function of $\hbar$ according to Eq. (2.19) in an implicitly way, in the convergence range of the series $u(\alpha)$ a plateau occurs. The number of such horizontal plateaus where $\delta(\hbar)$ becomes constant, gives the multiplicity of the solutions of problems (2.1) and (2.2), (we call this point as *rule of multiplicity of solutions*).

In the following, we give a proof to the above assertion. We notice that there is a direct connection between the existences of multiple solutions and the number of horizontal plateaus in the plot of $\delta(\hbar)$, so at first, let us discuss about the fundamental of $\hbar$-curve and its relation to the Taylor's series. The following theorem is basic and simple to understand.

Theorem 2.1. *If $f(x)$ is continuous on $[a, b]$ and is differentiable in (a, b) and $f(a) = f(b) = 0$, then there exists at least one point $\xi \in (a, b)$ such that $f'(\xi) = 0$.*

Proof. See Ref. [22]. □

Now, we discuss about occurrence of $\hbar$-curve for the arbitrary but smooth function $f(x)$.

Theorem 2.2. *Suppose that $g(\hbar)$ be a continuous function onto interval $[a, b]$ and all derivatives of $f : [a, b] \to \mathbb{R}$ exist and have a common $\mathbb{M}$ so that*

$$\max_{x\in[a,b]} |f^{(k)}(x)| \leq \mathbb{M}, \qquad \textit{for all } k. \tag{2.20}$$

Furthermore, assume that $G_n(x,\alpha)$ be the Taylor polynomial of degree n for $f(x)$ about some $\alpha \in (a, b)$, say $\alpha = g(\hbar)$, then $\forall \varepsilon > 0$ and $\gamma \in (a, b)$ there exists $N \in \mathbb{N}$ and interval (c, d) so that

$$\forall \hbar \in (c, d) \textit{ and } \; n \geq N : |f(\gamma) - G_n(\gamma, g(\hbar))| < \varepsilon. \tag{2.21}$$

Proof. Let $\beta \in [a, b]$ be the point at which we want to determine the error. We suppose (without loss of generality) that $\beta > \alpha$. Let

$$s(x) = f(\beta) - f(x) - \sum_{k=1}^{n} \frac{(\beta - x)^k}{k!} f^{(k)}(x), \tag{2.22}$$

then $s'(x)$ exists for $x \in (a, b)$ and

$$s'(x) = -\frac{(\beta - x)^n}{n!} f^{(n+1)}(x). \tag{2.23}$$

Now, consider the function

$$U(x) = s(x) - \left(\frac{\beta - x}{\beta - \alpha}\right)^{n+1} s(\alpha), \tag{2.24}$$

then

$$U(\alpha) = U(\beta) = 0. \tag{2.25}$$

From the differentiability of $s(x)$ and $\left(\frac{\beta - x}{\beta - \alpha}\right)^{n+1}$, it follows that $U(x)$ is differentiable on any subinterval of (a, b). Now, we apply Theorem 2.1 to $U(x)$ on the interval $[\alpha, \beta]$. Therefore there exists $\xi_\beta \in (\alpha, \beta)$ so that

$$U^{'}(\xi_\beta) = 0, \tag{2.26}$$

which implies

$$-\frac{(\beta - \xi_\beta)^n}{n!} f^{(n+1)}(\xi_\beta) + (n+1)\frac{(\beta - \xi_\beta)^n}{(\beta - \alpha)^{n+1}} s(\alpha) = 0, \tag{2.27}$$

then, since $\beta \neq \xi_\beta$,

$$s(\alpha) = \frac{(\beta - \alpha)^{n+1}}{(n+1)!} f^{(n+1)}(\xi_\beta). \tag{2.28}$$

Now, from Eq. (2.22), we have

$$s(\alpha) = f(\beta) - G_n(\beta, \alpha) = \frac{(\beta - \alpha)^{n+1}}{(n+1)!} f^{(n+1)}(\xi_\beta). \tag{2.29}$$

Since the parameter β has been chosen arbitrary then

$$\forall x \in [\alpha, b], \;\; \alpha \in [a, b]: \;\; f(x) - G_n(x, \alpha) = \frac{(x - \alpha)^{n+1}}{(n+1)!} f^{(n+1)}(\xi_x), \tag{2.30}$$

where $\xi_x \in (\alpha, x)$. Suppose that $\gamma \in (a, b)$ and $\varepsilon > 0$, let $\alpha \in (a, \gamma)$ then it is clear that there exists N such that

$$\forall n \geq N: \;\; \frac{(\gamma - \alpha)^{n+1}}{(n+1)!} < \frac{\varepsilon}{\mathbb{M}}, \tag{2.31}$$

then from Eqs. (2.20) and (2.30),

$$|f(\gamma) - G_n(\gamma, \alpha)| = \frac{(\gamma - \alpha)^{n+1}}{(n+1)!} |f^{(n+1)}(\xi_\gamma)| < \frac{\varepsilon}{\mathbb{M}} \cdot \mathbb{M} = \varepsilon. \tag{2.32}$$

Therefore, we have proved

$$\forall \gamma \in (a, b), \;\; \alpha \in (a, \gamma), \; \varepsilon > 0, \;\; \exists N \;\; \Rightarrow \;\; \forall n \geq N: |f(\gamma) - G_n(\gamma, \alpha)| < \varepsilon.$$

Since $g(\hbar)$ is continuous function onto interval $[a,b]$ then there exists interval (c,d) such that $g\{(c,d)\}=(a,\gamma)$. Hence the statement (2.32) is read equivalently as

$$\forall\gamma\in(a,b),\ \ \hbar\in(c,d),\ \ \varepsilon>0,\ \exists N\ \Rightarrow\ \forall n\geq N:|f(\gamma)-G_n(\gamma,g(\hbar))|<\varepsilon,$$

and the proof is completed. □

Corollary 2.1. *Suppose that $f(x)$ is sufficiently smooth on interval $[a,b]$, $g(x)$ be a continuous function onto interval $[a,b]$ and $G_n(x,g(\hbar))$ be the Taylor polynomial of degree n about $g(\hbar)$. Then, as n goes infinity, in the plot of $G_n(\gamma,g(\hbar))$, $a<\gamma<b$ versus $\hbar$ horizontal line occurs.*

Proof. The proof is straightforward from Theorem 2.2. □

Example 2.1. Consider $f(x)=exp(x)$, $g_1(\hbar)=\hbar^2$ and $g_2(\hbar)=\sinh(\hbar)$, the function $f(x)$ in $[a,b]$, $a,b\in\mathbb{R}$ with g_1 and g_2 satisfy all conditions in Theorem 2.2. For example, Taylor polynomials of degree three for $f(x)$ about $g_1(\hbar)$ and $g_2(\hbar)$ are given as follows

$$G_3(\gamma,g_1(\hbar))=\frac{1}{6}e^{\hbar^2}\left(\gamma-\hbar^2\right)^3+\frac{1}{2}e^{\hbar^2}\left(\gamma-\hbar^2\right)^2+e^{\hbar^2}\left(\gamma-\hbar^2\right)+e^{\hbar^2},\tag{2.33}$$

$$\begin{aligned}G_3(\gamma,g_2(\hbar))&=\frac{1}{6}e^{\sinh(\hbar)}(\gamma-\sinh(\hbar))^3+\frac{1}{2}e^{\sinh(\hbar)}(\gamma-\sinh(\hbar))^2\\&\quad+e^{\sinh(\hbar)}\left(\gamma-\sinh\left(\hbar\right)\right)+e^{\sinh(\hbar)}.\end{aligned}\tag{2.34}$$

We have shown $\hbar$-curve of $G_{10}(\gamma,g_1(\hbar))$ and $G_{10}(\gamma,g_2(\hbar))$ at $\gamma=-2,-1,$ $0,\frac{1}{4},\frac{1}{2},\frac{3}{4},1,\frac{5}{4}$ and $\frac{3}{2}$ in Figs. 2.1 and 2.2, respectively. As we see, it can be found horizontal line for each case.

Example 2.2. Let $f(x)=\cos(x)$, $g_1(\hbar)=\hbar^2$ and $g_2(\hbar)=1-\hbar$, the function $f(x)$ in $[a,b]$, $a,b\in\mathbb{R}$ with g_1 and g_2 satisfy all conditions in Theorem 2.2. They are obtained, for instance,

$$\begin{aligned}G_2(\gamma,g_1(\hbar))&=-\left(\gamma-\hbar^2\right)\sin\left(\hbar^2\right)-\frac{1}{2}\left(\gamma-\hbar^2\right)^2\cos\left(\hbar^2\right)\\&\quad+\cos\left(\hbar^2\right),\end{aligned}\tag{2.35}$$

$$\begin{aligned}G_2(\gamma,g_2(\hbar))&=-(\hbar+\gamma-1)\sin(1-\hbar)-\frac{1}{2}(\hbar+\gamma-1)^2\cos\left(1-\hbar\right)\\&\quad+\cos\left(1-\hbar\right).\end{aligned}\tag{2.36}$$

We have plotted $\hbar$-curve of $G_{25}(\gamma,g_1(\hbar))$ and $G_{20}(\gamma,g_2(\hbar))$ at $\gamma=0,1,2,3,$ $4,5,6,7$ and 8 in Figs. 2.3 and 2.4, respectively. As we see, it could be found horizontal line for each case.

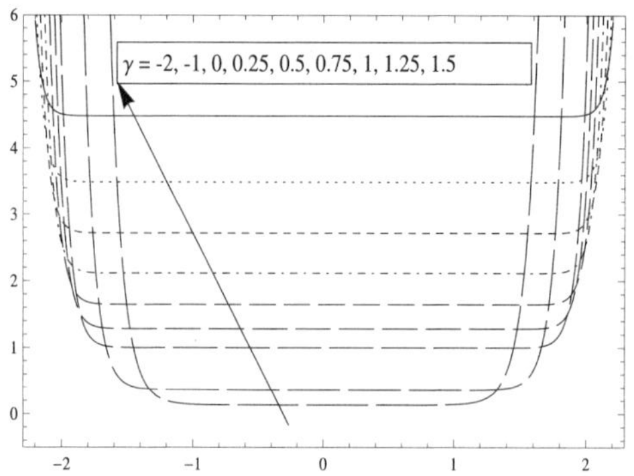

Fig. 2.1. Plot of $G_{10}(\gamma, g_1(\hbar))$ via $\hbar$ in Example 2.1.

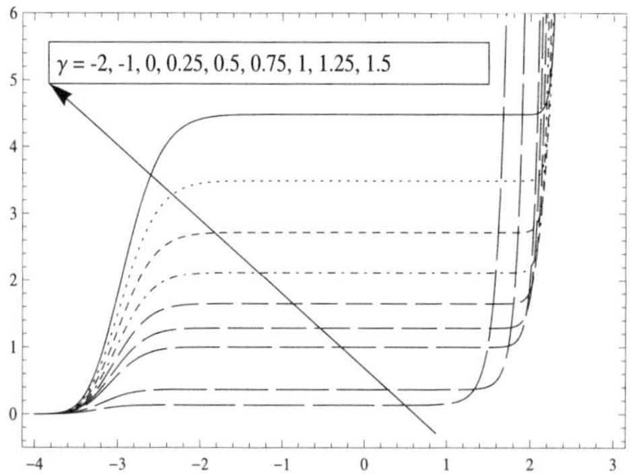

Fig. 2.2. Plot of $G_{10}(\gamma, g_2(\hbar))$ via $\hbar$ in Example 2.1.

The condition (2.20) is very severe in Theorem 2.2, for example, the Taylor series of the function $f(x) = \frac{1}{1+x}$ about zero has radius of convergence $R = 1$ and moreover the derivatives of $f(x)$ are unbounded but, as we will see in this section, Corollary 2.1 still holds for $\gamma \neq -1$. The following theorem describes that Theorem 2.2 still holds by a weakly conditions.

Theorem 2.3. *Suppose that $g(\hbar)$ be a continuous function onto interval $[a', b']$ and $G_n(x, \alpha) = \sum_{k=0}^{n} a_k(\alpha)(x-\alpha)^k$ be the Taylor polynomial of degree n for $f(x)$ about some $\alpha \in (a', b')$, say $\alpha = g(\hbar)$. Moreover, assume*

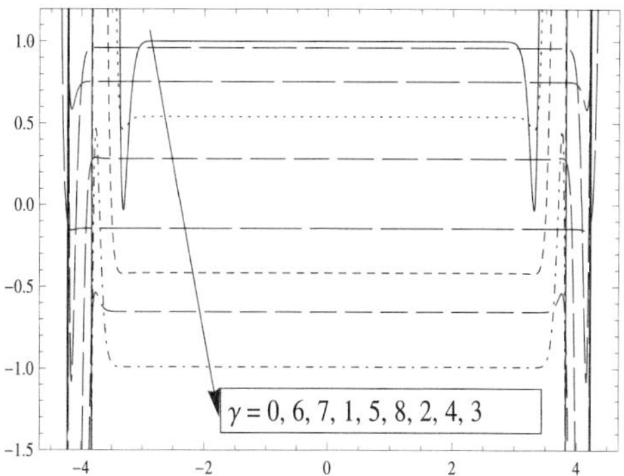

Fig. 2.3. Plot of $G_{25}(\gamma, g_1(\hbar))$ via $\hbar$ in Example 2.2.

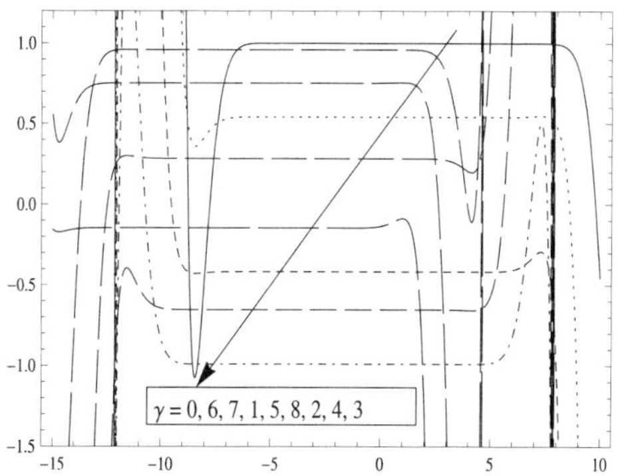

Fig. 2.4. Plot of $G_{25}(\gamma, g_2(\hbar))$ via $\hbar$ in Example 2.2.

that

$$\forall k \in \mathbb{N},\ \alpha \in [a', b'],\ x \in [a, b] : |x-\alpha| \leq |\frac{a_k(\alpha)}{a_{k+1}(\alpha)}|. \qquad (2.37)$$

Then for $\varepsilon > 0$ *and* $\gamma \in [a, b]$*, there exists* $\mathbf{N} \in \mathbb{N}$ *and interval* (c, d) *so that*

$$\forall \hbar \in (c, d) \text{ and } n \geq \mathbf{N}\text{: } |f(\gamma) - G_n(\gamma, g(\hbar))| < \varepsilon. \qquad (2.38)$$

Proof. From the statement (2.37), it follows that there exists $0 < \theta < 1$ so that

$$\forall k \in \mathbb{N},\ \alpha \in [a', b'],\ x \in [a, b] : |a_{k+1}(\alpha)||x-\alpha| \leq \theta |a_k(\alpha)|. \qquad (2.39)$$

To prove the sentence (2.38), we show that, for a fixed $\gamma \in [a,b]$, $G_n(\gamma,\alpha)$, $\alpha \in (a^{'},b^{'})$ is convergence sequence. We show that $\{G_n(\gamma,\alpha)\}_{n=0}^{\infty}$ is Cauchy sequence in the Hilbert space $\mathbb{R}$. From (2.39), it is obtained

$$\| G_{n+1}(\gamma,\alpha)-G_n(\gamma,\alpha) \| = |a_{n+1}(\alpha)| \; \| (\gamma-\alpha)\|^{n+1} \leq \theta|a_n(\alpha)| \; \| (\gamma-\alpha)\|^{n}$$

$$\leq \theta^2|a_{n-1}(\alpha)| \; \| (\gamma-\alpha)\|^{n-1} \leq \cdots \leq \theta^{n-p+1}|a_p(\alpha)| \; \| (\gamma-\alpha)\|^{p}, \qquad (2.40)$$

now, for every $n,m \in \mathbb{N}$, $n \geq m > p$, we have

$$\begin{aligned} &\| G_n(\gamma,\alpha)-G_m(\gamma,\alpha) \| = \| (G_n(\gamma,\alpha)-G_{n-1}(\gamma,\alpha)) + (G_{n-1}(\gamma,\alpha) \\ &- G_{n-2}(\gamma,\alpha)) + \cdots + (G_{m+1}(\gamma,\alpha)-G_m(\gamma,\alpha)) \| \\ &\leq \| (G_n(\gamma,\alpha)-G_{n-1}(\gamma,\alpha)) \| + \| (G_{n-1}(\gamma,\alpha)-G_{n-2}(\gamma,\alpha)) \| \\ &+ \cdots + \| G_{m+1}(\gamma,\alpha)-G_m(\gamma,\alpha) \| \leq \theta^{n-p}|a_p(\alpha)| \; \| (\gamma-\alpha)\|^{p} \\ &+ \theta^{n-p-1}|a_p(\alpha)| \; \| (\gamma-\alpha)\|^{p} + \cdots + \theta^{m-p+1}|a_p(\alpha)| \; \| (\gamma-\alpha)\|^{p} \\ &= \frac{1-\theta^{n-m}}{1-\theta}\theta^{m-p+1}|a_p(\alpha)| \; \| (\gamma-\alpha)\|^{p}. \end{aligned} \qquad (2.41)$$

Therefore, we arrive to

$$\lim_{m,n\to\infty} \| G_n(\gamma,\alpha)-G_m(\gamma,\alpha) \| = 0, \qquad (2.42)$$

then it follows that $\{G_n(\gamma,\alpha)\}_{n=0}^{\infty}$ converges. On the other hand $G_n(\gamma,\alpha)$ is the Taylor polynomial of $f(x)$ at $x=\gamma$. Hence for each $\varepsilon > 0$, there exists $\mathbf{N} \in \mathbb{N}$ so that

$$\forall n \geq \mathbf{N}\text{: } |f(\gamma)-G_n(\gamma,\alpha)| < \varepsilon. \qquad (2.43)$$

Now by this fact that α is chosen arbitrarily from $(a^{'},b^{'})$, we can equivalently say that for each $\varepsilon > 0$ there exists $\mathbf{N} \in \mathbb{N}$ and interval (c,d) so that

$$\forall \hbar \in (c,d) \text{ and } n \geq \mathbf{N}\text{: } |f(\gamma)-G_n(\gamma,g(\hbar))| < \varepsilon, \qquad (2.44)$$

and the proof is completed. □

Example 2.3. Consider $f(x) = \frac{1}{1+x}$ and $g(\hbar) = -1-\frac{1}{\hbar}$ ($g(\hbar)$ can be chosen other piecewise-continuous functions onto $\mathbb{R}$ except some points, we have chosen this function due to conforming aforesaid references. Let $[a,b] = [-6,-2]$ and $[a^{'},b^{'}] = [-8,-7]$ in Theorem 2.3, since $a_k(\alpha) = \frac{(-1)^k}{(1+\alpha)^{k+1}}$ for all $k \in \mathbb{N}$ then

$$\left|\frac{a_k(\alpha)}{a_{k+1}(\alpha)}\right| = \left|1+\alpha\right|. \qquad (2.45)$$

It is easy to see that

$$\forall\alpha \in [-8,-7] \ \text{ and } \ x \in [-6,-2] : |x-\alpha| \le |1+\alpha| = \left|\frac{a_k(\alpha)}{a_{k+1}(\alpha)}\right|, \tag{2.46}$$

then conditions of Theorem 2.3 hold. The Taylor polynomial of degree three for $f(x)$ about $g(\hbar)$ is given as follows

$$G_3(\gamma, g(\hbar)) = \hbar^4\left(-\left(\frac{1}{\hbar}+\gamma+1\right)^3\right) - \hbar^3\left(\frac{1}{\hbar}+\gamma+1\right)^2 - \hbar^2\left(\frac{1}{\hbar}+\gamma+1\right) - \hbar. \tag{2.47}$$

We have shown $\hbar$-curve of $G_{20}(\gamma, g(\hbar))$ at $\gamma = -6,-5,-4,-3,$ and -2 in Fig. 2.5. As it is seen, it can be found horizontal line for each case. In fact, if we let $a' < b' \le a < b$ in the interval $(-\infty,-1)$ and $a < b \le a' < b'$ in the interval $(-1,+\infty)$ then all conditions of Theorem 2.3 will hold for $\gamma \in \mathbb{R}$, $(\gamma \ne -1)$. Therefore we have horizontal line in the plot of $\hbar$-curve for each $\gamma \ne -1$.

Example 2.4. Consider the differential equation

$$f'(x)+f^2(x) = 1,\ f(0) = 0, \tag{2.48}$$

with the exact solution $f(x) = \tanh(x)$. Let $g(\hbar) = \frac{\pi}{2}\sqrt{-\frac{1}{\hbar(2+\hbar)}-1}$ and $[a,b] = [a',b'] = [0,1]$ in Theorem 2.3. It could be seen that all conditions of Theorem 2.3 hold. The Taylor polynomial of degree two for $f(x)$ about $g(\hbar)$ is given as follows

$$\begin{aligned} G_3(\gamma, g(\hbar)) &= \left(\gamma-\frac{1}{2}\pi\sqrt{-\frac{1}{\hbar(\hbar+2)}-1}\right)^2 \\ &\left(\tanh^3\left(\frac{1}{2}\pi\sqrt{-\frac{1}{\hbar(\hbar+2)}-1}\right) - \tanh\left(\frac{1}{2}\pi\sqrt{-\frac{1}{\hbar(\hbar+2)}-1}\right)\right) \\ &+ \left(\gamma-\frac{1}{2}\pi\sqrt{-\frac{1}{\hbar(\hbar+2)}-1}\right)\left(1-\tanh^2\left(\frac{1}{2}\pi\sqrt{-\frac{1}{\hbar(\hbar+2)}-1}\right)\right) \\ &+ \tanh\left(\frac{1}{2}\pi\sqrt{-\frac{1}{\hbar(\hbar+2)}-1}\right). \end{aligned} \tag{2.49}$$

If we get base functions as polynomials in the frame of HAM, we will obtain in fact the above generalized Taylor series as solution. We have plotted $\hbar$-curve of $G_{10}(\gamma, g(\hbar))$ at $\gamma = 0, \frac{1}{3}$ and $\frac{1}{2}$ in Fig. 2.6.

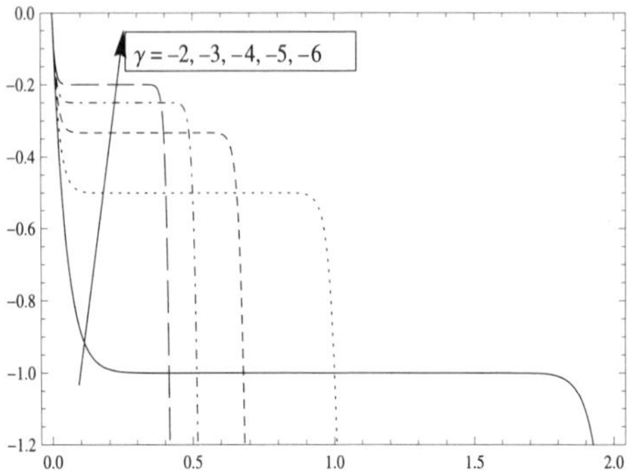

Fig. 2.5. Plot of $G_{20}(\gamma, g(\hbar))$ via $\hbar$ in Example 2.3.

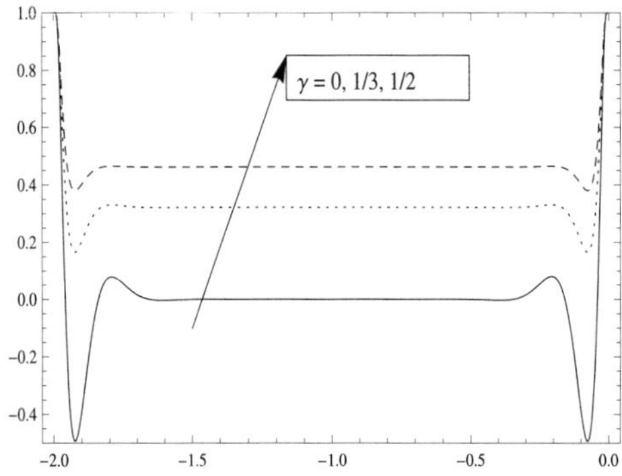

Fig. 2.6. Plot of $G_{10}(\gamma, g(\hbar))$ via $\hbar$ in Example 2.4.

2.3. Convergence analysis

Theorem 2.4. *Let* $0 < \gamma < 1$ *and the solution components* $u_0(r,\delta)$, $u_1(r,\delta)$, $u_2(r,\delta)$, ... *obtained by* (2.14) *satisfy the following condition*

$$\exists k_0 \in N, \ \ \forall k \geq k_0: \qquad \|u_{k+1}(r,\delta)\| \leq \gamma \|u_k(r,\delta)\|, \tag{2.50}$$

then the series solution $\sum_{k=0}^{+\infty} u_k(r,\delta)$ *is convergent.*

Proof. Define the sequence $\{S_n\}_{n=0}^{+\infty}$ as,

$$\begin{cases} S_0 = u_0(r,\delta) \\ S_1 = u_0(r,\delta) + u_1(r,\delta) \\ S_2 = u_0(r,\delta) + u_1(r,\delta) + u_2(r,\delta) \\ \vdots \\ S_n = u_0(r,\delta) + u_1(r,\delta) + \cdots + u_n(r,\delta) \end{cases} \tag{2.51}$$

and we show that $\{S_n\}_{n=0}^{+\infty}$ is a Cauchy sequence in the Hilbert space $\mathbb{R}$. For this purpose, consider,

$$\|S_{n+1} - S_n\| = \|u_{n+1}(r,\delta)\| \leq \gamma \|u_n(r,\delta)\| \leq$$

$$\gamma^2 \|u_{n-1}(r,\delta)\| \leq \ldots \leq \gamma^{n-k_0+1} \|u_{k_0}(r,\delta)\|. \tag{2.52}$$

For every $n, m \geq N$, $n \geq m > k_0$, we have

$$\begin{aligned} \|S_n - S_m\| &\leq \|(S_n - S_{n-1}) + (S_{n-1} - S_{n-2}) + \cdots + (S_{m+1} - S_m)\| \\ &\leq \|(S_n - S_{n-1})\| + \|(S_{n-1} - S_{n-2})\| + \cdots + \|(S_{m+1} - S_m)\| \\ &\leq \gamma^{n-k_0} \|u_{k_0}(r,\delta)\| + \gamma^{n-k_0-1} \|u_{k_0}(r,\delta)\| + \ldots \\ &\quad + \gamma^{m-k_0+1} \|u_{k_0}(r,\delta)\| \\ &= \frac{1-\gamma^{n-m}}{1-\gamma} \gamma^{m-k_0+1} \|u_{k_0}(r,\delta)\|, \end{aligned} \tag{2.53}$$

and since $0 < \gamma < 1$, we get,

$$\lim_{n,m \to \infty} \|S_n - S_m\| = 0. \tag{2.54}$$

Therefore, $\{S_n\}_{n=0}^{+\infty}$ is a Cauchy sequence in the Hilbert space $\mathbb{R}$ and it implies that the series solution defined in (2.18), converges. This completes the proof of Theorem 2.4. □

Theorem 2.5. *Assume that the series solution* $\sum_{k=0}^{+\infty} u_k(r,\delta)$ *defined in* (2.18), *is convergent to the solution* $u(r)$. *If the truncated series* $U_M(r,\delta,\hbar) = \sum_{m=0}^{M} u_m(r,\delta)$ *is used as an approximation to the solution* $u(r)$ *of the problem* (2.14), *then the maximum absolute truncated error is estimated as,*

$$\|u(r) - U_M(r,\delta,\hbar)\| \leq \frac{1}{1-\gamma} \gamma^{M-k_0+1} \|u_{k_0}(r,\delta)\|. \tag{2.55}$$

Proof. From Theorem 2.4, following inequality (2.53), we have

$$\|S_n - S_M\| \leq \frac{1-\gamma^{n-M}}{1-\gamma}\gamma^{M-k_0+1}\,\|u_{k_0}(r,\delta)\|\,, \tag{2.56}$$

for $n \geq M$. Now, as $n \to \infty$ then $S_n \to u(r)$ and $\gamma^{n-M} \to 0$. So,

$$\|u(r) - U_M(r,\delta,\hbar)\| \leq \frac{1}{1-\gamma}\gamma^{M-k_0+1}\,\|u_{k_0}(r,\delta)\|\,. \tag{2.57}$$

□

Theorems 2.3 and 2.4 together confirm that the convergence of series solution (2.18) leads to occurrence of horizontal plateaus in $\hbar$-curve, in which they give valid region for the convergence controller parameter $\hbar$, where $U_M(r,\delta,\hbar)$ converges. Now, we discuss about uniqueness of the solution of initial value problem (2.4)–(2.5) which plays fundamental rule in the existence of multiple solutions of the boundary value problem (2.1)–(2.2), for that we bring the below theorem.

Theorem 2.6. *Let f be a complex-valued continuous function defined on*

$$R : |x - x_0| \leq a,\ \ \|\mathbf{y} - \mathbf{y_0}\| \leq b,\ \ (a,\ b > 0), \tag{2.58}$$

such that $|f(x, \mathbf{y})| \leq N$, for all $(x,\ \mathbf{y})$ in R. Suppose there exists a constant $L > 0$ such that

$$|f(x, \mathbf{y}) - f(x, \mathbf{z})| \leq L\,\|\mathbf{y} - \mathbf{z}\|\,, \tag{2.59}$$

for all $(x,\ \mathbf{y})$ and $(x,\ \mathbf{z})$ in R. Then there exists one, and only one, solution φ of

$$y^{(n)} = f\left(x,\ y,\ y', \ldots, y^{(n-1)}\right), \tag{2.60}$$

on the interval

$$I : |x\ x_0| \leq \min\left\{a, \frac{b}{M}\right\}, \qquad (M = N + b + \|\mathbf{y_0}\|), \tag{2.61}$$

which satisfies

$$\varphi(x_0) = \alpha_1,\ \varphi'(x_0) = \alpha_2,\ \ldots, \varphi^{(n-1)}(x_0) = \alpha_n, \tag{2.62}$$

with $\mathbf{y_0} = (\alpha_1,\ \ldots, \alpha_n)$.

Proof. Please see Ref. [23]. □

Now, assuming the initial value problems (2.4)–(2.5) is in the form (2.60) and (2.61), we conclude that there exists one and only one solution to (2.4)–(2.5) for each value of the prescribed parameter δ.

Theorem 2.7. *Consider the boundary value problems* (2.1)–(2.2) *and suppose that the conditions of Theorem 2.6 hold for the initial value problem* (2.4)–(2.5) *and more, the series*

$$\sum_{m=0}^{+\infty} u_m(r,\delta), \quad \sum_{m=0}^{+\infty} \mathcal{L}[u_m(r,\delta)], \tag{2.63}$$

converge. If the number of K horizontal plateaus occur in the plane of $(\hbar,\delta)$ *where Eq.* (2.19) *is plotted implicitly, then the problems* (2.1)–(2.2) *admit the number of K multiple solutions in terms of the basis functions* (2.6).

Proof. Suppose that the number of horizontal plateaus occurring in the plane $(\hbar,\delta)$ is K namely

$$\delta_1(\hbar_1), \ \delta_2(\hbar_2), \ \ldots, \ \delta_K(\hbar_K), \tag{2.64}$$

where $(\hbar_j, \delta_j(\hbar_j))$, $j = 1, 2, \ldots, K$ are proper ordered pair which chosen from the plane $(\hbar,\delta)$. Since the conditions of Theorem 2.4 hold and by uniqueness of the Taylor's series we conclude all the series $\sum_{m=0}^{+\infty} u_m(r, \delta_j(\hbar_j))$, $j = 1, 2, \ldots, K$ converge. Suppose

$$s_1(r) = \sum_{m=0}^{+\infty} u_m(r, \delta_1(\hbar_1)),$$

$$s_2(r) = \sum_{m=0}^{+\infty} u_m(r, \delta_2(\hbar_2)),$$

$$\ldots \tag{2.65}$$

$$s_K(r) = \sum_{m=0}^{+\infty} u_m(r, \delta_K(\hbar_K)).$$

Now, we show all the above series are the solutions of the problem (2.1)–(2.2). To show this, it is sufficient to prove they are the solutions of problems (2.4)–(2.5) because the condition $u(\alpha) = \beta$ is automatically satisfied. Let us consider

$$s_j(r) = \sum_{m=0}^{+\infty} u_m(r, \delta_j(\hbar_j)), \tag{2.66}$$

by high-order deformation equations (2.14)–(2.16), we get

$$\begin{aligned}
&\hbar_j H(r)\sum_{m=1}^{+\infty} R_m\left(\vec{u}_{m-1},\ r,\delta_j(\hbar_j)\right)\\
&=\sum_{m=1}^{+\infty}\mathcal{L}\left[u_m(r,\delta_j(\hbar_j))-\chi_m u_{m-1}(r,\delta_j(\hbar_j))\right]\\
&=\mathcal{L}\left\{\sum_{m=1}^{+\infty}\left[u_m(r,\delta_j(\hbar_j))-\chi_m u_{m-1}(r,\delta_j(\hbar_j))\right]\right\}\\
&=\mathcal{L}\{u_1(r,\delta_j(\hbar_j))\}+\mathcal{L}\{u_2(r,\delta_j(\hbar_j))-u_1(r,\delta_j(\hbar_j))\}\\
&\quad+\mathcal{L}\{u_3(r,\delta_j(\hbar_j))-u_2(r,\delta_j(\hbar_j))\}+\ldots\\
&\quad+\mathcal{L}\{u_n(r,\delta_j(\hbar_j))-u_{n-1}(r,\delta_j(\hbar_j))\}+\ldots\\
&=\mathcal{L}\left\{\lim_{n\to\infty}u_n(r,\delta_j(\hbar_j))\right\}=\mathcal{L}\{0\}=0,
\end{aligned}\tag{2.67}$$

Notice that the linear operator $\mathcal{L}$ is Lipschitz in the above equations from the conditions of Theorem 2.6. Now, since $\hbar_j \neq 0$ and $H(r) \neq 0$ then

$$\sum_{m=1}^{+\infty} R_m\left(\vec{u}_{m-1},\ r,\delta_j(\hbar_j)\right)=0,\tag{2.68}$$

thus,

$$\begin{aligned}
&\sum_{m=0}^{+\infty}\frac{1}{m!}\frac{\partial^m\mathcal{N}[\Phi(r,\delta_j(\hbar_j);q)]}{\partial q^m}\bigg|_{q=0}\\
&=\sum_{m=1}^{+\infty}\frac{1}{(m-1)!}\frac{\partial^{m-1}\mathcal{N}[\Phi(r,\delta_j(\hbar_j);q)]}{\partial q^{m-1}}\bigg|_{q=0}=0.
\end{aligned}$$

Consider the residual of the original equation (2.1) as

$$\mathcal{R}(r,\delta_j(\hbar_j);q)=\mathcal{N}\left[\Phi(r,\delta_j(\hbar_j);q)\right].\tag{2.69}$$

Now, what it remains is to show $\mathcal{R}(r,\delta_j(\hbar_j);1)=0$, because then

$$\mathcal{N}\left[\Phi(r,\delta_j(\hbar_j);1)\right]=0,\tag{2.70}$$

and, on the other side we have $\Phi(r,\delta_j(\hbar_j);1)=s_j(r)$. The Taylor's series of $\mathcal{R}(r,\delta_j(\hbar_j);q)$ respect to q is

$$\mathcal{R}(r,\delta_j(\hbar_j);q)=\sum_{m=0}^{+\infty}\frac{1}{m!}\frac{\partial^m\mathcal{R}(r,\delta_j(\hbar_j);q)}{\partial q^m}\bigg|_{q=0}q^m,\tag{2.71}$$

and finally the above equation gives

$$\mathcal{R}(r, \delta_j(\hbar_j); 1) = \sum_{m=0}^{+\infty} \frac{1}{m!} \frac{\partial^m \mathcal{R}(r, \delta_j(\hbar_j); q)}{\partial q^m}\bigg|_{q=0}$$
$$= \sum_{m=0}^{+\infty} \frac{1}{m!} \frac{\partial^m \mathcal{N}[\Phi(r, \delta_j(\hbar_j); q)]}{\partial q^m}\bigg|_{q=0} = 0, \quad (2.72)$$

then the proof is completed. □

Example 2.5. Consider a one-dimensional semi-linear problem

$$u'' + u^3 = 0, x \in [0, \pi], u(0) = u(\pi) = 0. \quad (2.73)$$

Supposing $u'(0) = \delta$, where δ is a prescribed parameter, we can have the following initial value problem

$$u'' + u^3 = 0, u(0) = 0, \quad u'(0) = \delta, \quad (2.74)$$

with $u(\pi) = 0$, as forcing condition, attached to the above problem. Let us get the set of base functions as

$$\{x^n | n = 0, 1, 2, \dots\}. \quad (2.75)$$

Under the rule of solution expression and according to the initial conditions, it is easy to choose $u_0(x) = \delta x$ as initial guess of solution $u(x)$, $H(x) = 1$ as auxiliary function, and to choose auxiliary linear operator

$$\mathcal{L}[\varphi(x, \delta; p)] = \frac{\partial^2 \varphi(x, \delta; p)}{\partial x^2}, \quad (2.76)$$

with the property

$$\mathcal{L}[c_1 + c_2 x] = 0. \quad (2.77)$$

Thus, after two subsequent integrations, the Mth-order deformation equation yields for $m \geq 1$

$$u_m(x, \delta) = \chi_m u_{m-1}(x, \delta) + \hbar \int_0^x \int_0^s R_m(\overrightarrow{u}_{m-1}, \tau, \delta) d\tau ds$$
$$+ c_1 + c_2 x, \quad (2.78)$$

where from (2.74)

$$R_m(\overrightarrow{u}_{m-1}, \tau, \delta) = u''_{m-1}(x, \delta)$$
$$+ \sum_{j=0}^{m-1} u_{m-1-j}(x, \delta) \sum_{i=0}^{j} u_i(x, \delta) u_{j-i}(x, \delta), \quad (2.79)$$

and integration constants c_1 and c_2 are obtained from the boundary conditions

$$u_m(0) = 0, \quad u'_m(0) = 0. \tag{2.80}$$

In this way we obtain the functions $u_m(x, \delta)$ for $m = 1, 2, 3, \dots$ from Eq. (2.78) successively. Then, we can obtain Mth-order approximate solution

$$U_M(x, \delta, \hbar) = \sum_{m=0}^{M} u_m(x, \delta). \tag{2.81}$$

So Eq. (2.19), with the help of additional forcing condition $u(\pi) = 0$, becomes

$$u(\pi) \approx U_M(\pi, \delta, \hbar) = 0. \tag{2.82}$$

According to the above equation in Fig. 2.7, δ as a function of convergence controller parameter $\hbar$, for $M = 30$, has been plotted. The number of three δ-plateaus can be identified in this figure, namely $\delta = -0.9851$ in the range$[-0.35, -0.05]$, $\delta = 0$ in the range$[-0.5, 0]$ and $\delta = 0.9851$ in the range$[-0.35, -0.05]$ of $\hbar$. Accordingly, we conclude that the HAM furnishes triple solutions in terms of basis functions (2.75).

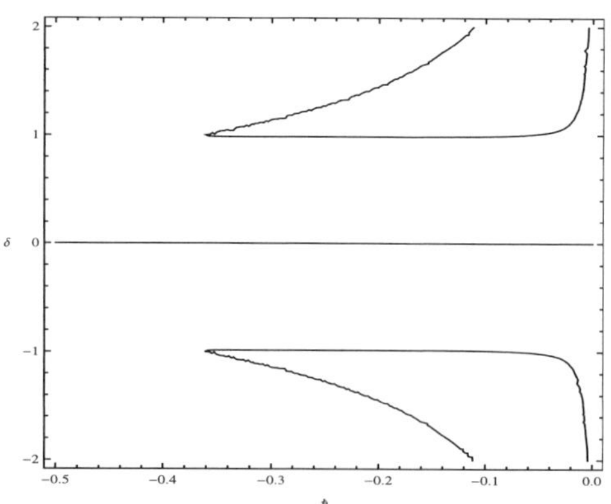

Fig. 2.7. The plot of δ as function of $\hbar$ through Eq. (2.82).

2.4. Some illustrative models

In this part, we apply aforesaid predictor homotopy analysis method to discover multiple solutions of some important nonlinear model and obtain approximations of these solutions simultaneously as well.

2.4.1. *Nonlinear problem arising in heat transfer*

2.4.1.1. *Model and exact solutions*

Consider a straight fin of length L with a uniform cross-section area A. The fin surface is exposed to a convective environment at temperature T_a and the local heat transfer coefficient h along the fin surface is assumed to exhibit a power-law-type dependence on the local temperature difference between the fin and the ambient fluid as

$$h = (T - T_a)^n, \tag{2.83}$$

where a is a dimensional constant defined by physical properties of the surrounding medium, T is the local temperature on the fin surface, and the exponent n depends on the heat transfer mode. The value of n can vary in a wide range between -4 and 5 [14]. For example, the exponent n may take the values -4, -0.25, 0, 2 and 3, indicating the fin subject to transition boiling, laminar film boiling or condensation, convection, nucleate boiling, and radiation into free space at zero absolute temperature, respectively. For one-dimensional steady state heat conduction, the equation in terms of dimensionless variables

$$x = \frac{X}{L}, \quad h = \frac{T - T_a}{T_b - T_a}, \tag{2.84}$$

can be written as

$$\frac{d^2\theta}{dx^2} - \psi^2 \theta^{n+1} = 0, \tag{2.85}$$

where the axial distance x is measured from the fin tip, T_b is the fin base temperature, and ψ is the convective-conductive parameter of the fin defined as

$$h = \left(\frac{h_b P L^2}{kA}\right)^{\frac{1}{2}} = \left(\frac{a P L^2}{kA}(T_b - T_a)^n\right)^{\frac{1}{2}}. \tag{2.86}$$

In the above equation h_b, P and k represent the heat transfer coefficient at fin base, the periphery of fin cross-section, and the conductivity of the fin, respectively. For simplicity, assume the fin tip is insulated and the boundary conditions to Eq. (2.85) can be expressed as

$$\frac{d\theta}{dx}(0) = 0, \quad \theta(1) = 1. \tag{2.87}$$

Equation (2.85) with boundary conditions (2.87) has been considered lately by researchers. In [24, 25], the authors have taken ψ, convective- conductive parameter, as a small parameter and gave approximate solutions by perturbation method and variational iteration method. M. S. H. Chowdhury and I. Hashim [14] transformed the boundary value problems (2.85) and (2.87) to an initial value problem with unknown parameter and then applied the perturbation method so that they could give approximate solutions for positive exponent of θ in Eq. (2.85). Abbasbandy [13] considered Eq. (2.85) with boundary conditions (2.87) in some special cases and, used homotopy analysis method in order to guarantee convergence of approximate series solutions by convergence-controller parameter.
It can be shown that the problems (2.85) and (2.87) admit unit solution for $-1 \leq n \leq 5$, both unit and dual solutions for $-2 < n < -1$, and only dual solutions for $-4 \leq n \leq -2$, therefore for our purpose during the present paper assume that $n = -3$, so Eq. (2.85) becomes

$$\frac{d^2\theta}{dx^2} - \psi^2\theta^{-2} = 0. \tag{2.88}$$

By transformation $\frac{d\theta}{dx} = y$, we have

$$\frac{d^2\theta}{dx^2} = \frac{dy}{dx} = \frac{dy}{d\theta}\frac{d\theta}{dx} = y\frac{dy}{d\theta}. \tag{2.89}$$

So, Eq. (2.88) is converted to $y\frac{dy}{d\theta} - \psi^2\theta^{-2} = 0$ which is the first-order ordinary differential equation of separable type so, by integration and replacing $\frac{d\theta}{dx} = y$, we reach to

$$\frac{1}{2}\left(\frac{d\theta}{dx}\right)^2 + \psi^2\theta^{-1} = C, \tag{2.90}$$

where C is integral constant, using the first boundary condition (2.87), Eq. (2.90) gives

$$C = \psi^2\delta^{-1}, \tag{2.91}$$

where $\delta = \theta(0)$ is unknown temperature at the fin tip ($x = 0$) which will be obtained later by second boundary condition (2.87). Using (2.91) and after simplification Eq. (2.90) becomes

$$dx = \frac{d\theta}{\sqrt{2\psi^2\left(\delta^{-1} - \theta^{-1}\right)}}. \tag{2.92}$$

Using $\delta = \theta(0)$ and by integration (2.92), the final solution of Eq. (2.88) in implicit form is given by

$$x = \int_{\delta}^{\theta} \frac{d\tau}{\sqrt{2\psi^2\left(\delta^{-1}-\tau^{-1}\right)}} = -\frac{\sqrt{\delta}\log\delta}{2\sqrt{\frac{2\psi^2}{\delta^2}}} + \frac{\delta}{\sqrt{2}\psi^2}\left(\theta\sqrt{\frac{(\theta-\delta)\,\psi^2}{\delta\theta}} - \delta^{\frac{3}{2}}\sqrt{\frac{\psi^2}{\delta^2}}\log 2 + \frac{\delta\sqrt{\theta}\sqrt{\frac{(\theta-\delta)\psi^2}{\delta\theta}}\log 2}{\sqrt{(\theta-\delta)}} + \frac{\delta\sqrt{\theta}\sqrt{\frac{(\theta-\delta)\psi^2}{\delta\theta}}\log\left(\sqrt{\theta}+\sqrt{(\theta-\delta)}\right)}{\sqrt{(\theta-\delta)}}\right). \tag{2.93}$$

The parameter δ can be easily obtained with the help of the boundary condition $\theta\,(1) = 1$ from (2.93) as follows

$$-\frac{\sqrt{\delta}\log\delta}{2\sqrt{\frac{2\psi^2}{\delta^2}}} + \frac{\delta}{\sqrt{2}\psi^2}\left(\sqrt{\frac{(1-\delta)\,\psi^2}{\delta}} - \delta^{\frac{3}{2}}\sqrt{\frac{\psi^2}{\delta^2}}\log 2 + \frac{\delta\sqrt{\frac{(1-\delta)\psi^2}{\delta}}\log 2}{\sqrt{(1-\delta)}} + \frac{\delta\sqrt{\frac{(1-\delta)\psi^2}{\delta}}\log\left(1+\sqrt{(1-\delta)}\right)}{\sqrt{(1-\delta)}}\right) = 1. \tag{2.94}$$

The above equation in implicit way, δ as function of ψ, has been plotted in Fig. 2.8.

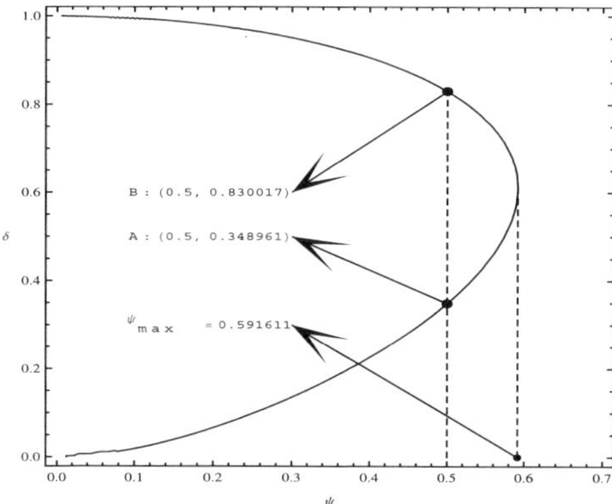

Fig. 2.8. The plot of δ as function of ψ through Eq. (2.94).

First insight to Fig. 2.8, reveals that corresponding to $0 \leq \psi \leq \psi_{max}= 0.591611$ there exist two δ so dual solutions occur, for example corresponding to $\psi = 0.5$, we have $\delta = 0.348961$ and $\delta = 0.830017$ as shown in Fig. 2.8 by points A and B, respectively.

2.4.1.2. *Prediction of dual solutions by the rule of multiplicity of solutions*

Considering $\psi = 0.5$, the two-point boundary value problem (2.85) and (2.87) can be replaced formally by the initial value problem

$$\frac{d^2\theta}{dx^2}-0.25\theta^{-2}= 0, \tag{2.95}$$

$$\theta(0)= \delta, \quad \frac{d\theta}{dx}(0)= 0, \tag{2.96}$$

subject to the additional forcing condition $\theta(1)= 1$, where δ denotes temperature of the fin tip and it is an unknown parameter of the problem and will be determined later by the rule of multiplicity of solutions. Now, HAM is applied on the problems (2.95) and (2.96) as follows:
It is straightforward to use the set of base functions

$$\left\{x^{2n} | n = 0, 1, 2, \ldots\right\}. \tag{2.97}$$

Under the rule of solution expression and according to the initial conditions, it is easy to choose $\theta_0(x) = x^2 + \delta$ as initial guess of solution $\theta(x)$, $H(x) = 1$ as auxiliary function, and to choose auxiliary linear operator

$$\mathcal{L}\left[\varphi(x,\delta;p)\right] = \frac{\partial^2\varphi(x,\delta;p)}{\partial x^2}, \tag{2.98}$$

with the property

$$\mathcal{L}\left[c_1 + c_2 x\right] = 0. \tag{2.99}$$

Thus, after two subsequent integrations, the Mth-order deformation equation (2.14) yields for $M \geq 1$

$$\theta_m(x,\delta) = \chi_m\theta_{m-1}(x,\delta) + \hbar\int_0^x\int_0^s R_m(\vec{\theta}_{m-1},\tau,\delta)\,d\tau\,ds + c_1 + c_2 x, \tag{2.100}$$

where

$$R_m\left(\vec{\theta}_{m-1},\tau,\delta\right) = \sum_{j=0}^{m-1}\theta''_{m-1-j}(x)\sum_{i=0}^{j}\theta_i(x)\,\theta_{j-i}(x) - 0.25\,(1-\chi_m), \tag{2.101}$$

and integration constants c_1 and c_2 are obtained from the boundary conditions

$$\theta_m(0) = 0, \quad \theta_m^{'}(0) = 0. \tag{2.102}$$

The above homogeneous boundary conditions imply that both the integration constants c_1 and c_2 occurring in Eq. (2.100) are zero. In this way we obtain the functions $\theta_m(x)$ for $m = 1, 2, 3, \ldots$ from Eq. (2.100) successively. In this way, we can obtain Mth-order approximate solution

$$\Theta_M(x,\delta,\hbar) = \sum_{m=0}^{M} \theta_m(x,\delta). \tag{2.103}$$

So Eq. (2.19), with the help of additional forcing condition $\theta(1) = 1$, becomes

$$\theta(1) \approx \Theta_M(1,\delta,\hbar) = 1. \tag{2.104}$$

According to the above equation in Fig. 2.9, δ as a function of convergence controller parameter $\hbar$, has been plotted in the $\hbar$-range $[-1.1, -0.1]$, for $M = 35$. Two δ-plateaus can be identified in this figure, namely $\delta = 0.3489$ in the range $[-0.95, -0.3]$ and $\delta = 0.8300$ in the range $[-0.45, -0.25]$ of $\hbar$. Accordingly, we conclude that the HAM furnishes dual solutions, in a full agreement with the exact result shown in Fig. 2.8.

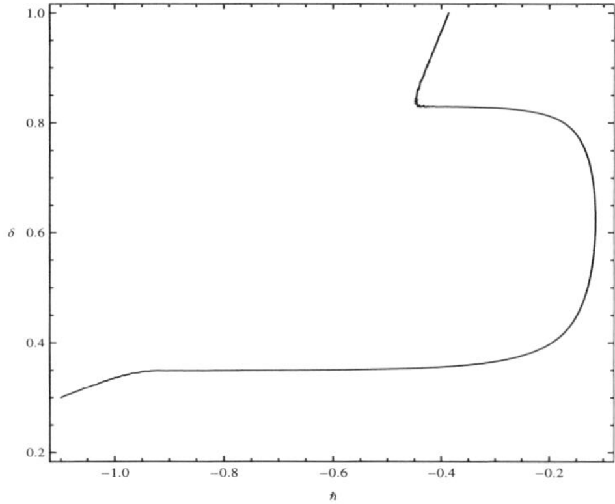

Fig. 2.9. The plot of δ as function of $\hbar$ through Eq. (2.104).

2.4.1.3. *Effective calculation of the two branches of solution*

As soon as the multiplicity of solutions (here dual solutions) in the parameter plane $(\hbar,\delta)$, have been identified, we may turn to calculate them explicitly to any desired order M of HAM-approximation according to Eq. (2.103). In the present section, we do this for both of the dual solutions corresponding to $\delta = 0.3489$ and $\delta = 0.8300$ as being identified in Fig. 2.9 and compare the HAM approximate series solutions given by Eq. (2.103) with the exact solutions (2.93) identified in the respective point pairs $(A,\ B)$ of Fig. 2.8. We remark here, as mentioned in introduction, both the lower branch and upper branch of solutions are calculated at the same time only by Eq. (2.103) with different δ and $\hbar$ which are specified from Fig. 2.9. Furthermore, we emphasis again that there is no need to use more than one initial approximation guess, one auxiliary linear operator, and one auxiliary function that is in a sharp contrast to all approximation methods which are used to converge to one solution. In the plot shown in Fig. 2.10(a), correspond to $\delta = 0.3489$ and $\hbar = -0.6$, the approximate HAM solutions $\Theta_3(x, 0.3489, -0.6)$, $\Theta_5(x, 0.3489, -0.6)$ and $\Theta_{10}(x, 0.3489, -0.6)$ given by Eq. (2.103) are compared to the exact lower branch solution $\theta(x)$ given by Eq. (2.93) for $\delta = 0.348961$ and $\psi = 0.5$ (Point A of Fig. 2.8). In a same action, in Fig. 2.10(b), correspond to $\delta = 0.8300$ and $\hbar = -0.4$, the approximate HAM solutions $\Theta_3(x, 0.8300, -0.4)$, $\Theta_5(x, 0.8300, -0.4)$ and $\Theta_{10}(x, 0.8300, -0.4)$ given by Eq. (2.103) are compared to the exact upper branch solution $\theta(x)$ given by Eq. (2.93) for $\delta = 0.830017$ and $\psi = 0.5$ (Point B of Fig. 2.8). One sees that with the increasing order M, the approximate solutions $\Theta_M(x)$ approach the exact solution smoothly. In Fig. 2.11 the dual HAM solutions $\Theta_M(x)$ of order $M = 35$ are compared to the exact dual solutions. To this order of approximation the HAM results (marked by bold red circle) and the exact results (solid blue lines), at the scale of Fig. 2.11 become undistinguishable.

2.4.2. *Strongly nonlinear Bratu's equation*

2.4.2.1. *Problem and exact solutions*

The nonlinear Bratu's problem has been lately investigated by researchers with various techniques [15, 16], [26, 27]. Shuicai Li and Shijun Liao [15] applied successfully the HAM in erudite way to obtain multiple branches of solutions of this nonlinear problem. The problem is given by

$$u'' + \lambda e^u = 0, \quad x \in (0,1), \tag{2.105}$$

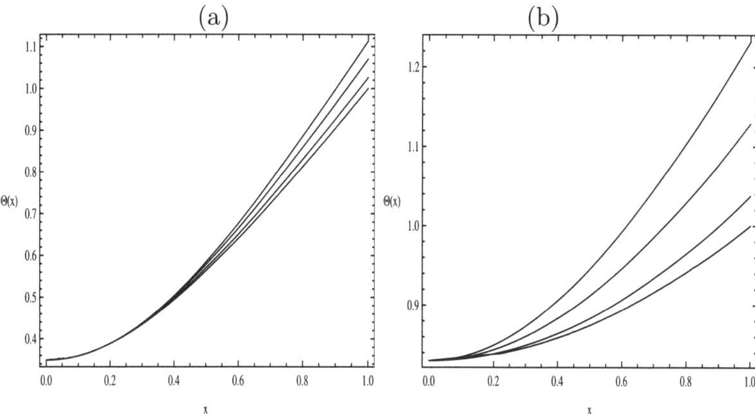

Fig. 2.10. (**a**) The comparison of approximate lower solutions with the exact one: $\Theta_3(x)$-Brown, $\Theta_5(x)$-Blue and $\Theta_{10}(x)$-Red; the exact lower solution-Black. (**b**) The comparison of approximate lower solutions with the exact one: $\Theta_3(x)$-Brown, $\Theta_5(x)$-Blue and $\Theta_{10}(x)$-Red; the exact lower solution-Black.

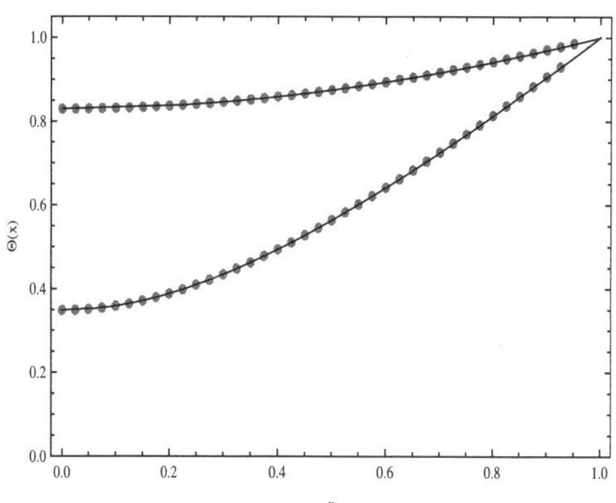

Fig. 2.11. The comparison of approximate dual solutions ($\Theta_{35}(x, 0.3489, -0.6)$ and $\Theta_{35}(x, 0.8300, -0.4)$) with exact dual solutions: Bold red circle-approximate solutions; Solid blue line-exact solutions.

with boundary conditions

$$u(0) = u(1) = 0. \tag{2.106}$$

By reduction order the exact solution of Eq. (2.105) is given by [50]

$$u\left(x\right) = \log\left[\frac{a^2}{2\lambda\cosh^2\left[-\frac{a}{2}\left(x+b\right)\right]}\right], \tag{2.107}$$

where a, b are two arbitrary constants. Introducing the boundary conditions to determine these constants (Eq. (108)), then $b = -\frac{1}{2}$ and a is determined such that

$$a^2 = 2\lambda\cosh^2\left[\frac{a}{4}\right]. \tag{2.108}$$

Suppose that $\alpha = \frac{a}{4}$, thus the solution of Bratu problem is given by

$$u\left(x\right) = 2\log\left[\frac{\cosh\alpha}{\cosh\left[\alpha\left(1-2x\right)\right]}\right], \tag{2.109}$$

where α satisfies

$$\cosh\alpha = \frac{4}{\sqrt{2\lambda}}\alpha. \tag{2.110}$$

The dependence of α on λ has been plotted in Fig. 2.12. It is easy to find out that, according to the value of λ relative to a maximum value λ_{max}, the problems (2.105) and (2.106) has no solution for $\lambda > \lambda_{max}$, one solution for $\lambda = \lambda_{max}$ and two solutions for $\lambda < \lambda_{max}$, for example, as indicated in Fig. 2.12, corresponding to $\lambda = 3$ there are two α ($\alpha = 0.84338$ and $\alpha = 1.64414$ (points C and D)) therefore there exist two solutions. Furthermore, differentiating (2.109) respect to x once time and setting $x = 0$ gives

$$u'\left(0\right) = 4\alpha\tanh\alpha\,. \tag{2.111}$$

In Fig. 2.13 also, for our future purpose, we have plotted $u'(0)$ as function of α according to Eq. (2.111). As we see, for those points C and D in Fig. 2.12, corresponding $u'(0)$ ($u'(0) = 2.3196$ and $u'(0) = 6.1034$ (Points E and F)) have been marked. In the other words, for $\lambda = 3$, dual solutions occur so that, we have $u'(0) = 2.3196$ for the first solution and $u'(0) = 6.1034$ for the second solution.

2.4.2.2. *Prediction of multiple solutions by the rule of multiplicity of solutions*

The purpose of this subsection is to forecast the existence of multiple solutions for the problem (2.105) and (2.106) in case $\lambda = 3$ by the rule of multiplicity of solutions. Assuming $\lambda = 3$, the two-point boundary value

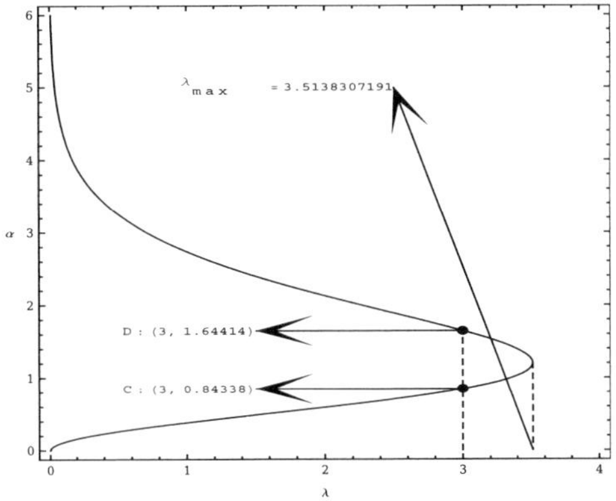

Fig. 2.12. α as function of λ according to Eq. (2.110).

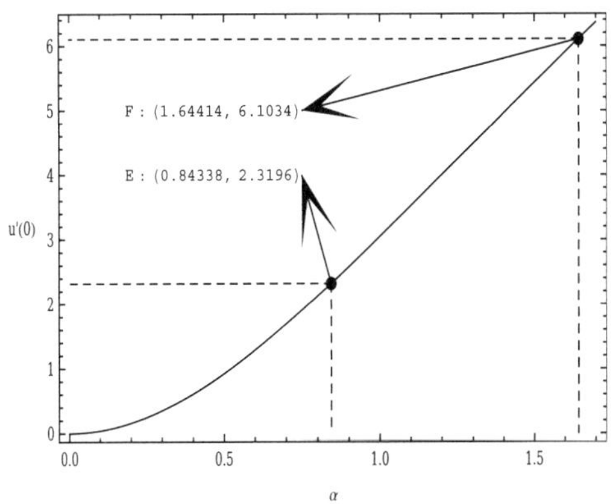

Fig. 2.13. $u'(0)$ as function of α according to Eq. (2.111).

problem (2.105) and (2.106) can be changed formally to the initial value problem as follows

$$u'' + 3e^u = 0, \quad x \in (0,1), \tag{2.112}$$

$$u(0) = 0, \quad u'(0) = \delta, \tag{2.113}$$

subject to the additional forcing condition

$$u(1) = 0. \tag{2.114}$$

In order to prevent suffering from the strongly nonlinear term e^u in the frame of HAM, we can transform the problem (2.112), (2.113) and additional condition (2.114) to an equivalent one. Suppose that

$$y\left(x\right)=e^{-u(x)} \quad or \quad u\left(x\right)=-\log\left[y\left(x\right)\right]. \tag{2.115}$$

Under the above transformation Eq. (2.112) is converted to

$$y\left(x\right)y^{''}\left(x\right)-\left[y^{'}\left(x\right)\right]^{2}-3y\left(x\right)=0, \quad x\in\left(0,1\right). \tag{2.116}$$

From Eq. (2.115) $y^{'}\left(x\right)=-u'(x)e^{-u(x)}$ so, the boundary conditions (2.113) become

$$y\left(0\right)=1, \;\; y^{'}\left(0\right)\;=\gamma=-\delta, \tag{2.117}$$

and additional condition by Eq. (2.115) that comes from (2.114), is

$$y\left(1\right)=1. \tag{2.118}$$

Now, instead of the problems (2.112)–(2.114), the technique is applied on the problems (2.116)–(2.118) and the parameter γ, which played an important role to realize about multiplicity of solutions, will be obtained with the help of rule of multiplicity of solutions.
It is straightforward to use the set of base functions

$$\left\{x^{n}|n=0,1,2,\ldots\right\}. \tag{2.119}$$

Under the rule of solution expression and according to the initial conditions (2.117), it is easy to choose $y_0\left(x\right)=\gamma x+1$ as initial guess of solution $y\left(x\right)$, $H\left(x\right)=1$ as auxiliary function, and to choose auxiliary linear operator

$$\mathcal{L}\left[\varphi(x,\gamma;p)\right]=\frac{\partial^{2}\varphi(x,\gamma;p)}{\partial x^{2}}, \tag{2.120}$$

with the property

$$\mathcal{L}\left[c_1+c_2x\right]=0. \tag{2.121}$$

Therefore, after two subsequent integrations, the Mth-order deformation equation (2.14) yields for $M\geq 1$

$$y_m\left(x,\gamma\right)=\chi_m y_{m-1}\left(x,\gamma\right)+\hbar\int_0^x\int_0^s R_m(\overrightarrow{y}_{m-1},\tau,\gamma)\,d\tau\,ds + c_1+c_2x, \tag{2.122}$$

where

$$R_m\left(\overrightarrow{y}_{m-1},\tau,\gamma\right)=\sum_{j=0}^{m-1}y^{''}_{m-1-j}\left(x\right)y_j\left(x\right)-\sum_{j=0}^{m-1}y^{'}_{m-1-j}\left(x\right)y^{'}_j\left(x\right) - 3y_{m-1}(x), \tag{2.123}$$

and integration constants c_1 and c_2 are obtained from the boundary conditions

$$y_m(0) = 0, \quad y_m^{'}(0) = 0. \tag{2.124}$$

The above homogeneous boundary conditions imply that both the integration constants c_1 and c_2 occurring in Eq. (2.122) are zero. In this way we obtain the functions $y_m(x)$ for $m = 1, 2, 3, \ldots$ from Eq. (2.122) successively. Finally, we can obtain Mth-order approximate solution

$$Y_M(x,\gamma,\hbar) = \sum_{m=0}^{M} y_m(x,\gamma). \tag{2.125}$$

So the Eq. (2.19), with the help of additional forcing condition $y(1) = 1$, becomes

$$y(1) \approx Y_M(1,\gamma,\hbar) = 1. \tag{2.126}$$

According to the above equation in Fig. 2.14, γ as a function of convergence controller parameter $\hbar$, has been plotted in the $\hbar$-range $[-2, 0]$, for $M = 40$. Two γ-plateaus can be identified in this figure, namely $\gamma = -6.1034$ ($\delta = 6.1034$) in the range $[-0.6, -0.4]$ and $\gamma = -2.3196$ ($\delta = 2.3196$) in the range $[-0.8, -0.3]$ of $\hbar$. Accordingly, we conclude that the HAM furnishes dual solutions, in a full agreement with the exact result shown in Figs. 2.12 and 2.13 (It is worth to remark that Figs. 2.12 and 2.13 indicate existence of two solutions for $\lambda = 3$ so that, $u^{'}(0) = 2.3196$ for the first solution and $u^{'}(0) = 6.1034$ for the second solution. On the other hand, Fig. 2.14 also indicates two solutions so that $u^{'}(0) = \delta = -\gamma = 2.3196$ for the first one and $u^{'}(0) = \delta = -\gamma = 6.1034$ for the other one).

2.4.2.3. *Effective calculation of the two branches of solution*

After that the multiplicity of solutions have been identified, we may turn to calculate them explicitly to any desired order M of HAM-approximation according to Eq. (2.125). In the present section, we do this for the two branches of solutions corresponding to $\delta = 2.3196$ and $\delta = 6.1034$ as being identified in Fig. 2.14 and compare the HAM approximate series solutions given by Eqs. (2.115) and (2.125) with the exact solutions (2.107) identified in the respective point pairs (C, D) of Fig. 2.12. The Mth-order approximate solution is given by equations (2.115) and (2.125) as follows:

$$U_M(x,\delta,\hbar) = -\log\left[Y_M(x,\gamma,\hbar)\right]. \tag{2.127}$$

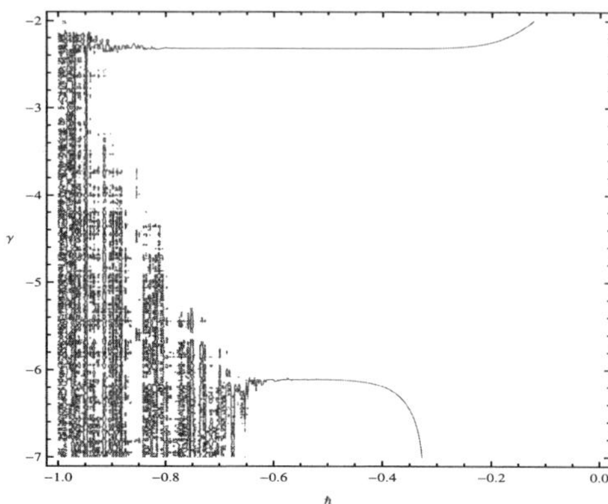

Fig. 2.14. The plot of γ as function of $\hbar$ through Eq. (2.126).

In the plot shown in Fig. 2.15, correspond to $\delta = 2.3196$ and $\hbar = -0.5$, the approximate HAM solutions $U_5(x, 2.3196, -0.5)$, $U_8(x, 2.3196, -0.5)$ and $U_{10}(x, 2.3196, -0.5)$ given by Eq. (2.127) are compared to the exact lower branch solution $u(x)$ given by Eq. (2.107) for $\alpha = 0.84338$ (Point C of Fig. 2.12). In a similar manner, in Fig. 2.16, correspond to $\delta = 6.1034$ and $\hbar = -0.5$, the approximate HAM solutions $U_{17}(x, 6.1034, -0.5)$, $U_{20}(x, 6.1034, -0.5)$ and $U_{22}(x, 6.1034, -0.5)$ given by Eq. (2.127) are compared to the exact upper branch solution $u(x)$ given by Eq. (2.107) for $\alpha = 1.64414$ (Point D of Fig. 2.12). Obviously, with the increasing order M, the approximate solutions $U_M(x)$ approach the exact solution smoothly. In Fig. 2.17 the both two branches of HAM solutions $U_M(x)$ (with $M = 20$ for the lower branch and $M = 30$ for the upper branch) are compared to the exact dual solutions. We observe again, as mentioned before, both the lower branch and upper branch of solutions are calculated at the same time only by Eq. (2.127) with different δ and $\hbar$ which are specified from Fig. 2.8. Moreover, we emphasize again that there is no need to use more than one initial approximation guess, one auxiliary linear operator, and one auxiliary function.

2.4.3. *Nonlinear reaction-diffusion model*

2.4.3.1. *Equation and exact solutions*

A nonlinear model of diffusion and reaction in porous catalysts has been investigated by approximate analytical methods [17, 28–30]. E. Magyari

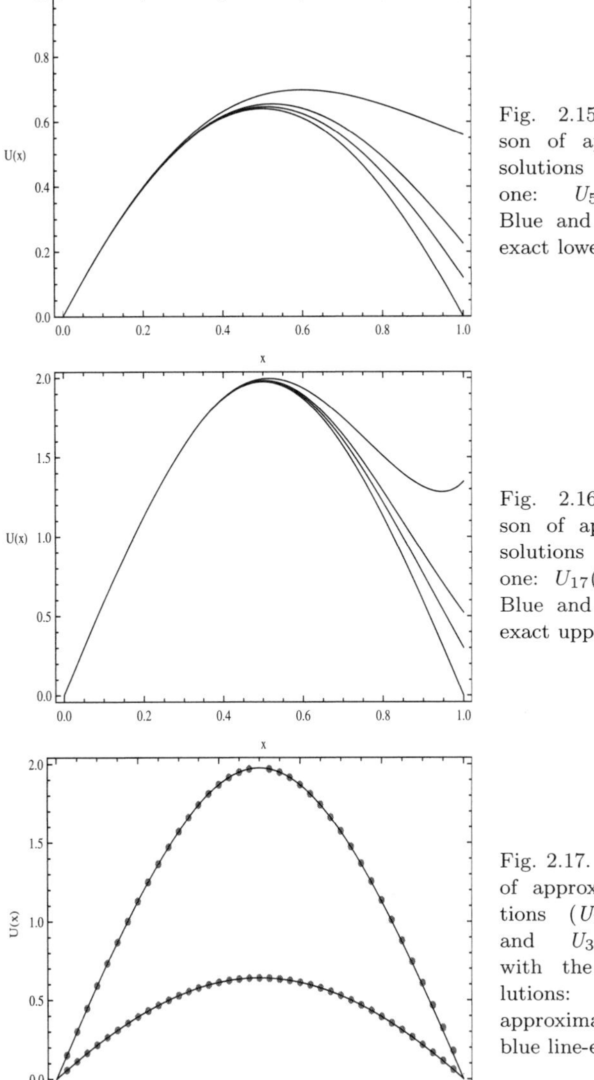

Fig. 2.15. The comparison of approximate lower solutions with the exact one: $U_5(x)$-Brown, $U_8(x)$-Blue and $U_{10}(x)$-Red; the exact lower solution-Black.

Fig. 2.16. The comparison of approximate upper solutions with the exact one: $U_{17}(x)$-Brown, $U_{20}(x)$-Blue and $U_{22}(x)$-Red; the exact upper solution-Black.

Fig. 2.17. The comparison of approximate dual solutions ($U_{20}(x, 2.3196, -0.5)$ and $U_{30}(x, 6.1034, -0.5)$) with the exact dual solutions: Bold red circle-approximate solutions; Solid blue line-exact solutions.

have considered this model [17] and given successfully exact analytical solutions in implicit form for all values of parameters of the problem. We just consider here a special case of that problem when the model takes -0.75

for reaction-order and 0.8 for Thiele modulus, as follows

$$u^{''}u^{0.75} - 0.64 = 0, \tag{2.128}$$

with the boundary conditions

$$u'(0) = 0, \quad u(1) = 1, \tag{2.129}$$

where u is the dimensionless concentration of the reactant, the primes denote differentiation with respect to the dimensionless transverse coordinate x, $0 \leq x \leq 1$.

In [17], it has been shown that the above problems (2.128) and (2.129) admit two following solutions

$$22.6539x = 2\sqrt{2 - 2\left(\frac{0.1836}{u}\right)^{\frac{1}{4}}}$$

$$\left(5 + 6\left(\frac{0.1836}{u}\right)^{\frac{1}{4}} + 8\sqrt{\frac{0.1836}{u}} + 16\left(\frac{0.1836}{u}\right)^{\frac{3}{4}}\right)\left(\frac{0.1836}{u}\right)^{\frac{1}{8}} u, \tag{2.130}$$

$$25.8821x = 2\sqrt{2 - 2\left(\frac{0.5330}{u}\right)^{\frac{1}{4}}}$$

$$\left(5 + 6\left(\frac{0.5330}{u}\right)^{\frac{1}{4}} + 8\sqrt{\frac{0.5330}{u}} + 16\left(\frac{0.5330}{u}\right)^{\frac{3}{4}}\right)\left(\frac{0.5330}{u}\right)^{\frac{1}{8}} u. \tag{2.131}$$

A simple inspection of the above two solutions gives

$$u(0) = 0.1836, \tag{2.132}$$

for the first solution, and

$$u(0) = 0.5330, \tag{2.133}$$

for the second one.

2.4.3.2. *Prediction of multiple solutions by the rule of multiplicity of solutions*

Let us consider the following initial value problem

$$u^{''}u^{0.75} - 0.64 = 0, \tag{2.134}$$

$$u(0) = \delta, \quad u'(0) = 0, \tag{2.135}$$

with additional forcing condition

$$u\left(1\right)=1, \tag{2.136}$$

where δ is an unknown parameter of the problem in this time and will be determined later by the *rule of multiplicity of solutions*. Now, HAM is applied on the problem (2.134) and (2.135) as follows: It is straightforward to use the set of base functions

$$\left\{x^{2n}|n=0,1,2,\ldots\right\}. \tag{2.137}$$

Under the rule of solution expression and according to the initial conditions, it is easy to choose $u_0\left(x\right)=\delta$ as initial guess of solution $u\left(x\right)$, $H\left(x\right)=1$ as auxiliary function, and to choose auxiliary linear operator

$$\mathcal{L}\left[\varphi(x,\delta;p)\right]=\frac{\partial^2\varphi(x,\delta;p)}{\partial x^2}, \tag{2.138}$$

with the property

$$\mathcal{L}\left[c_1+c_2x\right]=0. \tag{2.139}$$

Thus, the Mth-order deformation equation for $M\geq 1$ becomes

$$\mathcal{L}\left[u_m(x,\delta)-\chi_m u_{m-1}\left(x,\delta\right)\right]=\hbar R_m(\vec{u}_{m-1},x,\delta), \tag{2.140}$$

where

$$R_m\left(\vec{u}_{m-1},x,\delta\right)=\sum_{j=0}^{m-1}u''_{m-1-j}\left(x\right)z_j(x)-0.64\left(1-\chi_m\right), \tag{2.141}$$

with

$$\begin{aligned} z_n\left(x\right)&=\frac{1}{(n)!}\left.\frac{\partial^n[\varphi\left(x,\delta;p\right)]^{0.75}}{\partial p^n}\right|_{p=0}\\ &=\frac{1}{(n)!}\left.\frac{\partial^n\left[\sum_{k=0}^{k=+\infty}u_k(x,\delta)p^k\right]^{0.75}}{\partial p^n}\right|_{p=0}. \end{aligned} \tag{2.142}$$

For instance

$$z_0\left(x\right)=\left[u_0(x)\right]^{0.75}, \tag{2.143}$$

$$z_1\left(x\right)=\frac{0.75u_1(x)}{\left[u_0(x)\right]^{0.25}}, \tag{2.144}$$

$$z_2(x) = -\frac{0.09375[u_1(x)]^2}{[u_0(x)]^{1.25}} + \frac{0.75u_2(x)}{[u_0(x)]^{0.25}}. \quad (2.145)$$

The high-order deformation equation (2.140) with initial conditions

$$u_m(0) = 0, \quad u'_m(0) = 0, \quad (2.146)$$

which comes from (2.135), can be easily solved by starting $u_0(x,\delta) = \delta$, in this way we obtain the functions $u_m(x,\delta)$ for $m = 1, 2, 3, \ldots$ successively and Mth-order approximate solution

$$U_M(x,\delta,\hbar) = \sum_{m=0}^{M} u_m(x,\delta). \quad (2.147)$$

So the Eq. (2.19), with the help of additional forcing condition $u(1) = 1$, becomes

$$u(1) \approx U_M(1,\delta,\hbar) = 1. \quad (2.148)$$

According to the above equation in Fig. 2.18, δ as a function of convergence controller parameter $\hbar$, has been plotted in the $\hbar$-range $[-2, 0]$, for $M = 25$. Two δ-plateaus can be identified in this figure, namely $\delta = 0.1836$ in the range $[-1.3, -0.4]$ and $\delta = 0.5330$ in the range $[-1.7, -0.3]$ of $\hbar$. Accordingly, we conclude that the HAM furnishes dual solutions, in a full agreement with the exact result (2.130) and (2.131).

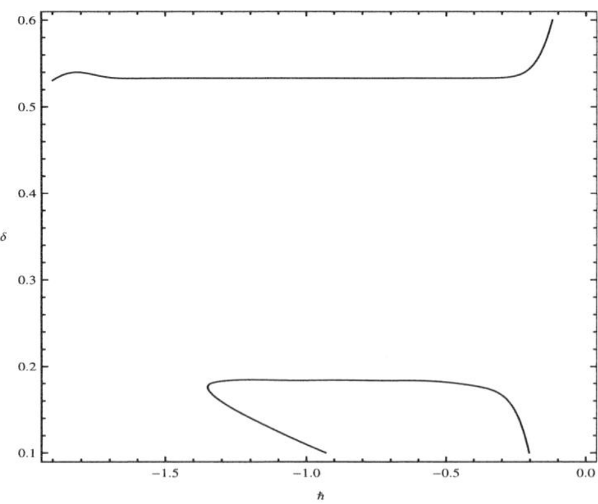

Fig. 2.18. The plot of δ as function of $\hbar$ through Eq. (2.148).

2.4.3.3. *Calculation of the two branches of solution*

Now, we jump to calculate explicitly to any desired order M of HAM-approximate solutions according to Eq. (2.147) for both of dual solutions corresponding to $\delta = 0.1836$ and $\delta = 0.5330$ as being identified in Fig. 2.18 and to compare them with the exact solutions (2.130)–(2.131). In the plot shown in Fig. 2.19, correspond to $\delta = 0.1836$ and $\hbar = -0.8$, the approximate HAM solutions $U_3(x, 0.1836, -0.8)$, $U_5(x, 0.1836, -0.8)$ and $U_7(x, 0.1836, -0.8)$ given by Eq. (2.147) are compared to the exact lower branch solution $u(x)$ given by Eq. (2.130). In a same action, in Fig. 2.20, correspond to $\delta = 0.5330$ and $\hbar = -1$, the approximate HAM solutions $U_0(x, 0.5330, -1) = 0.5330$, $U_1(x, 0.5330, -1)$ and $U_2(x, 0.5330, -1)$ given by Eq. (2.147) are compared to the exact upper branch solution $u(x)$ given by Eq. (2.131). One sees that with the increasing order M, the approximate solutions $U_M(x)$, rapidly approach to the exact solution smoothly. In Fig. 2.21 the dual HAM solutions $U_M(x)$ of order $M = 15$ are compared to the exact dual solutions. To this order of approximation the HAM results (marked by bold red circle) and the exact results (solid blue lines), at the scale of Fig. 2.21 become undistinguishable. We notice here, both the lower branch and upper branch of solutions are calculated at the same time only by Eq. (2.147) with different δ and $\hbar$ which are specified from the *rule of multiplicity of solutions* and, there is no need to use more than one initial approximation guess, one auxiliary linear operator, and one auxiliary function.

2.4.4. *Mixed convection flows in a vertical channel*

The aim of this section is to apply Predictor homotopy analysis method to analyze a kind of model in mixed convection flows namely combined forced and free flow in the fully developed region of a vertical channel with isothermal walls kept at the same temperature [31, 32]. In this model, the fluid properties are assumed to be constant and the viscous dissipation effect is taken into account. The set of governing balance equations for the velocity field is reduced to

$$\frac{d^4u}{dy^4} = \frac{\Xi}{16}\left(\frac{du}{dy}\right)^2, \tag{2.149}$$

with conditions

$$u'(0) = u'''(0) = u(1) = 0, \qquad \int_0^1 u(y)dy = 1, \tag{2.150}$$

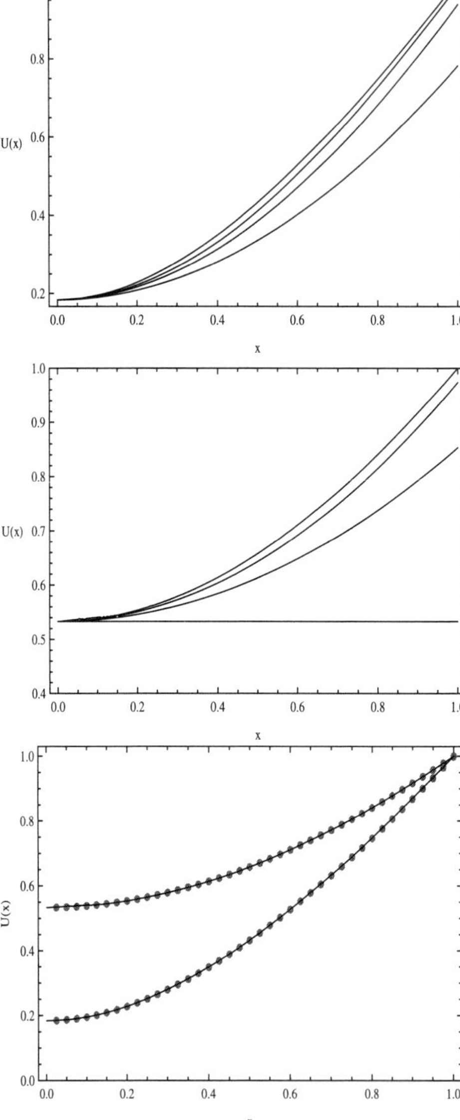

Fig. 2.19. The comparison of approximate lower solutions with the exact one: $U_3(x)$-Brown, $U_5(x)$-Blue and $U_7(x)$-Red; the exact lower solution-Black.

Fig. 2.20. The comparison of approximate upper solutions with the exact one: $U_0(x)$-Brown, $U_1(x)$-Blue and $U_2(x)$-Red; the exact upper solution-Black.

Fig. 2.21. The comparison of approximate dual solutions ($U_{15}(x, 0.1836, -0.8)$ and $U_{15}(x, 0.5330, -1)$) with the exact dual solutions: Bold red circle-approximate solutions; Solid blue line-exact solutions.

where u and y are dimensionless velocity and transversal coordinate, respectively and also

$$u = \frac{U}{U_m}, \; y = \frac{Y}{L}, \; Ge = \frac{4Lg\beta}{c_p}, Pr = \frac{\mu c_p}{k}, \quad Re = \frac{4LU_m}{\nu}, \; \Xi = GePrRe,$$

in which U_m, L, c_p, μ, k, ν, Ge, Pr and Re are mean fluid velocity, channel half-width, specific heat at constant pressure, dynamic viscosity, thermal conductivity, kinematic viscosity, Gebhart number, Prandtl number and Reynolds number, respectively.

In the case $\Xi = 0$, correspond either to a very small viscous dissipation heating or to negligible buoyancy effects, the Eqs. (2.149)–(2.150) is easily solved and admit the unique solution

$$u(y) = \frac{3}{2}\left(1-y^2\right). \tag{2.151}$$

It has been shown in [30, 31] by perturbation and numerical methods that Eqs. (2.1)–(2.2) admit dual solutions for any given Ξ in the interval $(-\infty, 0) \cup (0, \Xi_{max})$ in which $\Xi_{max} \cong 228.128$.

2.4.4.1. *Prediction of dual solutions by the rule of multiplicity of solutions*

The purpose of this subsection is to show how one can find out the existence of dual solutions for Eqs. (2.149)–(2.150) in aforesaid range for Ξ. Consider Eqs. (2.149)–(2.150) and suppose that $u^{''}(0) = \delta$, so the problem becomes

$$\frac{d^4u}{dy^4} = \frac{\Xi}{16}\left(\frac{du}{dy}\right)^2, \tag{2.152}$$

subject to boundary conditions

$$u^{'}(0) = u^{'''}(0) = u(1) = 0, \quad u^{''}(0) = \delta, \tag{2.153}$$

with additional forcing condition

$$\int_0^1 u(y)dy = 1. \tag{2.154}$$

Now, we apply Predictor homotopy analysis method on Eqs. (2.152)-(2.153) where *prescribed parameter* δ, which is played important role to realize about multiplicity of solutions, will be obtained with the help of *rule of multiplicity of solutions.*

It is straightforward to use the set of base functions

$$\{y^n, \ n = 0, 1, 2, ...\}. \tag{2.155}$$

Under the rule of solution expression and according to the initial conditions (2.153), it is easy to choose

$$u_0(y, \delta) = \frac{\delta}{2}\left(y^2-1\right),$$

as initial guess of solution $u(y)$, $H(y) = 1$ as auxiliary function, and to choose auxiliary linear operator

$$\mathcal{L}\left[\phi(y,\delta;p)\right] = \frac{\partial^4 \phi(y,\delta;p)}{\partial y^4}, \tag{2.156}$$

with the property

$$\mathcal{L}\left[c_1 + c_2 y + c_3 y^2 + c_4 y^3\right] = 0. \tag{2.157}$$

Therefore, after four subsequent integrations, the Mth-order deformation equation (2.14) yields for $M \geq 1$

$$\begin{aligned} u_m(y,\delta) &= \chi_m u_{m-1}(y,\delta) \\ &+ \hbar \int_0^y \int_0^{\eta_1} \int_0^{\eta_2} \int_0^{\eta_3} R_m(\vec{u}_{m-1},\tau,\delta)\, d\tau\, d\eta_3\, d\eta_2\, d\eta_1 \\ &+ c_1 + c_2 y + c_3 y^2 + c_4 y^3, \end{aligned} \tag{2.158}$$

where

$$R_m(\vec{u}_{m-1},\tau,\delta) = u''''_{m-1}(\tau,\delta) - \frac{\Xi}{16} \sum_{j=0}^{m-1} u'_j(\tau,\delta) u'_{m-1-j}(\tau,\delta), \tag{2.159}$$

and integration constants c_1, c_2, c_3, c_4 are obtained by the conditions

$$u_m(1,\delta) = u'_m(0,\delta) = u''_m(0,\delta) = u'''_m(0,\delta) = 0. \tag{2.160}$$

In this way we obtain the functions $u_m(y,\delta)$ for $m = 1, 2, 3, ...$ from Eq. (2.158) successively. Finally, we can obtain Mth-order approximate solution

$$U_M(y,\delta,\hbar) = \sum_{m=0}^{M} u_m(y,\delta), \tag{2.161}$$

we give below the series solution (2.161) from the order $M = 1$ until the order $M = 3$ in its form valid for any Ξ:

$$U_1(y,\delta,\hbar) = -\frac{y^6\delta^2\Xi\hbar}{5760} + \frac{1}{2}\left(y^2-1\right)\delta + \frac{\delta^2\Xi\hbar}{5760},$$

$$U_2(y,\delta,\hbar) = \frac{y^{10}\delta^3\Xi^2\hbar^2}{38707200} - \frac{y^6\delta^2\Xi\hbar^2}{5760} - \frac{y^6\delta^2\Xi\hbar}{2880} + \frac{1}{2}\left(y^2-1\right)\delta - \frac{\delta^3\Xi^2\hbar^2}{38707200}$$
$$+ \frac{\delta^2\Xi\hbar^2}{5760} + \frac{\delta^2\Xi\hbar}{2880},$$

$$U_3(y,\delta,\hbar) = -\frac{31y^{14}\delta^4\Xi^3\hbar^3}{7439214182400} + \frac{y^{10}\delta^3\Xi^2\hbar^3}{19353600} + \frac{y^{10}\delta^3\Xi^2\hbar^2}{12902400} - \frac{y^6\delta^2\Xi\hbar^3}{5760}$$
$$- \frac{y^6\delta^2\Xi\hbar^2}{1920} - \frac{y^6\delta^2\Xi\hbar}{1920} + \frac{1}{2}\left(y^2-1\right)\delta + \frac{31\delta^4\Xi^3\hbar^3}{7439214182400}$$
$$- \frac{\delta^3\Xi^2\hbar^3}{19353600} - \frac{\delta^3\Xi^2\hbar^2}{12902400} + \frac{\delta^2\Xi\hbar^3}{5760} + \frac{\delta^2\Xi\hbar^2}{1920} + \frac{\delta^2\Xi\hbar}{1920}.$$

So the Eq. (2.161), with the help of additional forcing condition (2.154), becomes

$$\Gamma_M\left(\delta,\hbar,\Xi\right) = \int_0^1 U_M(y,\Xi,\delta,\hbar)\,dy \approx \int_0^1 u(y,\Xi)\,dy = 1. \tag{2.162}$$

Now, to be specific, we consider two case consist of $\Xi = 20$ and $\Xi = -20$. According to the above equation in Fig. 2.22, δ (prescribed parameter) as a function of convergence controller parameter $\hbar$, has been plotted in the $\hbar$-range [–2.6, 0.6] implicitly, for $M = 25$ and $\Xi = 20$. Two δ-plateaus can be identified in this figure, namely $\delta = -3.08411$ in the range [–1.6,–0.4] of $\hbar$ and $\delta = -161.726$ in the range [–0.95,–0.55] of $\hbar$. It is noticeable that we have to magnify the Fig. 2.22 to obtain values of δ with high accuracy (Fig. 2.23). Consequently, we conclude that the PHAM furnishes dual solutions, in a full agreement with those obtained in [31] (It is worth mentioning here that Fig. 2.22 indicates existence of two solutions for $\Xi = 20$ so that, $u''(0) = -3.08411$ for the first branch solution and $u''(0) = -161.726$ for the second branch solution). Another technique to find out how many solutions the nonlinear problems (2.5)–(2.6) admit is to use this fact that cross point of residual of (2.15) (i.e. $\Gamma_M\left(\delta,\hbar,\Xi\right) - 1$) by horizontal axis dose not vary with the variation of $\hbar$. Figure 2.24 shows that there are two crosses with horizontal axis which do not vary with change of $\hbar$ so we turn out that there exist dual solutions.

The same procedure has been done for the case $\Xi = -20$. As we see in Figs. 2.25–2.26 or equivalently in Fig. 2.27, there exist dual solutions namely $u''(0) = -2.92300$ for the first branch solution and $u''(0) = 170.039$ for the second branch solution for the problems (2.152)–(2.153).

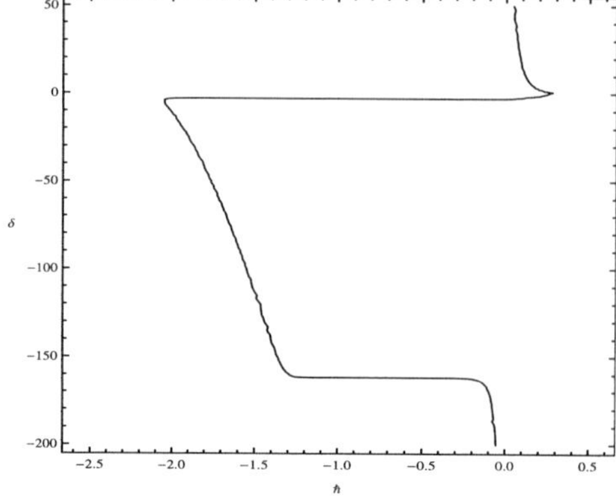

Fig. 2.22. Prescribed parameter δ via convergence controller parameter $\hbar$ in according to (2.162) with $M = 25$ for $\Xi = 20$.

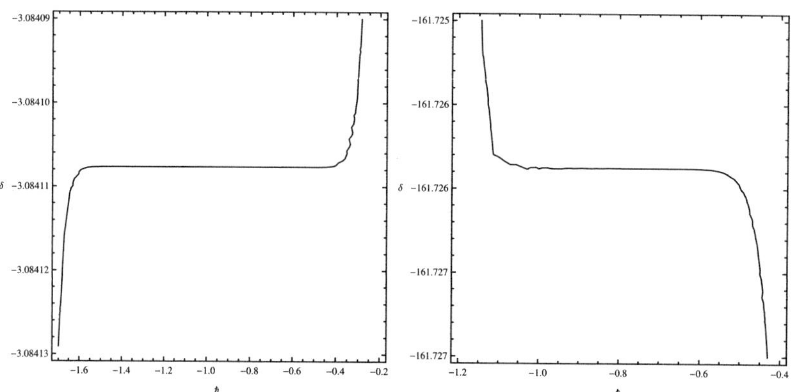

Fig. 2.23. Prescribed parameter δ via convergence controller parameter $\hbar$ in according to (2.162) with $M = 25$ for $\Xi = 20$. (Magnification of Fig. 2.22).

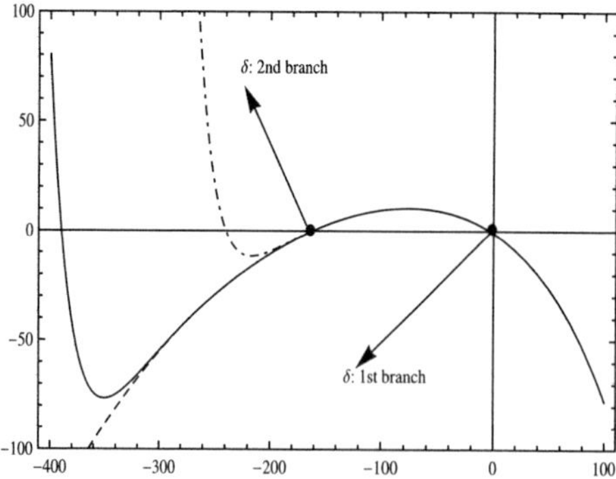

Fig. 2.24. The residual of (2.162) i.e. $\Gamma_M(\delta, \hbar, \Xi)$ with different values of $\hbar$ when $M = 20$ for $\Xi = 20$. Dashed line: $\hbar = 0.8$; bold line: $\hbar = 1$; dot-dashed line: $\hbar = 1.2$.

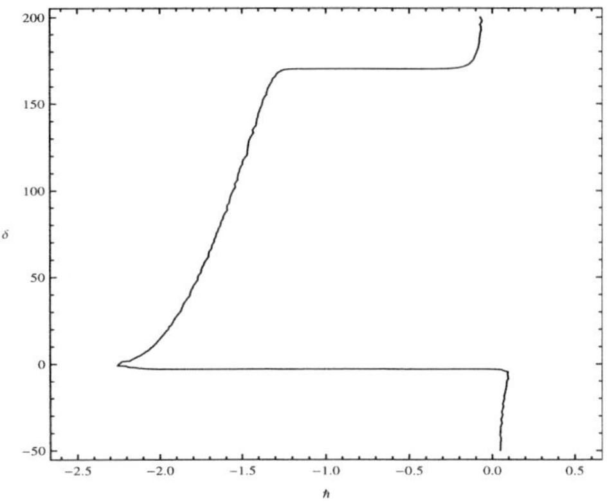

Fig. 2.25. Prescribed parameter δ via convergence controller parameter $\hbar$ in according to (2.162) with $M = 25$ for $\Xi = -20$.

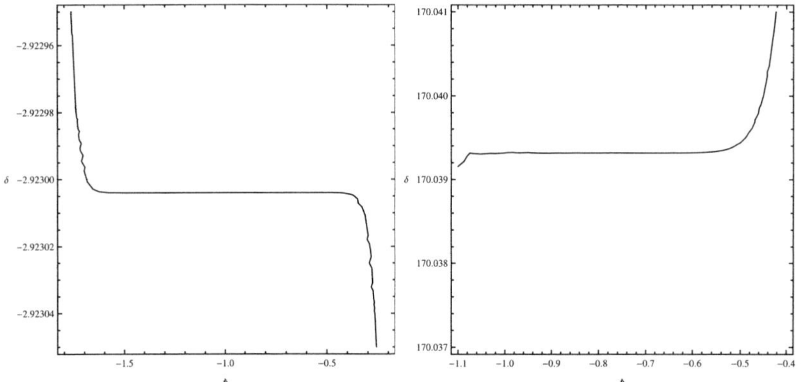

Fig. 2.26. Prescribed parameter δ via convergence controller parameter $\hbar$ in according to (2.162) with $M = 25$ for $\Xi = -20$. (Magnification of Fig. 2.25).

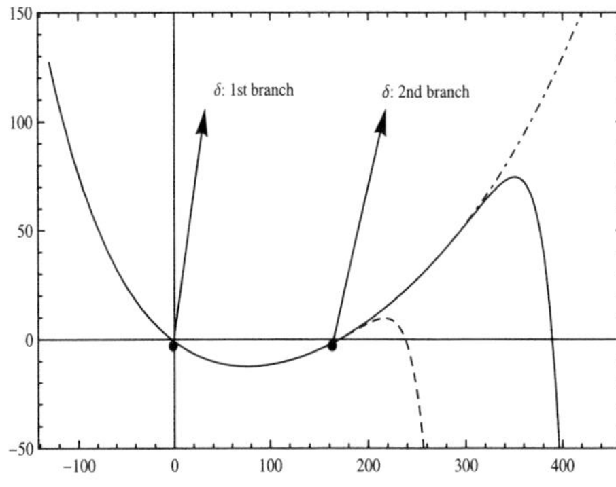

Fig. 2.27. The residual of (2.162) i.e. $\Gamma_M\,(\delta\,,\hbar\,,\Xi)$ with different values of $\hbar$ when $M = 20$ for $\Xi = -20$. Dashed line: $\hbar = -0.8$; bold line: $\hbar = -1$; dot-dashed line: $\hbar = -1.2$.

2.4.4.2. *Effective calculation of the two branches of solution*

As soon as the multiplicity of solutions (here dual solutions) of the problem (2.152)–(2.154) or equivalently problems (2.149)–(2.150), for $\Xi = -20$ and $\Xi = 20$ in the parameter plane ($\hbar$,δ), have been identified, we may turn to calculate them explicitly to any desired order M of PHAM-approximation according to Eq. (2.161). In the present section, we do this for the both of dual solutions corresponding to $\delta = -3.08411$ and $\delta = -161.726$ as being identified in Fig. 2.23. We remark here, as mentioned in introduction, both the first branch and second branch of solutions are calculated at the same time only by Eq. (2.161) with different δ and $\hbar$ which are specified from Fig. 2.23. Furthermore, we emphasize again that there is no need to use more than one initial approximation guess, one auxiliary linear operator, and one auxiliary function that is in a sharp contrast to all approximation methods which are used to converge to one solution.

In the plot shown in Fig. 2.28, correspond to $\delta = -3.08411$ and $\delta = -161.726$, the approximate PHAM solutions $U_{25}(y, -3.08411, -1)$ and $U_{25}(y, -161.726, -0.75)$ given by Eq. (2.161) have been normalized by dividing to $U_{25}(0, -3.08411, -1)$ and $U_{25}(0, -161.726, -0.75)$, respectively in order to have better view. In a same action, correspond to $\delta = -2.92300$ and $\delta = 170.039$, the approximate normalized PHAM solutions $U_{25}(y, -2.92300, -1)$ and $U_{25}(y, 170.039, -0.75)$ are shown in Fig. 2.29.

2.4.4.3. *Further results*

In two previous subsections, we applied Predictor homotopy analysis method on the problem (2.152)–(2.154) or equivalently problems (2.149)–(2.150), for $\Xi = -20$ and $\Xi = 20$. It has been shown that dual solutions, which are identified in the parameter plane ($\hbar$,δ), are calculated effectively only by putting ordered couples ($\hbar$,δ) in PHAM-series (2.161) to any desired of M, simultaneously. We have done a same procedure for other specific values of Ξ in this subsection.

To get a first insight into the Figs. 2.30 and 2.31 reveals that, for some other values of Ξ ($-200, -150, -100, -50, 50, 100, 150$ and 200), dual solutions occur for the velocity profile as well. Furthermore we have provided Table 2.1, by the same action in previous subsections, containing δ=$U''_{25}(0)$ and $U_{25}(0)$ for different values of Ξ in the valid region of $\hbar$.

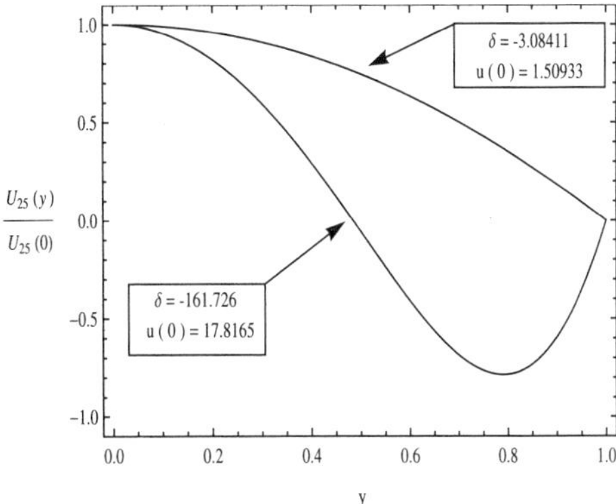

Fig. 2.28. Dual profile $\frac{U_{25}(y)}{U_{25}(0)}$ via dimensionless transversal coordinate y for $\Xi = 20$.

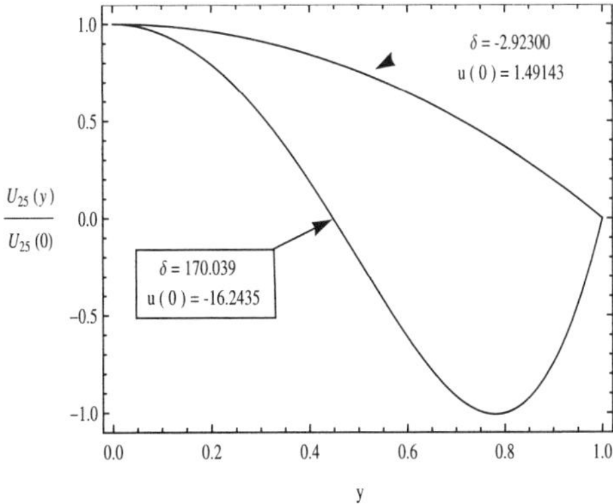

Fig. 2.29. Dual profile $\frac{U_{25}(y)}{U_{25}(0)}$ via dimensionless transversal coordinate y for $\Xi = -20$.

2.5. Concluding remarks

The purpose of this chapter is to introduce a method to predict the multiplicity of the solutions of the nonlinear boundary value problems so that it could be easily applied on nonlinear ordinary differential equations with boundary conditions. Our goal is to introduce a method not only to anticipate multiplicity of the solutions of the nonlinear differential equations but also to calculate effectively all branches of the solutions (on the condition that, there exist such solutions for the problem) analytically at the same

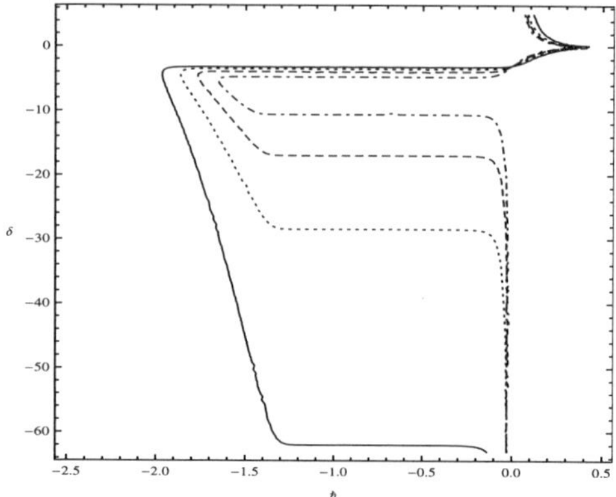

Fig. 2.30. Prescribed parameter δ via convergence controller parameter $\hbar$. Bold line: $\Xi = 50$; dotted line: $\Xi = 100$; dashed line: $\Xi = 150$; dot-dashed line: $\Xi = 200$.

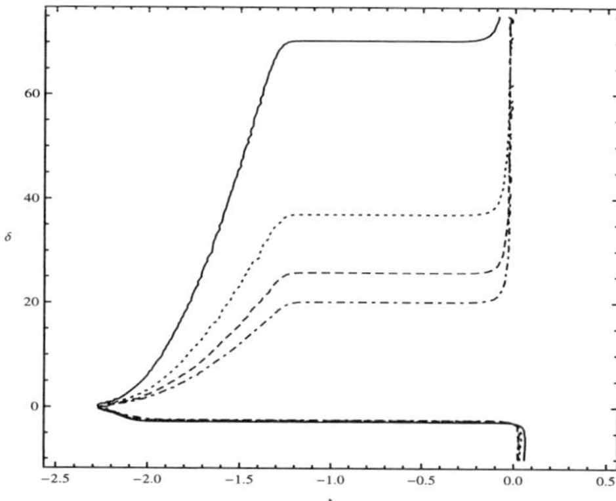

Fig. 2.31. Prescribed parameter δ via convergence controller parameter $\hbar$. Bold line: $\Xi = -50$; dotted line: $\Xi = -100$; dashed line: $\Xi = -150$; dot-dashed line: $\Xi = -200$.

time. In this manner, for practical use in science and engineering, such method might give new unfamiliar class of solutions which is of fundamental interest.

The PHAM has a new viewpoint to the homotopy analysis method by a new application of the convergence-controller parameter. Since this parameter plays important role to guarantee the convergence of the solutions of nonlinear differential equations. In the frame of the PHAM this parameter plays a fundamental role in the prediction of multiplicity of solutions. One can obtained all branches of solutions simultaneously by one initial approx-

Table 2.1. Calculation of $\delta = U_{25}''(0)$ and $U_{25}(0)$ for different values of Ξ according to Eq. (2.161)

	1st branch solution		2nd branch solution	
Ξ	$U_{25}''(0)$	$U_{25}(0)$	$U_{25}''(0)$	$U_{25}(0)$
–300	–2.24467	1.4142	14.4022	–0.265373
–280	–2.27829	1.41812	15.2299	–0.350282
–260	–2.31343	1.4222	16.1808	–0.447839
–240	–2.35019	1.42646	17.2853	–0.561177
–220	–2.38874	1.43091	18.5851	–0.694569
–200	–2.42923	1.43557	20.1383	–0.853988
–180	–2.47184	1.44046	22.0289	–1.04806
–160	–2.51678	1.4456	24.383	–1.28971
–140	–2.56429	1.45102	27.3982	–1.59926
–120	–2.61465	1.45675	31.4042	–2.01054
–100	–2.66818	1.46281	36.994	–2.58444
–80	–2.72528	1.46926	45.3536	–3.4427
–60	–2.78639	1.47614	59.2494	–4.86935
–40	–2.85208	1.4835	86.9805	–7.71641
–20	–2.923	1.49143	170.039	–16.2435
20	–3.08411	1.50933	–161.726	17.8165
40	–3.17666	1.51955	–78.6454	9.28697
60	–3.2794	1.53086	–50.8771	6.43596
80	–3.39469	1.54349	–36.9268	5.00354
100	–3.52581	1.55779	–28.493	4.1374
120	–3.67753	1.57427	–22.8046	3.55307
140	–3.85724	1.5937	–18.6688	3.12807
160	–4.07735	1.61737	–15.4805	2.80024
180	–4.36139	1.64775	–12.8869	2.53331
200	–4.7648	1.69064	–10.6349	2.3012
220	–5.50778	1.76902	–8.3785	2.06803
225	–5.90406	1.81059	–7.64578	1.99206
228	–6.50321	1.8732	–6.85183	1.90953

imation guess, one auxiliary linear operator and one auxiliary function.

References

[1] J.C. Butcher, *Numerical Methods for Ordinary Differential Equations* (2nd ed.), John Wiley & Sons, England, 2008.

[2] G.D. Smith, *Numerical Solution of Partial Differential Equations*, Oxford Univ. Press, 1985.

[3] C. Johnson, *Numerical solutions of partial differential equations by finite elements methods*, Cambridge Univ. Press, 1987.

[4] Y. Nath, M. Prithviraju, and A.A. Mufti, Nonlinear static and dynamics of antisymmetric composite laminated square plates supported on nonlinear

elastic subgrade, *Commun. Nonlinear. Sci. Numer. Simulat.* **11**: 340-354 (2006).
[5] O. Civalek, Harmonic differential quadrature-finite differences coupled approaches for geometrically nonlinear static and dynamic analysis of rectangular plates on elastic foundation, *J. Sound Vibration.* **294**: 966-980 (2006).
[6] O. Civalek, Nonlinear analysis of thin rectangular plates on winklerpasternak elastic foundations by dsc-hdq methods, *Appl. Math. Model.* **31**: 606-624 (2007).
[7] A.H. Nayfeh, *Perturbation Methods*, John Wiley & Sons, New York, 2000.
[8] J.D. Cole, *Perturbation methods in applied mathematics*, Blaisdell Publishing Company, Waltham, Massachusetts, 1968.
[9] A.M. Lyapunov, *General Problem on Stability of Motion* (English translation), Taylor and Francis, London, 1992.
[10] A.V. Karmishin, A.I. Zhukov, and V.G. Kolosov, *Methods of Dynamics Calculation and Testing for Thin-Walled Structures*, Mashinostroyenie, Moscow, 1990.
[11] G. Adomian, *Solving Frontier Problems of Physics: The Decomposition Method*, Kluwer Academic Publishers, Dordrecht, 1994.
[12] S. Abbasbandy and E. Shivanian, Application of variational iteration method for nth-order integro-differential equations, *Z. Naturforsch.* **64a**: 439-444 (2009).
[13] S. Abbasbandy, The application of the homotopy analysis method to nonlinear equations arising in heat transfer, *Phys. Lett. A.* **360**: 109-113 (2006).
[14] S. Abbasbandy and E. Shivanian, Prediction of multiplicity of solutions of nonlinear boundary value problems: Novel application of homotopy analysis method, *Commun. Nonlinear. Sci. Numer. Simula.* **15**: 3830-3846 (2010).
[15] S. Li and S.J. Liao, An analytic approach to solve multiple solutions of a strongly nonlinear problem, *Appl. Math. Comput.* **169**: 854-865 (2005).
[16] A. Mohsen, L.F. Sedeek, and S. A. Mohamed, New smoother to enhance multigrid-based methods for bratu problem, *Appl. Math. Comput.* **204**: 325-339 (2008).
[17] E. Magyari, Exact analytical solution of a nonlinear reaction-diffusion model in porous catalysts, *Chem. Eng. J.* **143**: 167-171 (2008).
[18] A. Barletta, Laminar convection in a vertical channel with viscous dissipation and buoyancy effects, *Int. Commun. Heat. Mass. Transfer.* **26**: 153-164 (1999).
[19] A. Barletta, E. Magyari, and B. Keller, Dual mixed convection flows in a vertical channel, *Int. J. Heat. Mass. Transfer.* **48**: 4835-4845 (2005).
[20] S.J. Liao, *On the proposed homotopy analysis techniques for nonlinear problems and its application*, PhD dissertation, Shanghai Jiao Tong University, 1992.
[21] S.J. Liao, *Beyond Perturbation: Introduction to the Homotopy Analysis Method*, Chapman & Hall/CRC Press, Boca Raton, 2003.
[22] G.B. Thomas and R.L. Finney, *Calculus and Analytic Geometry* (9th ed.), Addison Wesley, 1995.
[23] E.A. Coddington, *An introduction to ordinary differential equations*,

Prentice-Hall, Englewood Cliffs, New Jersey, 1961.

[24] D.D. Ganji, The application of Hes homotopy perturbation method to nonlinear equations arising in heat transfer, *Phys. Lett. A.* **355**: 337-341 (2006).

[25] H. Tari, D.D. Ganji, and H. Babazadeh, The application of Hes variational iteration method to nonlinear equations arising in heat transfer, *Phys. Lett. A.* **363**: 213-217 (2007).

[26] A.M. Wazwaz, Adomian decomposition method for a reliable treatment of the bratu-type equations, *Appl. Math. Comput.* **166**: 652-663 (2005).

[27] M.I. Syam and A. Hamdan, An efficient method for solving bratu equations, *Appl. Math. Comput.* **176**: 704-713 (2006).

[28] S. Abbasbandy, Approximate solution for the nonlinear model of diffusion and reaction in porous catalysts by means of the homotopy analysis method, *Chem. Eng. J.* **136**: 144-150 (2008).

[29] Y.P. Sun, S.B. Liu, and S. Keith, Approximate solution for the nonlinear model of diffusion and reaction in porous catalysts by decomposition method, *Chem. Eng. J.* **102**: 1-10 (2004).

[30] S. Abbasbandy, E. Magyari, and E. Shivanian, The homotopy analysis method for multiple solutions of nonlinear boundary value problems, *Commun. Nonlinear. Sci. Numer. Simulat.* **14**: 3530-3536 (2009).

[31] A. Barletta, Laminar convection in a vertical channel with viscous dissipation and buoyancy effects, *Int. Commun. Heat. Mass. Transfer.* **26**: 153-164 (1999).

[32] S. Abbasbandy and E. Shivanian, Predictor homotopy analysis method and its application to some nonlinear problems, *Commun. Nonlinear. Sci. Numer. Simulat.* **16**: 2456-2468 (2011).

Chapter 3

Spectral Homotopy Analysis Method for Nonlinear Boundary Value Problems

Sandile Motsa* and Precious Sibanda

School of Mathematics, Statistics and Computer Science
University of KwaZulu-Natal, Private Bag X01
Scottsville, Pietermaritzburg 3209, South Africa
**sandilemotsa@gmail.com*

In this chapter we provide a general review of the spectral homotopy analysis method (SHAM) for the solution of nonlinear boundary value problems. We demonstrate how the SHAM may be used to find multiple solutions of nonlinear boundary value problems (BVPs) and to solve nonlinear eigenvalue problems. Two approaches are suggested to determine the optimal convergence-control parameter $\hbar$, using respectively the so-called $\hbar$-curve and residual error analysis. We also introduce the iterative version of the SHAM, which leads to enhanced accuracy and efficiency, and accelerates convergence through systematically updating the initial approximation.

Contents

3.1. Introduction 86
3.2. Basic ideas of the spectral homotopy analysis method 86
3.3. Some applications of the spectral homotopy analysis method 89
3.3.1. Falkner–Skan boundary layer flow 89
3.3.2. Eigenvalue problems 95
3.3.3. Boundary value problems with multiple solutions 100
3.3.4. Coupled nonlinear boundary value equations 102
3.4. Convergence acceleration 112
3.4.1. Convergence acceleration through choice of linear operator 112
3.4.2. Convergence acceleration by iteration 116
3.5. Conclusion 119
References 120

3.1. Introduction

The spectral-homotopy analysis method (SHAM) was introduced in 2010 by Motsa *et al.* [20, 21], who used Chebyshev spectral collocation methods to solve the high-order deformation equations in the frame of the HAM. The initial approximation was also found systematically as the solution of the non-homogeneous linear part of the differential equation to be solved. The immediate benefit of these innovations was better accuracy and faster convergence of the solution series, requiring fewer iterations and less computational effort. Other benefits of the SHAM are that (i) the range of admissible $\hbar$ values is much wider in the spectral homotopy analysis method than in the original homotopy analysis method, (ii) the method allows for a much wider range of linear and nonlinear operators.

The use of the spectral homotopy analysis method has largely been restricted to the solution of nonlinear boundary value problems [2, 11, 20, 21, 23, 26, 27, 29] However, Atabakan *et al.* [3] recently used the method to solve Volterra and Fredholm integro-differential equations. A slightly different version of the SHAM that uses Chebyshev-Tau method to convert a BVP to algebraic equations is proposed in Kazem and Shaban [13].

In this chapter we show how the SHAM can used to solve BVPs with multiple solutions and eigenvalue problems. We propose two methods for identifying the optimal $\hbar$ for the SHAM using residual error analysis. Lastly, we present an iterated version of the SHAM which seeks to accelerate convergence of the SHAM through systematic updating of the initial approximation used at the start of the algorithm.

3.2. Basic ideas of the spectral homotopy analysis method

In this section we present the basic idea behind the development of the spectral homotopy analysis method (SHAM). For illustration purposes we describe the SHAM approach for the solution of general one-dimensional nonlinear differential equations. Extension to higher order systems of nonlinear BVPs can be done in a straight-forward manner.

Consider a non-linear ordinary differential equation of the form

$$\mathcal{L}[y(x)] + \mathcal{F}[y(x)] = \Phi(x) \tag{3.1}$$

where $\Phi(x)$ is a known function of the independent variable x and $y(x)$ is an unknown function. The functions $\mathcal{L}$ and $\mathcal{F}$ represent the linear and nonlinear components of the governing equation respectively. For illustrative purposes, we assume that equation (3.1) is to be solved in the domain $x \in [a, b]$ subject to the separated boundary conditions

$$B_a(y(a)) = 0, \quad B_b(y(b)) = 0, \tag{3.2}$$

where B_a and B_b are linear operators.

In the framework of the homotopy analysis method (HAM) [15, 16], we define the following zeroth-order deformation equations

$$(1-q)\mathcal{L}\left[Y(x;q) - y_0(x)\right] = q\hbar\left\{\mathcal{N}[Y(x;q)] - \Phi(x)\right\}, \tag{3.3}$$

where $q \in [0, 1]$ denotes an embedding parameter, $Y(x;q)$ is a kind of continuous mapping function of $y(x)$, $\hbar$ is the convergence-controlling parameter. The nonlinear operator $\mathcal{N}$ is defined from the governing equation (3.1) as

$$\mathcal{N}[Y(x;q)] = \mathcal{L}[Y(x;q)] + \mathcal{F}[Y(x;q)]. \tag{3.4}$$

By differentiating the zeroth-order equations (3.3) m times with respect to q, setting $q = 0$ and finally dividing the resulting equations by $m!$, we obtain the following mth-order deformation equations,

$$\mathcal{L}[y_m(x) - (\chi_m + \hbar)y_{m-1}(x)] = \hbar R_{m-1}[y_0, y_1, \ldots, y_{m-1}], \tag{3.5}$$

where

$$R_{m-1}[y_0, y_1, \ldots, y_{m-1}] = \frac{1}{(m-1)!}\left.\frac{\partial^{m-1}\left\{\mathcal{F}[Y(x;q)] - \Phi(x)\right\}}{\partial q^{m-1}}\right|_{q=0}, \tag{3.6}$$

and

$$\chi_m = \begin{cases} 0, & m \leqslant 1, \\ 1, & m > 1. \end{cases} \tag{3.7}$$

After obtaining solutions for equation (3.5), the approximate homotopy-series solution for $y(x)$ is determined as the series solution

$$y(x) = \sum_{k=0}^{+\infty} y_k(x). \tag{3.8}$$

A HAM solution is said to be of order M if the above series is truncated at $k = M$, that is, if

$$y(x) = \sum_{m=0}^{M} y_m(x). \tag{3.9}$$

In using the SHAM, the initial guess is obtained simply as a solution of the linear part of the governing equation (3.1) subject to the underlying boundary conditions (3.2). That is, we solve

$$\mathcal{L}[y_0] = \Phi(x). \tag{3.10}$$

In most cases, equation (3.10) together with the subsequent higher order deformation equations (3.5) cannot be solved exactly by means of analytical techniques. Numerical methods such as finite differences, finite element method, spectral method and many others can be used to solve equations of the form (3.5) and (3.10). Spectral methods, such as the Chebyshev pseudo-spectral method, have been found to be very convenient tools for solving the resulting HAM decomposed higher order deformation equations. It is for this reason that the method is referred to as the *spectral* homotopy analysis method. Spectral methods are now becoming the preferred tools for solving ordinary and partial differential equations because of their elegance and high accuracy in resolving problems with smooth functions.

For brevity, we omit the details of the spectral methods, and refer interested readers to Refs. [7, 30]. Before applying the spectral method, it is convenient to transform the domain on which the governing equation is defined to the interval $[-1, 1]$ where the spectral method can be implemented. We use the transformation $x = (b - a)(\tau + 1)/2$ to map the interval $[a, b]$ to $[-1, 1]$. The basic idea behind the spectral collocation method is the introduction of a differentiation matrix D which is used to approximate the derivatives of the unknown variables $y(x)$ at the collocation points (grid points) as the matrix vector product

$$\frac{dy}{dx} = \sum_{k=0}^{N} \mathbf{D}_{lk} y(x_k) = \mathbf{D}\mathbf{Y}, \qquad l = 0, 1, \ldots, N \tag{3.11}$$

where $N + 1$ is the number of collocation points, $\mathbf{D} = 2D/(b - a)$, and

$$\mathbf{Y} = [y(\tau_0), y(\tau_1), \ldots, y(\tau_N)]^T$$

is the vector function at the collocation points. Higher order derivatives are obtained as powers of $\mathbf{D}$, that is

$$y^{(p)} = \mathbf{D}^p \mathbf{Y}. \tag{3.12}$$

where p is the order of the derivative. We choose the Gauss-Lobatto collocation points to define the nodes in $[-1, 1]$ as

$$\tau_j = \cos\left(\frac{\pi j}{N}\right), \qquad j = 0, 1, \ldots, N. \tag{3.13}$$

The matrix D is of size $(N+1)\times(N+1)$ and its entries are defined [7, 30] as

$$\left.\begin{aligned} D_{jk} &= \frac{c_j}{c_k}\frac{(-1)^{j+k}}{\tau_j - \tau_k}, & & j \neq k; j,k = 0,1,\ldots,N, \\ D_{kk} &= -\frac{\tau_k}{2(1-\tau_k^2)}, & & k = 1,2,\ldots,N-1, \\ D_{00} &= \frac{2N^2+1}{6} = -D_{NN}, & & \end{aligned}\right\} \tag{3.14}$$

with

$$c_k = \begin{cases} 2 & k = 0, N \\ 1 & -1 \leq k \leq N-1 \end{cases}. \tag{3.15}$$

In the next section, we consider specific examples which highlight the main features of the SHAM and demonstrate the implementation of the SHAM algorithm.

3.3. Some applications of the spectral homotopy analysis method

3.3.1. *Falkner–Skan boundary layer flow*

In this section we demonstrate the application of the SHAM algorithm in solving the Falkner–Skan boundary layer flow equations in fluid mechanics, governed by the nonlinear differential equation

$$f'''(\eta) + \beta_0 f(\eta) f''(\eta) + \beta_1(1 - f'(\eta)^2) = 0, \quad \eta \in [0, \infty), \tag{3.16}$$

subject to the boundary conditions

$$f(0) = f'(0) = 0, \quad \lim_{\eta\to\infty} f(\eta) = 1. \tag{3.17}$$

where β_0 and β_1 are parameters whose values for certain special classes of flows are given as

(1) *Blasius Flow:* $\beta_0 = \dfrac{1}{2}, \quad \beta_1 = 0.$
(2) *Pohlhausen Flow:* $\beta_0 = 0, \quad \beta_1 = 1.$
(3) *Homann Flow:* $\beta_0 = 2, \quad \beta_1 = 1.$

In applying the SHAM, the linear operator is chosen to be the linear part of the governing equation. However, in the case of equation (3.16), we observe that the linear part is just f''' which suggests that the initial guess should be determined from solving

$$f''' = \beta_0, \quad f(0) = f'(0) = 0, \quad f(\infty) = 1. \tag{3.18}$$

The solution of (3.18) is not exponential and would not be a good initial approximation for the solution of (3.16) which is well known to have exponential solution profiles. To obtain the appropriate linear operator and initial guess to be used in the SHAM solution of (3.16), we set

$$f(\eta) = f_0(\eta) + g(\eta), \qquad f_0(\eta) = \eta - 1 + e^{-\eta} \tag{3.19}$$

where f_0 has been chosen as an exponential function that satisfies the boundary conditions. Substituting equation (3.19) in equation (3.16) gives,

$$g''' + \beta_0 f_0 g'' - 2\beta_1 f_0' g' + \beta_0 f_0'' g + \beta_0 g g'' - \beta_1 (g')^2 + \phi(\eta) = 0, \tag{3.20}$$

subject to the boundary conditions

$$g(0) = g'(0) = g'(\infty) = 0,$$

where

$$\phi(\eta) = f_0'''(\eta) + \beta_0 f_0(\eta) f_0''(\eta) + \beta_1 (1 - f_0'(\eta)^2).$$

By considering the linear part, the initial guess g_0 for solving the nonlinear equation (3.20) using the SHAM is obtained as a solution of

$$\begin{cases} g_0''' + \beta_0 f_0 g_0'' - 2\beta_1 f_0' g_0' + \beta_0 f_0'' g_0 + \phi(\eta) = 0, \\ g_0(0) = g_0'(0) = g_0'(\infty) = 0. \end{cases} \tag{3.21}$$

Using the linear part of (3.20), we choose the linear operator

$$\mathcal{L}(g) = g''' + \beta_0 f_0 g'' - 2\beta_1 f_0' g' + \beta_0 f_0'' g. \tag{3.22}$$

We remark that the linear operator is chosen in such a way that

$$\mathcal{L}(g_0) + \phi(\eta) = 0.$$

Based on the governing equation (3.20), the nonlinear operator is defined as

$$\mathcal{N}(g) = \mathcal{L}(g) + \mathcal{F}(g), \quad \mathcal{F}(g) = \beta_0 g g'' - \beta_1 (g')^2. \tag{3.23}$$

Thus, in the framework of the HAM, the zeroth-order deformation equation becomes

$$(1-q)\mathcal{L}\left[G(\eta; q) - g_0(\eta)\right] = q\hbar \left\{\mathcal{N}[G(\eta; q)] + \phi(\eta)\right\}, \tag{3.24}$$

where $q \in [0, 1]$ is the embedding parameter, $G(\eta; q)$ is a continuous mapping of $g(\eta)$ and $\hbar$ is the convergence-control parameter. The corresponding mth-order deformation equation reads

$$\mathcal{L}[g_m(\eta) - (\chi_m + \hbar)g_{m-1}(\eta)] = \hbar R_{m-1}[g_0, g_1, \ldots, g_{m-1}], \tag{3.25}$$

where

$$R_{m-1}[g_0, g_1, \ldots, g_{m-1}] = \frac{1}{(m-1)!} \frac{\partial^{m-1} \{\mathcal{N}[G(\eta; q)] + \phi(\eta)\}}{\partial q^{m-1}}\bigg|_{q=0}. \tag{3.26}$$

Thus, the homotopy series solution that approximates $f(\eta)$ reads

$$f(\eta) = f_0(\eta) + g_0(\eta) + \sum_{k=1}^{+\infty} g_k(\eta). \tag{3.27}$$

Using the definitions (3.23) and (3.26), the high-order deformation equations are

$$\begin{aligned}\mathcal{L}(g_m) = \hbar(1 - \chi_m)\phi(\eta) + (\chi_m + \hbar)\mathcal{L}(g_{m-1})& \\ + \hbar\beta_0 \sum_{n=0}^{m-1} g_m g''_{m-1-n} - \hbar\beta_1 \sum_{n=0}^{m-1} g'_m g'_{m-1-n},&\end{aligned} \tag{3.28}$$

subject to the boundary condition

$$g_m(0) = g'_m(0) = g'_m(\infty) = 0, \tag{3.29}$$

where $\mathcal{L}$ is defined by (3.22).

When $m = 1$, we have the 1st-order deformation equation

$$\begin{cases} g'''_1 + \beta_0 f_0 g''_1 - 2\beta_1 f'_0 g'_1 + \beta_0 f''_0 g_1 = \hbar\beta_0 g_0 g''_0 - \hbar\beta_1 (g'_0)^2, \\ g_1(0) = g'_1(0) = g'_1(\infty) = 0. \end{cases} \tag{3.30}$$

When $m > 1$, we have

$$\mathcal{L}(g_m) = (1+\hbar)\mathcal{L}(g_{m-1}) + \hbar\beta_0 \sum_{n=0}^{m-1} g_m g''_{m-1-n} - \hbar\beta_1 \sum_{n=0}^{m-1} g'_m g'_{m-1-n}, \tag{3.31}$$

subject to the boundary conditions

$$g_m(0) = g'_m(0) = g'_m(\infty) = 0.$$

Applying the spectral method on equations (3.21), (3.30) and (3.31) gives,

$$\mathbf{A}\mathbf{g}_m = \begin{cases} -\mathbf{\Phi}, & m = 0 \\ \hbar\beta_0 \mathbf{g}_0 \mathbf{g}_0'' - \hbar\beta_1 (\mathbf{g}_0')^2, & m = 1 \\ \mathbf{A}\mathbf{g}_{m-1} + \hbar\beta_0 \sum_{n=0}^{m-1} \mathbf{g}_m \mathbf{g}_{m-1-n}'' - \hbar\beta_1 \sum_{n=0}^{m-1} \mathbf{g}_m' \mathbf{g}_{m-1-n}', & m > 1 \end{cases} \tag{3.32}$$

subject to the boundary conditions

$$g_m(\xi_N) = 0, \quad \sum_{k=0}^{N} \mathbf{D}_{Nk} g_m(\xi_k) = 0, \quad \sum_{k=0}^{N} \mathbf{D}_{0k} g_m(\xi_k) = 0, \tag{3.33}$$

where $\xi = 2\eta/\eta_\infty - 1$ is a variable used to map the domain $[0, \eta_\infty]$ to $[-1, 1]$, η_∞ is a finite value used to numerically approximate the conditions at infinity, and

$$\begin{aligned} \mathbf{A} &= \mathbf{D}^3 + \beta_0 \operatorname{diag}(\mathbf{f}_0)\mathbf{D}^2 - 2\beta_1 \operatorname{diag}(\mathbf{f}_0')\mathbf{D} + \beta_0 \operatorname{diag}(\mathbf{f}_0'') \\ \mathbf{g}_m &= [g_m(\xi_0), g_m(\xi_1), \ldots, g_m(\xi_N)]^T, \\ \mathbf{\Phi} &= [\phi(\eta_0), \phi(\eta_1), \ldots, \phi(\eta_N)]^T, \\ \mathbf{f}_0 &= [f_0(\eta_0), f_0(\eta_1), \ldots, f_0(\eta_N)]^T. \end{aligned}$$

Here, diag() is a function that puts the vector () on the main diagonal. At each level m of the SHAM algorithm, the terms on the right-hand side of (3.32) are known from the previous $m - 1$ level. Thus, starting from the initial guess g_0, the solutions g_m can be obtained by recursively solving equation (3.32).

To improve computational efficiency and accuracy of the SHAM the optimal value of the convergence controlling parameter $\hbar$ must be carefully selected. In previous studies using the SHAM (see for example [11, 20–22]), admissible values of $\hbar$ were selected from a range of values that lie on a horizontal segment of the $\hbar$-curve, which is a plot of the derivative of an unknown function against $\hbar$. The disadvantage of this approach is that different values of $\hbar$ chosen from different locations of the flat segment of the $\hbar$-curve give different speeds of convergence and accuracy of the SHAM.

In this work, we propose two new approaches for identifying the optimal value of $\hbar$. The first approach is based on the observation that different $\hbar$-curves plotted at different SHAM orders seem to intersect at or near one point. We determined through numerical experimentation that this point of intersection of the different $\hbar$-curves gives the optimal $\hbar$. The second approach of choosing the optimal $\hbar$ is based on the residual of the governing equation. We define the maximum residual value on $(0, \infty)$ as

$$E_{max}^{\hbar} = \max\left|\sum_{k=0}^{m} \mathbf{f}_m''' + \beta_0 \sum_{k=0}^{m} \mathbf{f}_m \sum_{k=0}^{m} \mathbf{f}_m'' + \beta_1 \left[1 - \left(\sum_{k=0}^{m} \mathbf{f}_m'\right)^2\right]\right|, \quad (3.34)$$

where $\mathbf{f}$ is the approximate value of f at the collocation points. The optimal value of $\hbar$ is selected to be the value that corresponds to the minimum of the maximum residual curve. The $\hbar$-curve and maximum residual curves are shown in Figs. 3.1, 3.2 and 3.3 for the Blasius, Homann and Pohlhausen flows, respectively. It can be seen from Figs. 3.1, 3.2 and 3.3 that the optimal $\hbar$ obtained using the $\hbar$-curves and the maximum residual curves in each case are equal.

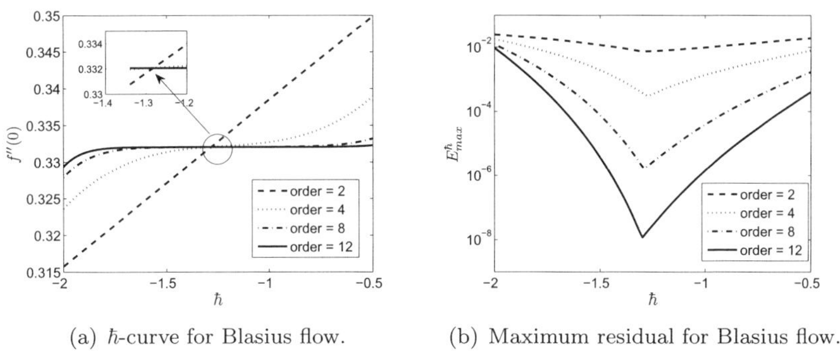

(a) $\hbar$-curve for Blasius flow. (b) Maximum residual for Blasius flow.

Fig. 3.1. Optimal $\hbar$ for the Blasius flow.

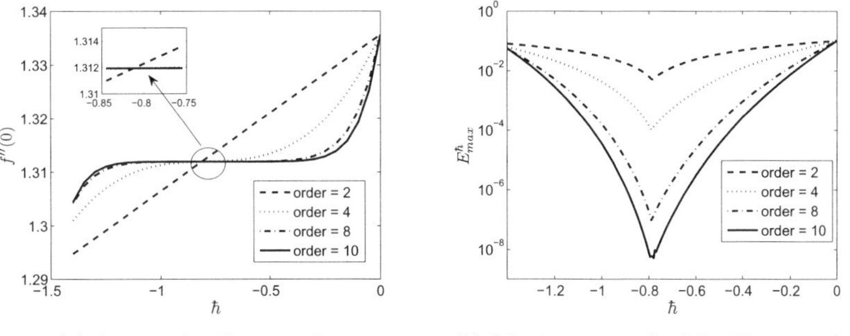

(a) $\hbar$-curve for Homann flow. (b) Maximum residual for Homann flow.

Fig. 3.2. Optimal $\hbar$ for the Homann flow.

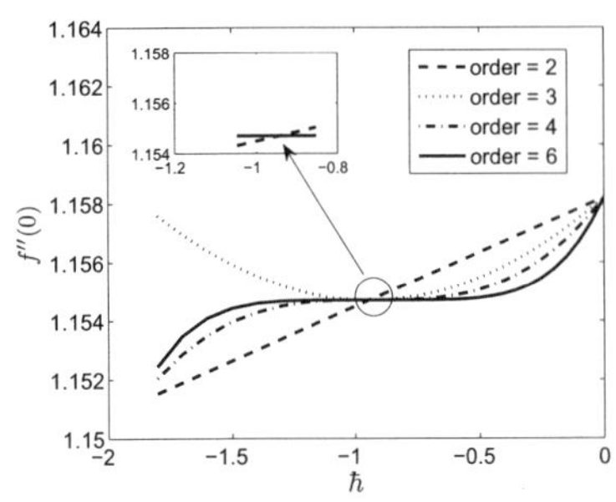

(a) $\hbar$-curve for Pohlhausen flow.

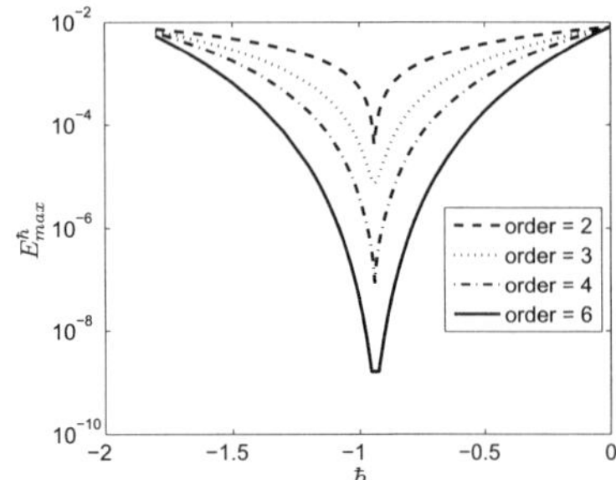

(b) Maximum residual for Pohlhausen flow.

Fig. 3.3. Optimal $\hbar$ for the Pohlhausen flow.

Table 3.1. Skin friction $f''(0)$.

Order	Blasius flow	Homann flow	Pohlhausen flow
2	0.3312181198	1.3125928575	1.1547124162
4	0.3320081757	1.3119529332	1.1547005593
6	0.3320539785	1.3119381208	1.1547005384
8	0.3320570908	1.3119377069	1.1547005384
10	0.3320573177	1.3119376943	1.1547005384
12	0.3320573348	1.3119376939	1.1547005384
14	0.3320573361	1.3119376939	1.1547005384
16	0.3320573362	1.3119376939	1.1547005384
Ref. [12]	0.3320573362	1.3119376939	1.1547005384
Optimal $\hbar$	−1.32	−0.79	−0.94

Table 3.1 presents the results for the wall skin friction rate (defined by $f''(0)$) at different orders of the SHAM for the Blasius, Homann and Pohlhausen flow. The SHAM results are compared against the recently reported results of Ganapol [12] who reported highly accurate results of between 10 and 30 decimal places using a robust algorithm based on Maclaurin series with convergence acceleration and analytical continuation techniques. We observe that all the iteration schemes rapidly converge to the results of cite GANAPOL to all 11 displayed decimal places. Full convergence is achieved after 16 iterations for the Blasius flow, 12 iterations for the Homann and after only 6 iterations in the case of the Pohlhausen flow. The results for Blasius and Homann flows were generated using $\eta_\infty = 30$ and $N = 100$ collocation points and the Pohlhausen flow results were generated using $\eta_\infty = 15$ and $N = 50$ collocation points.

3.3.2. *Eigenvalue problems*

In this section we illustrate how the spectral HAM can be used to solve nonlinear eigenvalue problems.

We consider as our first example the Lane–Emden equation described by the nonlinear singular initial value differential equation

$$y''(x) + \frac{2}{x}y'(x) + y^p = 0, \tag{3.35}$$

with initial conditions

$$y(0) = 1, \quad y'(0) = 0, \tag{3.36}$$

where $p \in [0, 5]$ is a constant parameter. This equation has applications in astrophysics in the study of polytropic models and stellar structures [9, 10]. For the special cases when $p = 0, 1, 5$ exact analytical solutions were obtained by Chandrasekhar [9]. For all other values of p approximate analytical methods and numerical methods are used to approximate the solution of the Lane–Emden equation. Here, we consider the case when $p = 2$. To solve equation (3.35) using the SHAM, it is convenient to recast the problem from an initial value problem to a boundary value problem by considering only the domain $x \in [0, \alpha]$ where α is the first zero of $y(x)$. In most practical applications of the Lane–Emden equation (3.35), the goal is to integrate the governing equation from 0 to α. Since α is an unknown parameter we rescale the problem by setting

$$x = \alpha z. \tag{3.37}$$

Substituting (3.37) in equation (3.35) and simplifying gives

$$y_{zz} + \frac{2}{z}y_z + \lambda y^2 = 0, \quad y(0) = 1, \ y(1) = 0, \ y_z(0) = 0, \tag{3.38}$$

which is a nonlinear eigenvalue problem with $\lambda = \alpha^2$ as the eigenvalue.

To obtain the linear operator and initial guess to be used in the SHAM solution of (3.38), we set

$$y(z) = y_0(z) + g(z), \qquad \lambda = \lambda_0 + \gamma, \quad y_0(z) = 1 - z^2 \tag{3.39}$$

where y_0 has been chosen to satisfy the boundary conditions and λ_0 is an initial approximation to λ. Substituting Eq. (3.39) in Eq. (3.38) gives

$$\begin{cases} g'' + \frac{2}{z}g' + 2\lambda_0 y_0 g + y_0^2\gamma + 2y_0\gamma g + \lambda_0 g^2 + \gamma g^2 + \phi(z) = 0, \\ g(0) = g'(0) = g(1) = 0, \end{cases} \tag{3.40}$$

where the primes denote differentiation with respect to z, and

$$\phi(z) = y_0'' + \frac{2}{z}y_0' + \lambda_0 y_0^2.$$

By considering the linear part, the initial guesses g_0 and γ_0 are obtained as solutions of

$$g_0'' + \frac{2}{z}g_0' + 2\lambda_0 y_0 g_0 + y_0^2\gamma_0 = -\phi, \quad g_0(0) = g_0'(0) = g_0(1) = 0. \tag{3.41}$$

Applying the Chebyshev spectral method to (3.41) and imposing the boundary conditions gives

$$\mathbf{A}\mathbf{G}_0 + \mathbf{y}_0^2\gamma_0 = -\mathbf{\Phi}, \quad g_0(\tau_N) = g_0(\tau_0) = 0, \ \sum_{k=0}^{N} \mathbf{D}_{Nk} g_0(\tau_k) = 0, \tag{3.42}$$

where $\tau = 2z - 1$ is used to transform the interval $z \in [0, 1]$ to $\tau \in [-1, 1]$. Thus, $\mathbf{D} = 2D$ where D is the Chebyshev derivative matrix defined by (3.14) and

$$\begin{aligned}
\mathbf{G}_0 &= [g_0(\tau_0), g_0(\tau_1), \ldots, g_0(\tau_{N-1}), g_0(\tau_N)]^T,\\
\mathbf{\Phi} &= [\phi(z_0), \phi(z_1), \ldots, \phi(z_{N-1}), \phi(z_N)]^T,\\
\mathbf{y}_0 &= [y_0(z_0), y_0(z_1), \ldots, y_0(z_{N-1}), y_0(z_N)]^T,\\
\mathbf{A} &= \mathbf{D}^2 + \text{diag}\left[\frac{2}{\mathbf{z}}\right]\mathbf{D} + 2\lambda_0 \text{diag}\,[\mathbf{y}_0],\\
\mathbf{z} &= [z_0, z_1, z_2, \ldots, z_{N-1}, z_N]^T.
\end{aligned}$$

The equation system (3.42) can be written as the following matrix equation

$$\left[\begin{array}{ccccc|c}
1 & 0 & \cdots & 0 & 0 & 0\\
 & & & & & y_0^2(\tau_1)\\
 & & \mathbf{A} & & & \vdots\\
 & & & & & y_0^2(\tau_{N-1})\\
0 & 0 & \cdots & 0 & 1 & 0\\
\hline
D_{N0} & D_{N1} & \cdots & D_{NN-1} & D_{NN} & 0
\end{array}\right]
\left[\begin{array}{c}
g_0(\tau_0)\\ g_0(\tau_1)\\ \vdots\\ g_0(\tau_{N-1})\\ g_0(\tau_N)\\ \hline \gamma_0
\end{array}\right]
=
\left[\begin{array}{c}
0\\ \phi(z_1)\\ \vdots\\ \phi(z_{N-1})\\ 0\\ \hline 0
\end{array}\right].$$

The linear operator for the SHAM algorithm is defined as

$$\mathcal{L}(g, \gamma) = g'' + \frac{2}{z}g' + 2\lambda_0 y_0 g + y_0^2\gamma. \tag{3.43}$$

We remark that the linear operator is chosen in such a way that

$$\mathcal{L}(g_0, \gamma_0) + \phi(z) = 0.$$

From the governing equation (3.40), the nonlinear operator is defined as

$$\mathcal{N}(g, \gamma) = \mathcal{L}(g, \gamma) + 2y_0\gamma g + \lambda_0 g^2 + \gamma g^2. \tag{3.44}$$

In the framework of the HAM, the zeroth-order deformation equation reads

$$(1-q)\mathcal{L}\left[\{G(z;q), \Gamma\} - \{g_0(z), \gamma_0\}\right] = q\hbar\left\{\mathcal{N}[G(z;q), \Gamma] + \phi(z)\right\}, \tag{3.45}$$

where $q \in [0,1]$ is the embedding parameter, $G(z;q)$, Γ are continuous mappings of $g(z)$ and γ, respectively, and $\hbar$ is the convergence controlling parameter. The mth order deformation equations reads

$$\begin{aligned}\mathcal{L}(g_m, \gamma_m) = {} & (\chi_m + \hbar)\mathcal{L}\left(g_{m-1}, \gamma_{m-1}\right) + \hbar(1-\chi_m)\phi(z) \\ & + 2\hbar y_0 \sum_{n=0}^{m-1} \gamma_n g_{m-1-n} + \hbar\lambda_0 \sum_{n=0}^{m-1} g_n g_{m-1-n} \\ & + \hbar \sum_{n=0}^{m-1} \gamma_{m-1-n} \sum_{i=0}^{n} g_i g_{n-i},\end{aligned} \tag{3.46}$$

subject to the boundary condition

$$g_m(0) = g_m(1) = g'_m(0) = 0, \tag{3.47}$$

where $\mathcal{L}(g, \gamma)$ is defined by (3.43).

Applying the spectral method to equation (3.46) and imposing the boundary conditions gives the following matrix equation,

$$\left[\begin{array}{ccccc|c} 1 & 0 & \cdots & 0 & 0 & 0 \\ & & & & & y_0^2(\tau_1) \\ & \mathbf{A} & & & & \vdots \\ & & & & & y_0^2(\tau_{N-1}) \\ 0 & 0 & \cdots & 0 & 1 & 0 \\ \hline D_{N0} & D_{N1} & \cdots & D_{NN-1} & D_{NN} & 0 \end{array}\right] \left[\begin{array}{c} g_m(\tau_0) \\ g_m(\tau_1) \\ \vdots \\ g_m(\tau_{N-1}) \\ g_m(\tau_N) \\ \hline \gamma_m \end{array}\right]$$

$$
= (\chi_m + \hbar) \left[\begin{array}{ccccc|c} 0 & 0 & \cdots & 0 & 0 & 0 \\ & & & & & y_0^2(\tau_1) \\ & & \mathbf{A} & & & \vdots \\ & & & & & y_0^2(\tau_{N-1}) \\ 0 & 0 & \cdots & 0 & 0 & 0 \\ \hline 0 & 0 & \cdots & 0 & 0 & 0 \end{array} \right] \left[\begin{array}{c} g_{m-1}(\tau_0) \\ g_{m-1}(\tau_1) \\ \vdots \\ g_{m-1}(\tau_{N-1}) \\ g_{m-1}(\tau_N) \\ \hline \gamma_{m-1} \end{array} \right]
$$

$$
+ \hbar(1 - \chi_m) \left[\begin{array}{c} 0 \\ \phi(z_1) \\ \vdots \\ \phi(z_{N-1}) \\ 0 \\ \hline 0 \end{array} \right] + \hbar \left[\begin{array}{c} 0 \\ Q(z_1) \\ \vdots \\ Q(z_{N-1}) \\ 0 \\ \hline 0 \end{array} \right], \tag{3.48}
$$

where

$$
Q(z) = 2\hbar y_0 \sum_{n=0}^{m-1} \gamma_n g_{m-1-n} + \hbar\lambda_0 \sum_{n=0}^{m-1} g_n g_{m-1-n} + \hbar \sum_{n=0}^{m-1} \gamma_{m-1-n} \sum_{i=0}^{n} g_i g_{n-1}.
$$

Thus, the homotopy series solution of $y(z)$ and λ read

$$
y(z) = y_0(z) + g_0(z) + \sum_{k=1}^{+\infty} g_k(z), \quad \lambda = \lambda_0 + \gamma_0 + \sum_{k=1}^{+\infty} \gamma_m. \tag{3.49}
$$

For a fixed λ_0, the minimum of the graph of the maximum residual against $\hbar$ gives the optimal $\hbar$. Optimal λ can be found by fixing $\hbar$ and locating the minimum of the graph of the maximum residual against λ_0. The maximum residual is defined as

$$
E_{max}^{\hbar,\lambda_0} = \max \left| \sum_{k=0}^{m} \mathbf{y}_m'' + \frac{2}{\mathbf{x}} \sum_{k=0}^{m} \mathbf{y}_m' + \left(\sum_{k=0}^{m} \mathbf{y}_m \right)^2 \right|, \tag{3.50}
$$

where $\mathbf{y}$ is the approximate value of y at the collocation points. The maximum residual curves for $\hbar$ and λ_0 are shown in Fig. 3.4. It can be seen from Fig. 3.4 that the minima of the maximum residuals are well defined. The value of the $\hbar$ (and λ_0) at which the minimum of the residual curve is located is the optimal value that gives the best convergence results for the SHAM algorithm.

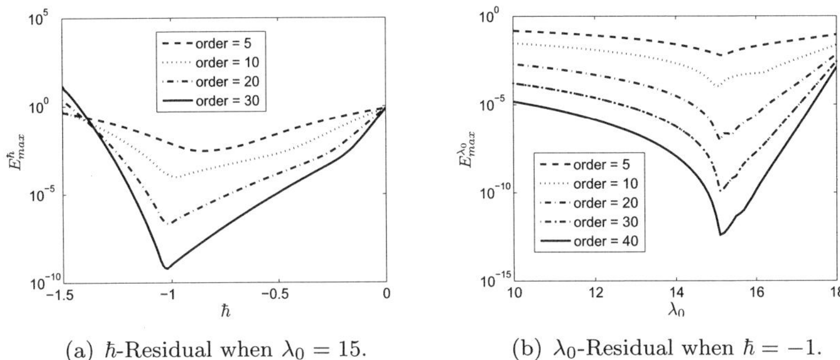

(a) $\hbar$-Residual when $\lambda_0 = 15$. (b) λ_0-Residual when $\hbar = -1$.

Fig. 3.4. Maximum residual curves for the SHAM solution of Lane–Emden equation.

In Table 3.2, we present the SHAM approximate solution of the Lane–Emden equation for the first zero α which is obtained using $\lambda_0 = 15$ and an optimal $\hbar = -1.03$. The results are compared with the recently reported accurate results of [6, 25]. We see that the SHAM results converge to the results of [6, 25].

Table 3.2. First zero α using $\lambda_0 = 15$, $N = 30$ and $\hbar = -1.03$.

Order	First zero α
0	4.3932120926
2	4.3550300471
4	4.3528974961
8	4.3528452341
16	4.3528742136
20	4.3528745567
24	4.3528745919
30	4.3528745958
32	4.3528745959
34	4.3528745959
Ref. [6, 25]	4.3528745959

3.3.3. *Boundary value problems with multiple solutions*

In this section, we illustrate the application of the SHAM in solving nonlinear boundary value problems with multiple solutions. We consider the following model of mixed convection in a porous medium with boundary conditions on the semi-infinite interval $[0, \infty)$ which admits multiple (dual) solutions [1, 18],

$$2f''' + f' - (f')^2 = 0, \quad f(0) = 0, \; f'(0) = 1 + b, \; f'(\infty) = 1, \tag{3.51}$$

where the primes denote differentiation with respect to a similarity variable η, f is a dimensionless stream function, and b is a constant. The nonlinear equation (3.51) was reported [18] to have multiple solutions

$$f'(\eta) = -\frac{1}{2} + \frac{3}{2}\tanh^2\left[\frac{\eta}{2\sqrt{2}} \pm \frac{1}{2}\ln\left(\frac{\sqrt{3}+\sqrt{3+2b}}{\sqrt{3}-\sqrt{3+2b}}\right)\right] \tag{3.52}$$

for any given value of $b \in [-3/2, 0)$.

To solve Eq. (3.51) using the SHAM, it is convenient to reduce the order of the differential equation by introducing the transformation $f' = u$. This results in the reduced equation,

$$2u'' + u - u^2 = 0, \quad u(0) = 1 + b, \; u(\infty) = 1. \tag{3.53}$$

In the context of the SHAM, we begin by choosing the initial guess

$$u_0(\eta) = 1 + (b + \sigma\eta)e^{-\eta}, \tag{3.54}$$

where σ is an unknown constant. We remark that choosing different values of σ leads to different initial guesses. In particular, we observe that when varying σ between negative and positive values, the concavity of the profile of u_0 changes in the region near $\eta = 0$. By fixing the value of the convergence-control parameter $\hbar$ and κ, and varying the values of σ, in the SHAM implementation, optimal values of σ can be identified from the residual of (3.53).

To obtain the linear operator and initial guess to be used in the SHAM solution of (3.53), we set

$$u(\eta) = u_0(\eta) + w(\eta). \tag{3.55}$$

Substituting Eq. (3.55) in Eq. (3.53) gives

$$w'' + \frac{1}{2}w - u_0 w - \frac{1}{2}w^2 + \phi(\eta) = 0, \; w(0) = w(\infty) = 0, \tag{3.56}$$

where

$$\phi(\eta) = u_0'' + \frac{1}{2}u_0 - \frac{1}{2}u_0^2.$$

By considering the linear part, the initial guess w_0 for solving the nonlinear equation (3.56) using the SHAM is obtained as a solution of

$$w_0'' + \left(\frac{1}{2} - u_0\right) w_0 + \phi(\eta) = 0, \qquad w_0(0) = w_0(\infty) = 0. \tag{3.57}$$

The linear operator, for developing the SHAM algorithm is defined as

$$\mathcal{L}(w) = w'' + \left(\frac{1}{2} - u_0\right) w. \tag{3.58}$$

The linear operator is chosen in such a way that

$$\mathcal{L}(w_0) + \phi(\eta) = 0.$$

Thus, the corresponding nonlinear operator becomes

$$\mathcal{N}(w) = \mathcal{L}(w) - \frac{1}{2}w^2. \tag{3.59}$$

In the framework of the HAM, the zeroth-order deformation equation becomes

$$(1-q)\mathcal{L}\left[\tilde{W}(\eta;q) - w_0(\eta)\right] = q\hbar\left\{\mathcal{N}[\tilde{W}(\eta;q)] + \phi(\eta)\right\}, \tag{3.60}$$

where $q \in [0,1]$ is the embedding parameter, $\tilde{W}(\eta;q)$ is the continuous mapping of $w(\eta)$ and $\hbar$ is the convergence controlling parameter. The mth-order deformation equations reads

$$\begin{aligned} w_m'' &+ \left(\frac{1}{2} - u_0\right) w_m \\ &= (\chi_m + \hbar)\left[w_{m-1}'' + \left(\frac{1}{2} - u_0\right) w_{m-1}\right] \\ &\quad + \hbar(1-\chi_m)\phi(\eta) - \frac{1}{2}\hbar\sum_{n=0}^{m-1} w_n w_{m-1-n}, \end{aligned} \tag{3.61}$$

subject to the boundary conditions

$$w_m(0) = w_m(\infty). \tag{3.62}$$

Applying the SHAM in (3.61) it can easily be shown that the SHAM recursive scheme is obtained as

$$W_m = (\chi + \hbar)W_{m-1} + E_c^{-1}\left[\hbar(1-\chi_m)\phi(\eta) - \frac{1}{2}\hbar\sum_{n=0}^{m-1} W_m W_{m-1-n}\right],$$

where

$$E_c = \mathbf{D}^2 - \frac{1}{2}\mathbf{I} - \text{diag}[\mathbf{u}_0],$$

$\mathbf{D}$ is the differentiation matrix, $\mathbf{I}$ is the identity matrix and $\mathbf{u}_0$ is the initial guess evaluated at the collocation points.

Figure 3.5 shows the maximum residual curve for a fixed $\hbar = -1$ plotted against varying values of σ. The graph has two local minima at which the maximum residual is very small and decreases with an increase in the order of the SHAM approximation. Figure 3.6 gives the two solutions of the velocity profile generated using $\sigma = 0$ and $\sigma = -1.3$. The results are compared with the exact solution (3.52) and good agreement is observed between the two results.

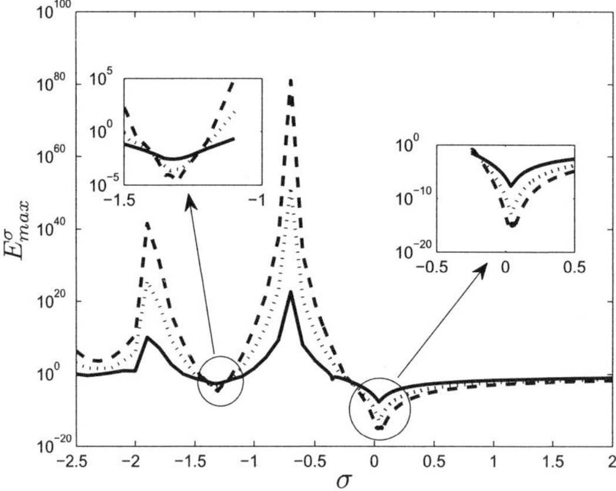

Fig. 3.5. Maximum residual curve when $b = -1$, $\hbar = -1$ order 4 (solid line), order 8 (dotted line), order 12 (dashed line).

3.3.4. *Coupled nonlinear boundary value equations*

In this section we discuss the extension of the SHAM algorithm to coupled systems of nonlinear boundary value problems. For illustration purposes we consider systems of two and three coupled nonlinear differential equations. We begin with the original von Kármán equations for the steady, laminar, axially-symmetric viscous flow induced by an infinitely rotating disk. The governing equations for the problem are given [15, 17, 19] in similarity

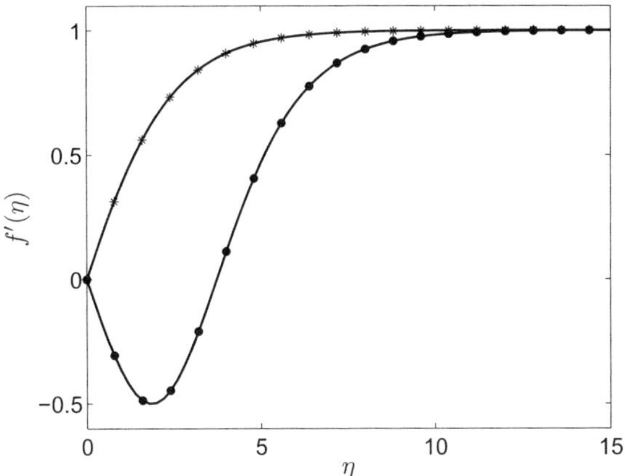

Fig. 3.6. Comparison between exact and SHAM solution when $b = -1$, $\hbar = -1$, $\kappa = 0.5$ using $\sigma = 0$ (upper branch) and $\sigma = -1.3$ (lower branch).

variable form, by

$$H'''(\eta) - H''(\eta)H(\eta) + \frac{1}{2}H'(\eta)H'(\eta) - 2G(\eta)^2 = 0, \tag{3.63}$$

$$G''(\eta) - H(\eta)G'(\eta) + H'(\eta)G(\eta) = 0, \tag{3.64}$$

subject to the boundary conditions

$$H(0) = H'(0) = H'(\infty) = 0, \quad G(0) = 1, \quad G(\infty) = 0, \tag{3.65}$$

where G is the azimuthal velocity and H is the axial velocity.

For effective application of the SHAM on (3.63)–(3.65), we homogenize the system by introducing the following transformations

$$H(\eta) = h(\eta) + H_0(\eta), \quad G(\eta) = g(\eta) + G_0(\eta), \tag{3.66}$$

where H_0 and G_0 are initial approximations that are chosen to satisfy the boundary conditions (3.65). Following [17], we choose, as initial guesses, the following functions

$$H_0(\eta) = -1 + e^{-\eta} + \eta e^{-\eta}, \quad G_0(\eta) = e^{-\eta}. \tag{3.67}$$

Substituting equation (3.66) in (3.63)–(3.65) gives,

$$h''' - H_0 h'' + H_0' h' - H_0'' h - 4G_0 g - h''h + \frac{1}{2}h'^2 - 2g^2 = \phi_1(\eta), \tag{3.68}$$

$$g'' - H_0 g' + H_0' g - G_0' h + G_0 h' - hg' + h'g = \phi_2(\eta), \tag{3.69}$$

subject to the boundary conditions

$$h(0) = h'(0) = h'(\infty) = 0, \quad g(0) = 0, \quad g(\infty) = 0, \tag{3.70}$$

where

$$\phi_1(\eta) = -H_0''' + H_0 H_0'' - \frac{1}{2} H_0' H_0' + 2G_0^2,$$
$$\phi_2(\eta) = -G_0'' + H_0 G_0' - H_0' G_0.$$

The initial approximation to be used in the SHAM algorithm is obtained by solving the linear part of equations (3.68)–(3.70), namely

$$h_0''' - H_0 h_0'' + H_0' h_0' - H_0'' h_0 - 4G_0 g_0 = \phi_1(\eta), \tag{3.71}$$
$$g_0'' - H_0 g_0' + H_0' g_0 - G_0' h_0 + G_0 h_0' = \phi_2(\eta), \tag{3.72}$$

subject to the boundary conditions

$$h_0(0) = h_0'(0) = h_0'(\infty) = 0, \quad g_0(0) = 0, \quad g_0(\infty) = 0. \tag{3.73}$$

The Chebyshev spectral collocation method is then applied to solve h_0 and g_0 of Eqs. (3.71)–(3.73). This gives,

$$\mathbf{A}\mathbf{F}_0 = \mathbf{\Phi}, \tag{3.74}$$

subject to the boundary conditions

$$\sum_{k=0}^{N} \mathbf{D}_{0k} h_0(\xi_k) = 0 \quad , \sum_{k=0}^{N} \mathbf{D}_{Nk} h_0(\xi_k) = 0, \quad h_0(\xi_N) = 0, \tag{3.75}$$
$$g_0(\xi_0) = 0, \quad g_0(\xi_N) = 0, \tag{3.76}$$

where $\xi = 2\eta/\eta_\infty - 1$ is a variable used to map the domain $[0, \eta_\infty]$ to $[-1, 1]$, η_∞ is a finite value used to numerically approximate the conditions at infinity, N is the number of collocation points, and

$$\mathbf{A} = \begin{pmatrix} \mathbf{D}^3 - \mathbf{H}_0\mathbf{D}^2 + \mathbf{H}_0'\mathbf{D} - \mathbf{H}_0'' & -4\mathbf{G}_0 \\ \mathbf{G}_0\mathbf{D} - \mathbf{G}_0' & \mathbf{D}^2 - \mathbf{H}_0\mathbf{D} + \mathbf{H}_0' \end{pmatrix}, \tag{3.77}$$

$$\begin{aligned} \mathbf{F}_0 &= [h_0(\xi_0), h_0(\xi_1), \ldots, h_0(\xi_N), g_0(\xi_0), g_0(\xi_1), \ldots, g_0(\xi_N)]^T, \\ \mathbf{\Phi} &= [\phi_1(\eta_0), \phi_1(\eta_1), \ldots, \phi_1(\eta_N), \phi_2(\eta_0), \phi_2(\eta_1), \ldots, \phi_2(\eta_N)]^T, \\ \mathbf{H}_0 &= diag[H_0(\eta_0), H_0(\eta_1), \ldots, H_0(\eta_N)], \\ \mathbf{G}_0 &= diag[G_0(\eta_0), G_0(\eta_1), \ldots, G_0(\eta_N)]. \end{aligned} \tag{3.78}$$

To obtain the SHAM solution of (3.68) and (3.69) we begin by defining the linear operators

$$\mathcal{L}_h[\tilde{h}, \tilde{g}] = \tilde{h}''' - H_0\tilde{h}'' + H_0'\tilde{h}' - H_0''\tilde{h} - 4G_0\tilde{g}, \tag{3.79}$$

$$\mathcal{L}_g[\tilde{h}, \tilde{g}] = \tilde{g}'' - H_0\tilde{g}' + H_0'\tilde{g} - G_0'\tilde{h} + G_0\tilde{h}', \tag{3.80}$$

where $q \in [0, 1]$ is the embedding parameter, and $\tilde{h}(\xi; q)$ and $\tilde{g}(\xi; q)$ are unknown functions. We observe that the linear operators (3.79–3.80) are coupled. This is one of the main features of the SHAM when applied to nonlinear systems of BVP governing by two or more coupled equations.

The zeroth-order deformation equations read

$$(1-q)\mathcal{L}_h[\{\tilde{h}(\eta; q), \tilde{g}(\eta; q)\} - h_0] = q\hbar\left\{\mathcal{N}_h[\tilde{h}(\eta; q), \tilde{g}(\eta; q)] - \phi_1\right\}, \tag{3.81}$$

$$(1-q)\mathcal{L}_g[\{\tilde{h}(\eta; q), \tilde{g}(\eta; q)\} - g_0] = q\hbar\left\{\mathcal{N}_g[\tilde{h}(\eta; q), \tilde{g}(\eta; q)] - \phi_2\right\}, \tag{3.82}$$

where

$$\mathcal{N}_h[h, g] = \mathcal{L}_h[h, g] - h''h + \frac{1}{2}h'^2 - 2g^2, \tag{3.83}$$

$$\mathcal{N}_g[h, g] = \mathcal{L}_g[h, g] - hg' + h'g. \tag{3.84}$$

From (3.81)–(3.84), it can be shown that the high-order deformation equations are given by

$$\mathcal{L}_h[h_m, g_m] = (\hbar + \chi_m)\mathcal{L}_h[h_{m-1}, g_{m-1}] - \phi_1(\eta)\hbar(1 - \chi_m) + \hbar\sum_{n=0}^{m-1}\left(\frac{1}{2}h_n'h_{m-1-n}' - h_nh_{m-1-n}'' - 2g_ng_{m-1-n}\right), \tag{3.85}$$

$$\mathcal{L}_g[h_m, g_m] = (\hbar + \chi_m)\mathcal{L}_g[h_{m-1}, g_{m-1}] - \phi_2(\eta)\hbar(1 - \chi_m) + \hbar\sum_{n=0}^{m-1}(h_n'g_{m-1-n} - g_n'h_{m-1-n}), \tag{3.86}$$

subject to the boundary conditions

$$h_m(0) = h_m'(0) = h_m'(\infty) = 0, \quad g_m(0) = g_m(\infty) = 0. \tag{3.87}$$

Note that the high-order deformation equations (3.85)–(3.87) are coupled.

Applying the Chebyshev pseudo-spectral method to Eqs. (3.85)–(3.87) gives

$$\mathbf{AF}_m = (\chi_m + \hbar)\mathbf{AF}_{m-1} - \hbar(1 - \chi_m)\mathbf{\Phi} + \hbar\mathbf{Q}_{m-1}, \tag{3.88}$$

subject to the boundary conditions

$$\sum_{k=0}^{N} \mathbf{D}_{0k} h_m(\xi_k) = 0, \quad \sum_{k=0}^{N} \mathbf{D}_{Nk} h_m(\xi_k) = 0, \quad h_m(\xi_N) = 0, \tag{3.89}$$

$$g_m(\xi_0) = 0, \quad g_m(\xi_N) = 0, \tag{3.90}$$

where $\mathbf{A}$ and $\mathbf{\Phi}$ are respectively defined by (3.77) and (3.78), and

$$\mathbf{F}_m = [h_m(\xi_0), h_m(\xi_1), \ldots, h_m(\xi_N), g_m(\xi_0), g_m(\xi_1), \ldots, g_m(\xi_N)]^T,$$

$$\mathbf{Q}_{m-1} = \begin{pmatrix} \sum_{n=0}^{m-1} \left[\frac{1}{2}(\mathbf{Dh}_n)(\mathbf{Dh}_{m-1-n}) - \mathbf{h}_n(\mathbf{D}^2\mathbf{h}_{m-1-n}) - 2\mathbf{g}_n\mathbf{g}_{m-1-n} \right] \\ \sum_{n=0}^{m-1} \left[(\mathbf{Dh}_n)\mathbf{g}_{m-1-n} - (\mathbf{Dg}_n)\mathbf{h}_{m-1-n} \right] \end{pmatrix}.$$

Thus, starting from the initial approximation, which is obtained from (3.74), higher order approximations $F_m(\xi)$ for $m \geq 1$, can be obtained through the recursive formula (3.88).

In Table 3.3 we give the SHAM computed values of $H(\infty)$, $H''(0)$ and $G'(0)$ at different orders of approximation. The results are compared against results generated using MATLAB's `bvp4c` routine for solving boundary value problems. It can be seen that full convergence to the `bvp4c` results is achieved after ten or twelve iterations.

Table 3.3. SHAM values of $H(\infty)$, $H''(0)$, $G'(0)$ at different orders of approximation when $\hbar = -1$, $N = 100$ and $\eta_\infty = 22$.

Order	$H(\infty)$	$H''(0)$	$G'(0)$
2	−0.88424618	−1.02167240	−0.61499561
4	−0.88447172	−1.02040332	−0.61594691
6	−0.88447466	−1.02046906	−0.61592106
8	−0.88447401	−1.02046498	−0.61592205
10	−0.88447410	−1.02046526	−0.61592201
12	−0.88447409	−1.02046524	−0.61592201
14	−0.88447409	−1.02046524	−0.61592201
`bvp4c`	−0.88447409	−1.02046524	−0.61592201

In Fig. 3.7 we give a comparison between the SHAM and `bvp4c` generated results for the velocity profiles $H(\eta)$ and $G(\eta)$. We observe that there is good agreement between the two results.

Next, we consider the SHAM application on a three-equation system that models the problem of unsteady free convective heat and mass transfer

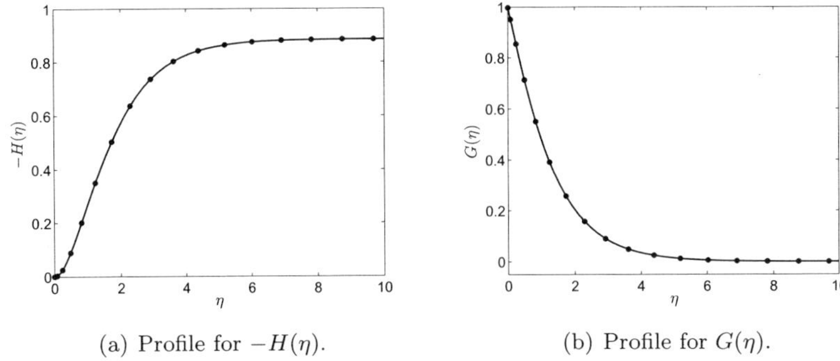

(a) Profile for $-H(\eta)$.

(b) Profile for $G(\eta)$.

Fig. 3.7. Comparison between SHAM (solid line) and `bvp4c` results for the velocity profiles.

on a stretching surface in a porous medium in the presence of a chemical reaction. The governing equations [8, 24] for this problem are given as the following dimensionless system of equations

$$f''' + ff'' - (f')^2 - Kf' - A\left(f' + \frac{\eta}{2}f''\right) + Gr\theta + Gc\phi = 0, \tag{3.91}$$

$$\frac{1}{Pr}\theta'' - f'\theta + f\theta' - A\left(\theta + \frac{1}{2}\eta\theta'\right) = 0, \tag{3.92}$$

$$\frac{1}{Sc}\phi'' - f'\phi + f\phi' - A\left(\phi + \frac{1}{2}\eta\phi'\right) - \gamma\phi = 0, \tag{3.93}$$

subject to the boundary conditions

$$f(0) = f_w, \quad f'(0) = 1, \quad \theta(0) = 1, \quad \phi(0) = 1, \tag{3.94}$$

$$f'(\infty) = 0\ , \quad \theta(\infty) = 0, \quad \phi(\infty) = 0, \tag{3.95}$$

where $f(\eta)$, $\theta(\eta)$ and $\phi(\eta)$ are, respectively, the dimensionless velocity, temperature and concentration, f_w is the suction/injection parameter, γ is the chemical reaction constant, Pr is the Prandtl number, Sc is the Schmidt number, K is the permeability parameter, Gr and Gc are the temperature and concentration dependent Grashof numbers respectively.

To apply the SHAM on (3.91)–(3.95), we begin by homogenizing the system by introducing the following transformations

$$f(\eta) = F(\eta) + f_0(\eta), \quad \theta(\eta) = G(\eta) + \theta_0(\eta), \quad \phi(\eta) = H(\eta) + \phi_0(\eta) \tag{3.96}$$

where f_0, θ_0 and ϕ_0 are initial approximations that are chosen to satisfy the boundary conditions (3.94)–(3.95). The appropriate initial guesses are chosen as

$$f_0(\eta) = f_w + 1 - e^{-\eta}, \quad \theta_0(\eta) = e^{-\eta}, \quad \phi_0(\eta) = e^{-\eta}. \tag{3.97}$$

Substituting equation (3.96) in (3.91)–(3.95) gives,

$$F''' - KF' - A\left(F' + \frac{\eta}{2}F''\right) + f_0F'' - 2f_0'F' + f_0''F + FF'' - (F')^2 + GrG + GcH = \phi_1, \tag{3.98}$$

$$\frac{1}{Pr}G'' - A\left(G + \frac{1}{2}\eta G'\right) - f_0'G + f_0G' - \theta_0F' + \theta_0'F - F'G + FG' = \phi_2, \tag{3.99}$$

$$\frac{1}{Sc}H'' - A\left(H + \frac{1}{2}\eta H'\right) - \gamma H - f_0'H + f_0H' - \phi_0F' + \phi_0'F - F'H + FH' = \phi_3, \tag{3.100}$$

subject to the boundary conditions

$$F(0) = 0, \quad F'(0) = 0, \quad G(0) = 0, \quad H(0) = 0, \tag{3.101}$$

$$F'(\infty) = 0\ , \quad G(\infty) = 0, \quad H(\infty) = 0, \tag{3.102}$$

where

$$\phi_1(\eta) = -\left(f''' + ff'' - (f')^2 - Kf' - A\left(f' + \frac{\eta}{2}f''\right) + Gr\theta + Gc\phi\right),$$

$$\phi_2(\eta) = -\left(\frac{1}{Pr}\theta'' - f'\theta + f\theta' - A\left(\theta + \frac{1}{2}\eta\theta'\right)\right),$$

$$\phi_3(\eta) = -\left(\frac{1}{Sc}\phi'' - f'\phi + f\phi' - A\left(\phi + \frac{1}{2}\eta\phi'\right) - \gamma\phi\right).$$

The initial approximation to be used in the SHAM algorithm is obtained by solving the linear part of equations (3.98)–(3.100), subject to the boundary conditions (3.101)–(3.102), that is, we solve

$$F_0''' - KF_0' - A\left(F_0' + \frac{\eta}{2}F_0''\right) + f_0F_0'' - 2f_0'F_0' + f_0''F_0 + GrG_0 + GcH_0 = \phi_1,$$

$$\frac{1}{Pr}G_0'' - A\left(G_0 + \frac{1}{2}\eta G_0'\right) - f_0'G_0 + f_0G_0' - \theta_0F_0' + \theta_0'F_0 = \phi_2,$$

$$\frac{1}{Sc}H_0'' - A\left(H_0 + \frac{1}{2}\eta H_0'\right) - \gamma H_0 - f_0'H_0 + f_0H_0' - \phi_0F_0' + \phi_0'F_0 = \phi_3,$$

subject to the boundary conditions

$$F_0(0) = 0, \quad F_0'(0) = 0, \quad G_0(0) = 0, \quad H_0(0) = 0, \tag{3.103}$$

$$F_0'(\infty) = 0\ , \quad G_0(\infty) = 0, \quad H_0(\infty) = 0. \tag{3.104}$$

The Chebyshev spectral collocation method is then applied to solve the above linear equations about F_0, G_0 and H_0. This gives,

$$\mathbf{B\Psi}_0 = \mathbf{\Phi}, \tag{3.105}$$

subject to the boundary conditions

$$\sum_{k=0}^{N} \mathbf{D}_{0k} F_0(\xi_k) = 0\,, \quad \sum_{k=0}^{N} \mathbf{D}_{Nk} F_0(\xi_k) = 0, \quad F_0(\xi_N) = 0, \tag{3.106}$$

$$G_0(\xi_0) = 0, \quad G_0(\xi_N) = 0, \quad H_0(\xi_0) = 0, \quad H_0(\xi_N) = 0, \tag{3.107}$$

where

$$\begin{cases} \mathbf{\Psi}_0 = [F_0(\xi_0), \ldots, F_0(\xi_N), G_0(\xi_0), \ldots, G_0(\xi_N), H_0(\xi_0), \ldots, H_0(\xi_N)]^T, \\ \mathbf{\Phi} = [\phi_1(\eta_0), \ldots, \phi_1(\eta_N), \phi_2(\eta_0), \ldots, \phi_2(\eta_N), \phi_3(\eta_0), \ldots, \phi_3(\eta_N)]^T, \\ \mathbf{F}_0 = diag[F_0(\eta_0), F_0(\eta_1), \ldots, F_0(\eta_N)], \\ \mathbf{H}_0 = diag[H_0(\eta_0), H_0(\eta_1), \ldots, H_0(\eta_N)], \\ \mathbf{G}_0 = diag[G_0(\eta_0), G_0(\eta_1), \ldots, G_0(\eta_N)], \end{cases} \tag{3.108}$$

and

$$\mathbf{B} = \begin{pmatrix} B_{11} & B_{12} & B_{13} \\ B_{21} & B_{22} & B_{23} \\ B_{31} & B_{32} & B_{33} \end{pmatrix} \tag{3.109}$$

with

$$\begin{aligned} B_{11} &= \mathbf{D}^3 - K\mathbf{D}' - A\left(\mathbf{D} + diag\left(\frac{\eta}{2}\right)\mathbf{D}^2\right) + diag(\mathbf{f}_0)\mathbf{D}^2 \\ &\quad - 2\,diag(\mathbf{f}_0')\mathbf{D} + diag(\mathbf{f}_0''), \\ B_{12} &= Gr\mathbf{I}, \quad B_{13} = Gc\mathbf{I}, \quad B_{21} = -diag(\theta_\mathbf{0})\mathbf{D} + diag(\theta_0'), \\ B_{22} &= \frac{1}{Pr}\mathbf{D}^2 - A\left(\mathbf{I} + \frac{1}{2}diag(\eta)\mathbf{D}\right) - diag(\mathbf{f}_0')G_0 + diag(\mathbf{f}_0)\mathbf{D}, \\ B_{23} &= \mathbf{O}, \quad B_{31} = -diag(\phi_\mathbf{0})\mathbf{D} + diag(\phi_0'), \quad B_{32} = \mathbf{O}, \\ B_{33} &= \frac{1}{Sc}\mathbf{D}^2 - A\left(\mathbf{I} + \frac{1}{2}diag(\eta)\mathbf{D}\right) - \gamma\mathbf{I} - diag(\mathbf{f}_0')G_0 + diag(\mathbf{f}_0)\mathbf{D}. \end{aligned}$$

The linear operator to be used in the SHAM solution of (3.91)–(3.95) is given by

$$\mathcal{L}_1[F,G,H] = F''' - KF' - A\left(F' + \frac{\eta}{2}F''\right) + f_0F'' - 2f_0'F' \\ + f_0''F + GrG + GcH, \tag{3.110}$$

$$\mathcal{L}_2[F,G,H] = \frac{1}{Pr}G'' - A\left(G + \frac{1}{2}\eta G'\right) - f_0'G + f_0G' - \theta_0F' \\ + \theta_0'F, \tag{3.111}$$

$$\mathcal{L}_3[F,G,H] = \frac{1}{Sc}H'' - A\left(H + \frac{1}{2}\eta H'\right) - \gamma H - f_0'H + f_0H' \\ - \phi_0F' + \phi_0'F. \tag{3.112}$$

The corresponding zeroth-order deformation equations are given by

$$(1-q)\mathcal{L}_1[\{\tilde{F}(\eta;q), \tilde{G}(\eta;q), \tilde{H}(\eta;q)\} - \{F_0, G_0, H_0\}] \\ = q\hbar\left\{\mathcal{N}_1[\tilde{F}(\eta;q), \tilde{G}(\eta;q), \tilde{H}(\eta;q)] - \phi_1\right\}, \tag{3.113}$$

$$(1-q)\mathcal{L}_2[\{\tilde{F}(\eta;q), \tilde{G}(\eta;q), \tilde{H}(\eta;q)\} - \{F_0, G_0, H_0\}] \\ = q\hbar\left\{\mathcal{N}_2[\tilde{F}(\eta;q), \tilde{G}(\eta;q), \tilde{H}(\eta;q)] - \phi_2\right\}, \tag{3.114}$$

$$(1-q)\mathcal{L}_3[\{\tilde{F}(\eta;q), \tilde{G}(\eta;q), \tilde{H}(\eta;q)\} - \{F_0, G_0, H_0\}] \\ = q\hbar\left\{\mathcal{N}_3[\tilde{F}(\eta;q), \tilde{G}(\eta;q), \tilde{H}(\eta;q)] - \phi_3\right\}, \tag{3.115}$$

where

$$\mathcal{N}_1[F,G,H] = \mathcal{L}_1[F,G,H] + FF'' - (F')^2, \tag{3.116}$$

$$\mathcal{N}_2[F,G,H] = \mathcal{L}_2[F,G,H] - F'G + FG', \tag{3.117}$$

$$\mathcal{N}_3[F,G,H] = \mathcal{L}_3[F,G,H] - F'H + FH'. \tag{3.118}$$

From (3.113)–(3.118), it can be shown that the high-order deformation equations read

$$\mathcal{L}_1[F_m, G_m, H_m] = (\hbar + \chi_m)\mathcal{L}_1[F_{m-1}, G_{m-1}, H_{m-1}] - \phi_1(\eta)\hbar(1 - \chi_m) + \hbar \sum_{n=0}^{m-1} \left(F_n F''_{m-1-n} - F'_n F'_{m-1-n}\right), \tag{3.119}$$

$$\mathcal{L}_2[F_m, G_m, H_m] = (\hbar + \chi_m)\mathcal{L}_2[F_{m-1}, G_{m-1}, H_{m-1}] - \phi_2(\eta)\hbar(1 - \chi_m) + \hbar \sum_{n=0}^{m-1} (F_n G'_{m-1-n} - F'_n G_{m-1-n}), \tag{3.120}$$

$$\mathcal{L}_3[F_m, G_m, H_m] = (\hbar + \chi_m)\mathcal{L}_3[F_{m-1}, G_{m-1}, H_{m-1}] - \phi_3(\eta)\hbar(1 - \chi_m) + \hbar \sum_{n=0}^{m-1} (F_n H'_{m-1-n} - F'_n H_{m-1-n}), \tag{3.121}$$

subject to the boundary conditions

$$\begin{cases} F_m(0) = F'_m(0) = F'_m(\infty) = 0, \quad G_m(0) = G_m(\infty) = 0, \\ H_m(0) = H_m(\infty) = 0. \end{cases} \tag{3.122}$$

Applying the Chebyshev pseudo-spectral transformation to equations (3.119)–(3.122) gives

$$\mathbf{B}\mathbf{P}_m = (\chi_m + \hbar)\mathbf{B}\mathbf{P}_{m-1} - \hbar(1 - \chi_m)\mathbf{\Phi} + \hbar\mathbf{Q}_{m-1}, \tag{3.123}$$

subject to the boundary conditions

$$\sum_{k=0}^{N} \mathbf{D}_{0k} F_m(\xi_k) = 0, \quad \sum_{k=0}^{N} \mathbf{D}_{Nk} F_m(\xi_k) = 0, \quad F_m(\xi_N) = 0, \tag{3.124}$$

$$G_m(\xi_0) = 0, \quad G_m(\xi_N) = 0, \quad H_m(\xi_0) = 0, \quad H_m(\xi_N) = 0 \tag{3.125}$$

where $\mathbf{B}$ and $\mathbf{\Phi}$ are defined by (3.108) and (3.109), respectively, and

$$\mathbf{P}_m = [F_m(\xi_0), \ldots, F_m(\xi_N), G_m(\xi_0), \ldots, G_m(\xi_N), H_m(\xi_0), \ldots, H_m(\xi_N),]^T,$$

$$\mathbf{Q}_{m-1} = \begin{pmatrix} \sum_{n=0}^{m-1} \left[\mathbf{F}_n(\mathbf{D}^2\mathbf{F}_{m-1-n}) - (\mathbf{D}\mathbf{F}_n)(\mathbf{D}\mathbf{F}_{m-1-n})\right] \\ \sum_{n=0}^{m-1} [(\mathbf{D}\mathbf{G}_n)\mathbf{F}_{m-1-n} - (\mathbf{D}\mathbf{F}_n)\mathbf{G}_{m-1-n}] \\ \sum_{n=0}^{m-1} [(\mathbf{D}\mathbf{H}_n)\mathbf{F}_{m-1-n} - (\mathbf{D}\mathbf{F}_n)\mathbf{H}_{m-1-n}], \end{pmatrix}.$$

Thus, starting from the initial approximation, which is obtained as a solution of equations (3.105)–(3.106), the SHAM approximate solutions for f, θ, ϕ are obtained by recursively solving equations (3.123)–(3.124).

In Table 3.4 we present the SHAM computations of flow properties, namely the skin friction $f''(0)$, surface heat transfer rate at the $\theta'(0)$ and mass transfer rate at the wall $\phi'(0)$. The accuracy of the SHAM results are verified by comparing with the MATLAB in-built routine `bvp4c`. It can be seen from the results that the SHAM results converge to the `bvp4c` results.

Table 3.4. SHAM values of $f''(0)$, $\theta'(0)$, $\phi'(0)$ at different orders of approximation when $\hbar = -1$, $N = 100$ and $\eta_\infty = 30$.

Order	$f''(0)$	$\theta'(0)$	$\phi'(0)$
2	−1.58973019	−1.89527636	−2.24305196
4	−1.59345387	−1.89839841	−2.24540546
6	−1.59362599	−1.89853714	−2.24548919
8	−1.59363732	−1.89854621	−2.24549407
10	−1.59363819	−1.89854691	−2.24549442
12	−1.59363826	−1.89854697	−2.24549445
14	−1.59363827	−1.89854697	−2.24549446
16	−1.59363827	−1.89854697	−2.24549446
`bvp4c`	−1.59363827	−1.89854697	−2.24549446

3.4. Convergence acceleration

3.4.1. *Convergence acceleration through choice of linear operator*

In the framework of the HAM, there is great freedom to choose the linear operator for the zeroth-order and higher order deformation equations. In this section, we suggest a general approach that can be used to accelerate the convergence of the SHAM. We illustrate how using a different linear operator can significantly improve convergence of the SHAM by considering the following Darcy–Brinkman–Forchheimer equation that models the steady state pressure driven fully-developed parallel flow through a horizontal channel that is filled with porous media [2, 20, 28],

$$\frac{d^2y}{dx^2} - s^2y - Fsy^2 + \frac{1}{M} = 0, \quad y(-1) = 0, \quad y(1) = 0, \tag{3.126}$$

where F is the dimensionless the Forchheimer number and s is the porous media shape parameter. This problem was previously solved using the

SHAM in [20] In the context of the SHAM, the initial guess is chosen to be

$$y_0(x) = \frac{1}{s^2 M}\left(1 - \frac{\cosh(sx)}{\cosh(s)}\right), \tag{3.127}$$

which is the solution of the linear part of (3.126). Using the linear part of (3.126) to form the linear operator for the zeroth-order deformation equations, it can be shown that the high-order deformation scheme (see [20] for details) is given by

$$\begin{aligned} y_m'' - s^2 y_m &= (\chi_m + \hbar)(y_{m-1}'' - s^2 y_{m-1}) + \frac{\hbar}{M}(1 - \chi_m) \\ &\quad - \hbar F s \sum_{n=0}^{m-1} y_n y_{m-1-n}, \end{aligned} \tag{3.128}$$

subject to the boundary conditions

$$y_m(-1) = y_m(1) = 0. \tag{3.129}$$

Starting from the initial approximation (3.127), the high-order deformation equations (3.128) can be solved iteratively for y_m, $m \geq 1$ and the approximate solution $y(x)$ is given, in series form, as

$$y(x) = \sum_{m=0}^{+\infty} y_m(x). \tag{3.130}$$

The modified linear operator is obtained by introducing the transformation

$$y(x) = y_0(x) + u(x), \tag{3.131}$$

where $y_0(x)$ is the initial approximation given by (3.127). Substituting (3.131) into the governing equation (3.126) gives,

$$u'' - s^2 u - 2Fsy_0 u - Fsu^2 + \phi(x) = 0, \quad u(1) = u(-1) = 0, \tag{3.132}$$

where

$$\phi(x) = y_0'' - s^2 y_0 + \frac{1}{M} - Fsy_0^2.$$

Thus, the initial guess for solving the nonlinear equation (3.132) using the SHAM is

$$u_0'' - s^2 u_0 - 2Fsy_0 u_0 + \phi(x) = 0, \quad u_0(1) = u_0(-1) = 0. \tag{3.133}$$

The linear operator for is chosen as function of $y_0(x)$ as

$$\mathcal{L}(u) = u'' - s^2 u - 2Fsy_0 u. \tag{3.134}$$

We remark that the linear operator is chosen in such a way that

$$\mathcal{L}(u_0) + \phi(x) = 0.$$

Based on the governing equation (3.132), the nonlinear operator is defined as

$$\mathcal{N}(u) = \mathcal{L}(u) + \mathcal{N}_1(u), \quad \mathcal{N}_1(u) = -Fsu^2. \tag{3.135}$$

Thus, in the framework of the HAM, the zeroth-order deformation equation becomes

$$(1-q)\mathcal{L}\left[U(x;q) - u_0(x)\right] = q\hbar\left\{\mathcal{N}[U(x;q)] + \phi(x)\right\}, \tag{3.136}$$

where $q \in [0,1]$ is the embedding parameter, $U(x;q)$ is a continuous mapping of $u(x)$ and $\hbar$ is the convergence controlling parameter. The mth-order deformation equations corresponding to (3.136) are given by

$$\mathcal{L}[u_m(x) - (\chi_m + \hbar)u_{m-1}(x)] = \hbar R_{m-1}[u_0, u_1, \ldots, u_{m-1}], \tag{3.137}$$

where

$$R_{m-1}[u_0, u_1, \ldots, u_{m-1}] = \frac{1}{(m-1)!}\frac{\partial^{m-1}\left\{\mathcal{N}_1[U(x;q)] + \phi(x)\right\}}{\partial q^{m-1}}\bigg|_{q=0}.$$

Thus, the homotopy series solution that approximates $y(x)$ reads

$$y(x) = y_0(x) + u_0(x) + \sum_{k=1}^{+\infty} u_k(x). \tag{3.138}$$

Using the definitions (3.134) and (3.135), the high-order deformation equations are given as

$$\begin{aligned} u_m'' - s^2 u_m - 2Fsy_0u_m &= (\chi_m + \hbar)(u_{m-1}'' - s^2 u_{m-1} - 2Fsy_0u_{m-1}) \\ &\quad + \hbar(1-\chi_m)\phi(x) - \hbar Fs\sum_{n=0}^{m-1} u_m u_{m-1-n}. \end{aligned} \tag{3.139}$$

From equation (3.139) it can easily be shown that the SHAM recursive scheme is obtained as

$$U_m = (\chi + \hbar)U_{m-1} + E_s^{-1}\left[\hbar(1-\chi_m)\phi(x) - \hbar Fs\sum_{n=0}^{m-1} U_m U_{m-1-n}\right],$$

where

$$E_s = \mathbf{D}^2 - s^2 I - 2Fs\,\mathrm{diag}[y_0].$$

Table 3.5 gives a comparison between the results obtained using equations (3.130) and (3.138) for the maximum residual obtained using the optimal $\hbar$, in each case, at different orders of approximation. The results for the basic SHAM implementation (3.130) are labelled as SHAM1 and the results for the version of the SHAM that incorporates the initial guess y_0 in the linear operator (3.138) are labelled as SHAM2. It can be seen from the Table 3.5 that the convergence in the SHAM2 results is significantly faster than that of SHAM1. This shows that the convergence can be accelerated by incorporating the initial guess y_0 and the parameter F into the linear operator. In general, using different linear operators in the SHAM implementation is expected to give results with different levels of accuracy.

Table 3.5. Maximum residual at selected optimal values of $\hbar$ when $M = s = 1$, $F = 2$, $N = 101$.

	SHAM 1		SHAM 2	
Order	Optimal $\hbar$	$E^{\hbar}_{\max}$	Optimal $\hbar$	$E^{\hbar}_{\max}$
4	−0.91	7.92E-04	−1	3.60E-07
6	−0.97	3.97E-04	−1	1.04E-09
8	−0.99	3.68E-04	−1	1.05E-10
10	−1.00	6.75E-04	−1	1.05E-10
12	−1.00	2.66E-04	−1	1.05E-10
14	−1.00	1.09E-04	−1	1.05E-10
16	−1.00	4.59E-05	−1	1.05E-10

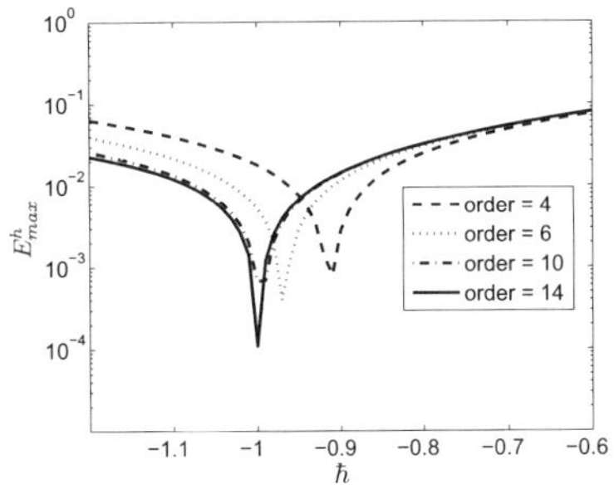

(a) Residual curve for SHAM1.

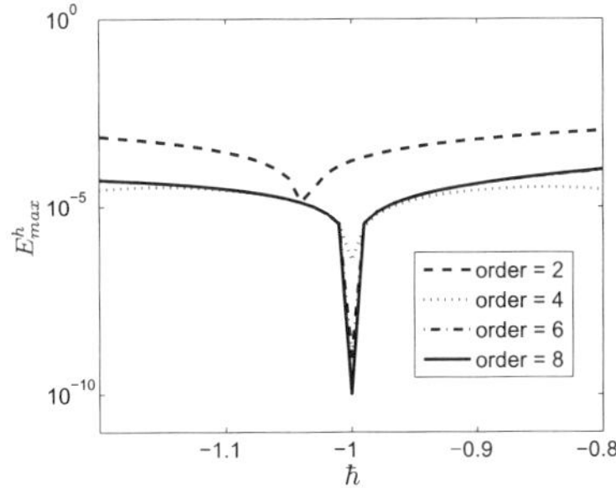

(b) Residual curve for SHAM2.

Fig. 3.8. Maximum residual curve for $s = M = 1$, $F = 2$, $N = 101$.

Figures 3.8 show the maximum residual curves, that can be used to locate the optimal value of the convergence controlling parameter $\hbar$ in both the SHAM1 and SHAM2 approaches.

3.4.2. *Convergence acceleration by iteration*

In this section we discuss a modification of the spectral (or compact finite difference) homotopy analysis method that is based on successively updating the initial approximation through iteration. The algorithm for the proposed method is outlined below for a second order differential equation for $y(x)$ with known boundary conditions at $y(a)$ and $y(b)$.

Consider the nonlinear differential equations

$$y'' + p(x)y' + q(x)y(x) + F(x, y, y') = 0, \quad y(a) = y_a, \ \ y(b) = y_b, \tag{3.140}$$

where $F(x, y, y')$ is a nonlinear function, $p(x)$ and $q(x)$ are known functions of x, and y_a and y_b are known constants.

1. Starting from a given initial guess y_0, which is chosen to satisfy the boundary condition, define $u(x)$ such that

$$y(x) = u(x) + y_0(x) \tag{3.141}$$

and substitute in (3.140) to obtain the following equation for $u(x)$,

$$\begin{cases} u'' + \tilde{p}(x)u' + \tilde{q}(x)u(x) + F(x, u, u') + \phi(x) = 0, \\ u(a) = 0, \ u(b) = 0, \end{cases} \tag{3.142}$$

where

$$\phi(x) = y_0'' + p(x)y_0' + q(x)y_0(x) + F(x, y_0, y_0'),$$
$$\tilde{p}(x) = p(x) + \frac{\partial F}{\partial y'}(y_0, y_0'), \quad \tilde{q}(x) = q(x) + \frac{\partial F}{\partial y}(y_0, y_0').$$

2. At the current iteration, r, choose a linear operator from the linear part of (3.142),

$$\mathcal{L}(u_r) = u_r'' + \tilde{p}(x)u_r' + \tilde{q}(x)u_r(x). \tag{3.143}$$

The nonlinear operator is obtained by adding the nonlinear part of (3.142) to the linear operator, that is

$$\mathcal{N}(u_r) = \mathcal{L}(u_r) + F(x, u_r, u_r'). \tag{3.144}$$

3. Obtain the initial guess $u_{r,0}(x)$ for solving (3.142) as the solution of

$$\mathcal{L}(u_{r,0}) + \phi(x) = 0. \tag{3.145}$$

4. Define the zeroth-order deformation equation as

$$(1-q)\mathcal{L}\left[U_r(x;q)-u_{r,0}(x)\right]=q\hbar\left\{\mathcal{N}[U_r(x;q)]+\phi(x)\right\}, \tag{3.146}$$

where $q\in[0,1]$ is the embedding parameter, $U_r(x;q)$ is a continuous mapping of $u_r(x)$ and $\hbar$ is the convergence-control parameter. The mth-order deformation equations corresponding to (3.146) are given by

$$\mathcal{L}[u_{r,m}(x)-(\chi_m+\hbar)u_{r,m-1}(x)]=\hbar R_{m-1}[u_{r,0},u_{r,1},\ldots,u_{r,m-1}], \tag{3.147}$$

where

$$R_{m-1}=\frac{1}{(m-1)!}\frac{\partial^{m-1}\left\{\mathcal{F}[U_r(x;q)]+\phi(x)\right\}}{\partial q^{m-1}}\bigg|_{q=0}.$$

Thus, the mth-order homotopy series solution of $u_r(x)$ reads

$$u_r(x)=u_{r,0}(x)+\sum_{k=1}^{m}u_{r,k}(x). \tag{3.148}$$

The current estimate for the solution $y(x)$ is

$$y_r(x)=y_0(x)+u_r(x). \tag{3.149}$$

Clearly, the mth-order approximation, after r iterations given by equation (3.149) satisfies the problem's underlying boundary conditions and can be used as the initial approximation at the next iteration, $r+1$.
5. Replace y_0 in Step 1, by the current estimate for $y_r(x)$, and repeat Steps 1–4.

The above algorithm provides us with an iteration scheme in the framework of the SHAM. For an m homotopy series solution, when using r iterations, the method is called the $[m,r]$ iterated spectral homotopy analysis method (iSHAM).

In Table 3.6 we present the iSHAM results for the solution of the Lane–Emden equation. The table demonstrates the significant improvement in the convergence of the SHAM when the iteration approach is used. Using an $m=5$ SHAM series, convergence to 13 decimal places is achieved after only 3 iterations. We remark that the results corresponding to $r=1$ are the original SHAM results.

Table 3.7 gives the results for the maximum residual of the Darcy–Brinkman–Forchheimer problem (3.126) using the $[m,r]$ iSHAM. Again, it

Table 3.6. First zero α of the Lane–Emden equation using $\lambda_0 = 15$, $N = 30$ and $\hbar = -1$.

Order $m \setminus r$	First zero, α 1	2
1	4.3932120925622	4.3528749640728
2	4.3359411795877	4.3528745959903
3	4.3538878282128	4.3528745959461
4	4.3500472931029	4.3528745959461
5	4.3525587860375	4.3528745959461
Ref. [6, 25]		4.3528745959461

Table 3.7. Maximum SHAM residual for the Darcy–Brinkman–Forchheimer problem when $\hbar = -1$, $M = 1$, $s = 1$, $F = 2$ and $N = 30$.

$m \setminus r$	1	2
1	4.6948e-003	5.1289e-014
2	1.6722e-004	3.6195e-014
3	7.3501e-006	3.6195e-014
4	3.6032e-007	3.6195e-014

can be seen that the iSHAM approach results in accelerated convergence of the original SHAM approach.

Tables 3.8–3.10 give the results of the iSHAM evaluation of the skin friction $f''(0)$ using $r = 1, 2$ for the Blasius, Pohlhausen and Homann flow respectively. Again, it can be seen from the tables that the iSHAM significantly improves the convergence to the benchmark numerical results of [12].

Table 3.8. Skin friction $f''(0)$ for Blasius flow using $\hbar = -1.32$.

Order $m \setminus r$	$f''(0)$ 1	2
1	0.3612452750	0.3320573363
2	0.3312181198	0.3320573362
3	0.3329709291	0.3320573362
4	0.3320081757	0.3320573362
5	0.3321112010	0.3320573362
6	0.3320539785	0.3320573362
Ref. [12]		0.3320573362

Table 3.9. Skin friction $f''(0)$ for Pohlhausen flow using $\hbar = -0.94$.

Order	$f''(0)$	
$m \setminus r$	1	2
1	1.1581939047	1.1547005384
2	1.1547124162	1.1547005384
3	1.1547040340	1.1547005384
Ref. [12]		1.1547005384

Table 3.10. Skin friction $f''(0)$ for Homann flow using $\hbar = -0.79$.

Order	$f''(0)$	
$m \setminus r$	1	2
1	1.3356339199	1.3119376940
2	1.3125928575	1.3119376939
3	1.3122405473	1.3119376938
4	1.3119529332	1.3119376938
Ref. [12]		1.3119376938

3.5. Conclusion

In this chapter, the basic idea of the Spectral homotopy analysis method (SHAM) for the solution of boundary value problems is described through the solution of Falkner–Skan boundary layer equations. In particular, the Blasius, Pohlhausen and Homann flows are discussed. The SHAM uses the Chebyshev spectral collocation method to solve the linearized higher order deformation equations which are developed using the concept of the original HAM. In this work, we present a general approach of implementing the SHAM and introduce two methods of identifying the optimal convergence-control parameter $\hbar$ that controls and adjusts accuracy and convergence of the SHAM. In the first approach, it is suggested that the optimal $\hbar$ is the value that lies at the intersection of $\hbar$-curves plotted at different orders of the SHAM approximation. In the second approach, the optimal $\hbar$ is identified to be the value at which the minimum of the maximum residual curve is located. The application of the SHAM is also extended to solve a nonlinear eigenvalue problem derived from the Lane–Emden equation.

The chapter also demonstrates that the convergence of the SHAM can be accelerated by using a linear operator that includes the initial guess and as many of the governing physical constants of the governing equations as possible. This can be achieved by homogenization of the boundary conditions using a function described in terms of the initial approximation of the solution.

The study also illustrates how the SHAM can be used to solve nonlinear equations with multiple (dual) solutions that arise in fluid mechanics applications. This is achieved through the introduction of an unknown parameter into the initial guess which when varied changes the concavity of the initial guess profile. An iterated version of the SHAM is also suggested. This approach involves successive modification of the initial guess and linear operator used in the SHAM algorithm. The iterated version, called iSHAM, is shown to be significantly more accurate and robust than the standard SHAM.

Lastly, the extension of the SHAM to nonlinear systems of two and three equations is demonstrated. It is noted that for systems of two or more equations, the linear operators used in the SHAM implementation may be coupled.

Thus, the SHAM have great potential to be applicable in more complicated problems of science and engineering, including some nonlinear partial differential equations.

References

[1] S. Abbasbandy and E. Shivanian, Multiple solutions of mixed convection in a porous medium on semi-infinite interval using pseudo-spectral collocation method, *Communications in Nonlinear Science and Numerical Simulation*, **16**, 2745–2752 (2011).

[2] S.M. Rassoulinejad-Mousavi, S. Abbasbandy. Analysis of forced convection in a circular tube filled with a Darcy-Brinkman-Forchheimer porous medium using spectral homotopy analysis method. *J. Fluid Eng.* **133** (2011) 101207.

[3] Z.P. Atabakan, A. Kiliçman, A.K. Nasab, On spectral homotopy analysis method for solving linear Volterra and Fredholm integrodifferential equations, *Abstract and Applied Analysis*, Vol. **2012**, Article ID 960289, 16 pages doi:10.1155/2012/960289 (2012).

[4] W. Auzinger, E. Karner, O. Koch and E. Weinmüller, Collocation methods for the solution of eigenvalue problems for singular ordinary differential equations, *Opuscula Math.* **26 (2)**, 229–241 (2006).

[5] R.E. Bellman and R.E. Kalaba, *Quasilinearization and nonlinear boundary-value problems*, Elsevier, New York, (1965).

[6] J.P. Boyd, Chebyshev Spectral Methods and the Lane-Emden Problem, *Numer. Math. Theor. Meth. Appl.* **4**, 142–157 (2011).

[7] C. Canuto, M.Y. Hussaini, A. Quarteroni, and T.A. Zang, *Spectral Methods in Fluid Dynamics*, Springer-Verlag, Berlin, (1988).

[8] A.J. Chamkha, A.M. Aly, M.A. Mansour, Similarity solution for unsteady heat and mass transfer from a stretching surface embedded in a porous medium with suction/injection and chemical reaction effects, *Chem. Eng. Comm.*, **197**, 846–858 (2010).

[9] S. Chandrasekhar, *An Introduction to the Study of Stellar Structure*, Dover, New York, (1958).

[10] H.T. Davis, *Introduction to nonlinear differential and integral equations*, Dover, New York, 371–394, (1962).

[11] R. Ellahi, E. Shivanian, S. Abbasbandy, S.U. Rahman, T. Hayat, Analysis of steady flows in viscous fluid with heat/mass transfer and slip effects, *Int. J of Heat and Mass Trans*, **55**, 6384–6390 (2012).

[12] B.D. Ganapol, Highly Accurate Solutions of the Blasius and Falkner-Skan Boundary Layer Equations via Convergence Acceleration, `arXiv:1006.3888` (June 2010).

[13] S. Kazem, M. Shaban, Tau-homotopy analysis method for solving micropolar flow due to a linearly stretching of porous sheet, *Communications in Numerical Analysis*, **2012** Article ID cna-00114, doi: 10.5899/2012/cna-00114.

[14] S.K. Lele, Compact finite difference schemes with spectral-like resolution, *J. Comp. Phys.* **103**, 16–42 (1992).

[15] S.J. Liao, *Beyond Perturbation: Introduction to the Homotopy Analysis Method*, Boca Raton: Chapman & Hall/CRC Press (2003).

[16] S.J. Liao, *Homotopy Analysis Method in Nonlinear Differential Equations*, Springer Berlin Heidelberg, (2012).

[17] C. Yang, S.J. Liao, On the explicit, purely analytic solution of Von Karman swirling viscous flow, *Communications in Nonlinear Science and Numerical Simulation.* **11**, 83–93 (2006).

[18] E. Magyari, I. Pop and B. Keller, Exact dual solutions occurring in the Darcy mixed convection flow, *Int. J. Heat Mass Transfer*, **44**, 4563–6 (2001).

[19] Z. Makukula, P. Sibanda and S.S. Motsa, A Note on the Solution of the Von Kármán Equations Using Series and Chebyshev Spectral Methods, *Boundary Value Problems*, Vol. **2010**, Article ID 471793, 17 pages doi:10.1155/2010/471793 (2010)

[20] S.S. Motsa, P. Sibanda and S. Shateyi, A new spectral-homotopy analysis method for solving a nonlinear second order BVP, *Communications in Nonlinear Science and Numerical Simulation* **15**, 2293–2302 (2010).

[21] S.S. Motsa, P. Sibanda, F.G. Awad, S. Shateyi, A new spectral-homotopy analysis method for the MHD Jeffery-Hamel problem, *Computer & Fluids* **39** 1219–1225 (2010).

[22] P. Sibanda, S.S. Motsa, A spectral-homotopy analysis method for heat transfer flow of a third grade fluid between parallel plates, *International Journal of Numerical Methods for Heat & Fluid Flow*, **22(1)** 4–23 (2012).

[23] S.S. Motsa, Application of the new spectral homotopy analysis method (sham) in the non-linear heat conduction and convective fin problem with variable thermal conductivity, *International Journal of Computational Meth-*

ods **9** 1250039 (2012), DOI: 10.1142/S0219876212500399.

[24] S.S. Motsa, S. Shateyi, Successive Linearisation Analysis of Unsteady Heat and Mass Transfer From a Stretching Surface Embedded in a Porous Medium With Suction/Injection and Thermal Radiation Effects, *Can. J. Chem. Eng.* **9999**, 1–13, (2011).

[25] S.S. Motsa, S. Shateyi, A successive linearization method approach to solving Lane-Emden type of equations, *Mathematical Problems in Engineering*, **vol. 2012**, Article ID 280702, 14 pages, 2012. doi:10.1155/2012/280702.

[26] Z.G. Makukula, P. Sibanda, S.S. Motsa, S. Shateyi, On new numerical techniques for the MHD flow past a shrinking sheet with heat and mass transfer in the presence of a chemical reaction, *Mathematical Problems in Engineering*, **2011**, Article ID 489217, 19 pages doi:10.1155/2011/489217.

[27] A.A. Khidir, P. Sibanda, On spectral-homotopy analysis solutions of steady magnetohydrodynamic (MHD) flow and heat transfer from a rotating disk in a porous medium, *Scientific Research and Essays*, **7** 2770–2780 (2012).

[28] S.M. Rassoulinejad-Mousavi, S. Abbasbandy, Analysis of forced convection in a circular tube filled with a Darcy-Brinkman-Forchheimer porous medium using spectral homotopy analysis method, *Journal of Fluids Engineering* **133** 101207-1 (2011).

[29] P. Sibanda, S.S. Motsa and Z.G. Makukula, A spectral-homotopy analysis method for heat transfer flow of a third grade fluid between parallel plates, *International Journal of Numerical Methods for Heat & Fluid Flow* **22**, 4–23 (2012).

[30] L.N. Trefethen, *Spectral Methods in MATLAB*, SIAM, (2000).

Chapter 4

Stability of Auxiliary Linear Operator and Convergence-Control Parameter in the Homotopy Analysis Method

Robert A. Van Gorder*

Department of Mathematics, University of Central Florida
Orlando, Florida 32816 USA
rav@knights.ucf.edu

We consider the stability of the homotopy analysis method under the choice of both linear operator and convergence-control parameter. In particular, through several examples, we determine how changes in the linear operator can influence the convergence properties of homotopy solutions. It is seen that there is often a best way to pick the linear operator, but this can change for each problem. We consider various linear operators for some ordinary differential operators, and also discuss the method of selection for some nonlinear evolution PDEs. Throughout this chapter, we consider the optimal homotopy analysis method, which permits us to select a convergence-control parameter that minimizes residual errors. It is natural to ask whether the optimal value of the convergence-control parameter varies much as we change the number of iterations taken. For computational efficiency, we would like to take as few terms as possible in order to guarantee a low error of approximation, so learning when the optimal convergence-control parameter stabilizes could help us in knowing when to truncate our approximation. We then turn our attention to other properties of the homotopy analysis method. Through applications, we study the effect of homotopies which are nonlinear in the embedding parameter, q. In another application, we show that the auxiliary function $H(x)$, which is often taken to unity, can be useful in a more general form. Finally, we present an application of the homotopy analysis method to a highly singular problem, and we demonstrate how to get accurate approximate solutions for such problems.

*This work supported in part by NSF grant number 1144246.

Contents

4.1. Overview 124
4.2. Ordinary auxiliary differential operators 129
4.2.1. Painlevé I equation 130
4.2.2. Lane–Emden equation 134
4.2.3. Flow over a nonlinearly stretching sheet 138
4.3. Time evolution PDEs and auxiliary linear operators 141
4.3.1. Simple polynomial evolution 141
4.3.2. Evolution and exponential temporal decay 142
4.3.3. Nonlinear Klein–Gordon equation 143
4.3.4. Zakharov system with dissipation 146
4.4. The convergence control parameter 150
4.4.1. Lane–Emden equation under $\mathcal{L}_1 = y''$ 150
4.4.2. Lane–Emden equation under $\mathcal{L}_3 = y'' + \frac{2}{x}y'$ 151
4.4.3. Flow over a nonlinearly stretching sheet 153
4.5. Modifying the homotopy 154
4.5.1. The general homotopy 154
4.5.2. Standard homotopy analysis method 155
4.5.3. A homotopy quadratic in q 157
4.6. What about the auxiliary function $H(x)$? 158
4.7. Avoiding singularities 165
4.8. Conclusions 169
References 171

4.1. Overview

The homotopy analysis method (HAM) [1–10] has recently been applied to the study of a number of non-trivial and traditionally hard to solve nonlinear differential equations, for instance nonlinear equations arising in heat transfer [11–14], fluid mechanics [15–22], solitons and integrable models [23–27], nanofluids [28, 29], the Lane–Emden equation which appears in stellar astrophysics [30–33], and models frequently used in mathematical physics [34–36], to name a few areas.

For those unfamiliar with the method, the homotopy analysis method is an analytical technique which may be used to solve complicated nonlinear problems when other approaches, such as perturbation or numerical analysis, fail to provide desirable results. Consider, for instance, the perturbed oscillator equation

$$y'' + y + \epsilon y^3 = 0, \; y(0) = a, y'(0) = b,$$

where ϵ, a and b are constants with *physical* meanings. When $|\epsilon| << 1$, then we are in the perturbative regime, and we may express the solution as a function of ϵ in the standard way:

$$y(x;\epsilon) = y_0(x) + \epsilon y_1(x) + \epsilon^2 y_2(x)^2 + \cdots .$$

The reason for such an expansion is that, when $|\epsilon|$ is sufficiently small, and the terms $y_k(x)$ are sufficiently bounded, this representation exhibits a geometric rate of convergence. Hence, in the perturbative regime, the standard perturbation solution is rather good. However, what happens if ϵ is not small enough, or if there is no such small *physical* parameter in the problem whatsoever?

Note that for perturbation problems with a small physical parameter ϵ, we often have an equation of the form

$$\mathcal{A}[u] + \epsilon\, \mathcal{B}[u] = 0\,, \tag{4.1}$$

where $\mathcal{A}$ is a linear operator and $\mathcal{B}$ is either linear or nonlinear. When $\mathcal{B}$ is nonlinear, the perturbation assumption

$$y(x;\epsilon) = u_0(x) + \epsilon u_1(x) + \epsilon^2 u_2(x) + \cdots \tag{4.2}$$

effectively reduces this nonlinear equation into infinitely many linear equations, indexed by powers of ϵ, of the form

$$\mathcal{A}[u_k] = f[u_0, \ldots, u_{k-1}].$$

In this way, we obtain the u_k's successively. Often we then calculate the first few terms of a perturbation expansion, say

$$u(x;\epsilon) = u_0(x) + \epsilon u_1(x) + \cdots \epsilon^\ell u_\ell(x). \tag{4.3}$$

If ϵ is small enough, then this expansion often is a good approximation to the true solution.

In many nonlinear problems arising in mathematics, physics, engineering, economics, finance, biology, fluid dynamics, and the like, there is no such a natural small parameter ϵ. If we introduce such a parameter, we risk obtaining solutions which are not interesting or relevant. However, without such small parameters, analytical solutions can be challenging to obtain, and numerical methods could be considered. Even then, we would prefer analytical solutions in many cases, since numerical results, while useful, do not give the whole picture.

In a way, this is one reason to consider the homotopy analysis method (HAM) [1–10]. This method maintains one fundamental aspect of perturbation theory, the fact that we may iteratively solve linear equations, while featuring an improvement over perturbation in the fact that it does not require a small parameter. In the frame of the HAM, we link the original nonlinear equation, say

$$\mathcal{N}[u] = 0,$$

with a linear problem

$$\mathcal{L}[u] = 0$$

(which is easier to solve, since it is free for us to select) by means of enforcing a homotopy

$$\mathcal{H}(q)[u] : (1-q)\mathcal{L}[u] - h\, q\, \mathcal{N}[u] \tag{4.4}$$

to be zero, i.e.

$$(1-q)\mathcal{L}[u] = h\, q\, \mathcal{N}[u], \tag{4.5}$$

where $q \in [0,1]$ is the embedding parameter, $h \neq 0$ is called *the convergence-control parameter*, and $\mathcal{L}$ is called *the auxiliary linear operator*, respectively. Note that we have large freedom to choose the auxiliary linear operator $\mathcal{L}$ and the value of the convergence-control parameter h. Besides, the convergence-control parameter h has *no* phyical meaning at all. Observe that when $q = 0$, $\mathcal{L}[u] = 0$ while, when $q = 1$ we have the original nonlinear problem, namely $\mathcal{N}[u] = 0$. So, equation (4.5) constructs a continuous variation, or deformation, from the solution of the linear equation $\mathcal{L}[u] = 0$ to that of the original nonlinear equation $\mathcal{N}[u] = 0$. For this reason, equation (4.5) is called *the zeroth-order deformation equation* in the frame of the HAM. If we can solve this equation, and then localize at $q = 1$, we shall indeed have a solution to the original nonlinear equation.

If we assume

$$u(x;q) = u_0(x) + u_1(x)q + u_2(x)q^2 + \cdots$$

and then match powers of q in the zeroth-order deformation equation (4.5), we find that we may iteratively solve equations of the form

$$\mathcal{L}[u_k] = F[u_0, \ldots, u_{k-1}] \tag{4.6}$$

where F depends on the lower order terms. Again, this is just like in the perturbation case. The zeroth-order solution, $u_0(x)$, satisfies an equation of the form $\mathcal{L}[u_0] = 0$, and we call it the zeroth-order deformation equation. In this case, the zeroth-order approximation is the solution to the linearized problem. In order to determine the contribution of the nonlinear terms, we need higher order corrections. By solving the so-called *higher-order deformation equation*

$$\mathcal{L}[u_k - \chi_k\, u_{k-1}] = h\, R_{k-1}[u_0, \dots, u_{k-1}], \tag{4.7}$$

we determine these corrections, where

$$R_n = \frac{1}{n!} \left. \frac{\partial^n \mathcal{N}[\phi(x;q)]}{\partial q^n} \right|_{q=0}$$

and

$$\chi_k = \begin{cases} 0, \text{ when } k \le 1, \\ 1 \text{ when } k > 1. \end{cases}$$

Comparing (4.7) with (4.6) gives

$$\mathcal{L}[u_k] = F[u_0, \dots, u_{k-1}] = \chi_k\, \mathcal{L}[u_{k-1}] + h\, R_{k-1}[u_0, \dots, u_{k-1}].$$

The convergence-control parameter $h \neq 0$ is selected in a way that will improve the convergence properties of solutions. Indeed, as in standard perturbation, we assume a solution

$$u(x;q) = u_0(x) + u_1(x)q + u_2(x)q^2 + \cdots . \tag{4.8}$$

However, unlike in standard perturbation, we must have $q = 1$ (that is, the solution to the nonlinear problem $\mathcal{N}[u]$ is $u(x;1)$), so q is not at all a small parameter. To remedy this, note that the solutions depend implicitly on the convergence-control parameter h, so that really $u(x;1) = U(x;h)$. Then, we have some hope to find a proper value of h for which $U(x;h)$ is a solution to the original nonlinear problem. Determining the infinite sum of terms in

$$U(x;h) = u_0(x) + u_1(x;h) + u_2(x;h) + \cdots \tag{4.9}$$

is often difficult or even impossible in closed-form, so we must truncate this expression to obtain a sort of approximate analytical solution, say after ℓ terms:

$$\hat{U}(x;h) = u_0(x) + u_1(x;h) + u_2(x;h) + \cdots + u_\ell(x;h). \tag{4.10}$$

This introduces a number of questions. First, how do we best pick the auxiliary linear operator $\mathcal{L}$ in order to iteratively solve for the solution of the higher order deformation equation (4.7)? After this, how do we pick the approximate solution (4.10)? If we pick to few terms, we will likely not get a good approximation. If we pick too many terms, then the computations may not be efficient. Even then, how do we pick the convergence-control parameter h so that the convergence rate of the solutions is good enough? Is it possible to pick the convergence-control parameter h so that the error inherent in our approximation (4.10) is the lowest possible? And, how much does the solution method depend on the problem at hand? Must we consider everything on a case-by-case basis, or can we make some general observations?

In the present chapter, we shall consider some of these questions. In particular, we shall take a look at some of the features of the homotopy analysis method, in order to determine the influence of selecting certain quantities, such as the auxiliary linear operator $\mathcal{L}$ and the convergence control parameter h. While there have been improvements on how the homotopy analysis method has been applied in recent years, there is still much work to be done in answering the above questions. We shall outline a number of issues which must be considered when applying the homotopy analysis method to nonlinear differential equations.

In the first section, we shall study the effect of changing auxiliary operators when solving a nonlinear ordinary differential equation. We show that while many options are available, there is often one choice of auxiliary linear operator which is best. So, it appears that the homotopy analysis method is not always stable under the choice of auxiliary linear operator: rather, in some cases, certain auxiliary linear operators can contribute to large error. For each choice of an auxiliary linear operator, we might obtain rather different values of the optimal convergence control parameter. We shall refer to optimal convergence control parameters, throughout. Recall that the so-called convergence control parameter h, appears in the zeroth-order deformation equation linking the linear and an original nonlinear problem, namely

$$(1-q)\mathcal{L}[u] = h\, q\, \mathcal{N}[u]\,,$$

where $q \in [0,1]$ is the embedding parameter, $\mathcal{N}$ is the nonlinear operator describing the original problem, $\mathcal{L}$ is the auxiliary linear operator. If $\hat{u}(x;h)$ is an approximate solution obtained by the homotopy analysis method, we may define the accumulated L^2 norm of the residual error by

$$E(h) = \int_{\mathcal{D}} \left(\mathcal{N}[\hat{u}(x;h)]\right)^2 dx\,,$$

where $\mathcal{D}$ is the problem domain. By construction, $E(h)$ is positive definite, so there exists a global minimum, say h^*. Then, such an h^* is a minimizer for the accumulated L^2 norm of the residual error, and we refer to it as optimal. Then, we refer to $\hat{u}(x; h^*)$ as an optimal homotopy analysis solution. This method has been employed to study optimal approximations for a number of nonlinear problems [37–44].

In the second section, we then consider partial differential equations, in particular nonlinear time evolution equations, and show that when selecting the auxiliary linear operator, it can be important to choose an operator which permits the proper type of time-evolution. We show that operators involving $u_t + u$ permit small-time convergence due to the appearance of terms of the form e^{-t}, whereas operators simply of the form u_t can lead to blow-up in solutions, owing to terms of the form t^α.

In the third section, we study the behavior of the convergence control parameter as a function of the number of iterations. The value of the optimal convergence control parameter seems to stabilize after repeated iterations, provided the auxiliary linear operator is appropriately chosen.

In the next three sections, we consider additional features of the homotopy analysis method. In the fourth section, we show that the homotopy analysis method appears stable under modifications to the form of the homotopy. To demonstrate this, we construct a nonlinear homotopy in q, and demonstrate that the residual error in the two approaches are almost equivalent. Another way to modify the homotopy is to include the auxiliary function $H(x)$ as a multiplier for the nonlinear operator in the homotopy. We highlight a case in the literature where this approach was useful. Finally, in the sixth section, we demonstrate how the method can be used on nonlinear problems with strong singularities when we appropriately select the auxiliary linear operator.

Each of these topics demonstrates the versatility of the homotopy analysis method, which makes it a highly useful tool in the study of nonlinear phenomenon, in particular for obtaining approximate solutions for the nonlinear ordinary and partial differential equations governing such phenomenon.

4.2. Ordinary auxiliary differential operators

Here we shall study the influence of selecting different auxiliary linear operators in order to arrive at HAM solutions. In order to best compare the different solutions, we shall always truncate the infinite HAM expansion to arrive at an accurate approximation. We pick the convergence-control parameter, h, in order to minimize residual errors.

For each linear operator given, we shall determine the corresponding error-minimizing value of h. We shall then be able to compare the choices of linear operators, to determine which is the best fit for the problem. Three distinct problems are given to demonstrate the method. In each of these problems, we see that there is indeed a best choice for the linear operator. Hence, some operators can provide better convergence and control of error than others.

4.2.1. *Painlevé I equation*

The Painlevé I equation reads

$$y'' = y^2 + x\,, \tag{4.11}$$

with associated initial conditions

$$y(0) = 1 \qquad \text{and} \qquad y'(0) = 0\,. \tag{4.12}$$

Here prime denotes differentiation with respect to x. We shall restrict our attention to the interval $x \in [0, 1]$. For more information on the Painlevé equations, refer to [45] and references therein. There are six Painlevé transcendents, corresponding to six second-order ordinary differential equations whose only movable singularities are ordinary poles (this characteristic is known as the Painlevé property) and which cannot be integrated in terms of other known functions or transcendents; see the original works on Painlevé transcendents [46–54], or any modern textbook covering the theory of nonlinear ordinary differential equations (e.g., Ince [55]). As the solutions to the six Painlevé equations cannot be obtained exactly, one may resort to series or perturbation solutions.

For this problem, we define a nonlinear operator

$$\mathcal{N}[y] = y'' - y^2 - x \tag{4.13}$$

and construct the zeroth-order deformation equation

$$(1-q)\mathcal{L}[y] = h\, q\, \mathcal{N}[y]\,, \tag{4.14}$$

subject to the initial condition

$$y = 1, y' = 0, \ \text{at } x = 0, \tag{4.15}$$

where $\mathcal{L}$ is a linear auxiliary operator, h is the convergence-control parameter, and x is the inhomogeneity present in the original equation.

4.2.1.1. $\mathcal{L}_1[y] = y'' - y$

Using the linear operator

$$\mathcal{L}_1[y] = y'' - y\,,$$

the zeroth-order deformation equation becomes

$$(1-q)\mathcal{L}_1[y] = h\, q\mathcal{N}[y],\ y(0) = 1,\ y'(0) = 0. \tag{4.16}$$

Assuming a three-term solution

$$F_1(x) = y_0(x) + y_1(x)q + y_2(x)q^2\,,$$

and then balancing powers of q in (4.16) (after which we set $q = 1$, to recover the three-term approximate solution to the nonlinear problem), we obtain

$$\begin{aligned}
\mathcal{L}_1[y_0] &= 0\,,\ y_0(0) = 1,\ y_0'(0) = 0\,,\\
\mathcal{L}_1[y_1] &= \mathcal{L}_1[y_0] + h\left(y_0'' - y_0^2 - x\right), y_1(0) = 0,\ \ y_1'(0) = 0\,,\\
\mathcal{L}_1[y_2] &= \mathcal{L}_1[y_1] + h\left(y_1'' - 2y_0y_1\right), y_2(0) = 0,\ y_2'(0) = 0\,,\\
&\ \ \vdots
\end{aligned}$$

We find successively that

$$\begin{aligned}
y_0(x) &= \cosh(x)\,,\\
y_1(x) &= \frac{h}{2} + hx - \frac{h}{6}\cosh(2x) + \frac{h}{2}\sinh(x) - \frac{2h}{3}e^{x} + \frac{h}{3}e^{-x}\,,
\end{aligned}$$

and so on.

In order to determine the accuracy of this three-term approximation, we compute the accumulated L^2 norm of the residual error of $F_1(x)$ over $x \in [0,1]$:

$$E_1(h) = \int_0^1 (\mathcal{N}[F_1(x;h)])^2 dx = \mu_0^1 + \mu_1^1 h + \cdots + \mu_8^1 h^8\,,$$

where μ_0^1 through μ_8^1 are constants. $E_1(h)$ is positive definite, so a global minimum exists. We find that

$$h_1^* = \operatorname*{argmin}_{h\in\mathbb{R}} E_1(h) = -1.09958\,,$$

which gives minimal error of $E_1(h_1^*) = 1.15996 \times 10^{-5}$. This error is very good, and would improve with the addition of higher order terms. For our sake, it is sufficient, since we desire only to compare the error between solutions involving different linear operators.

4.2.1.2. $\mathcal{L}_2[y] = y'' - y'$

Using the linear operator

$$\mathcal{L}_2[y] = y'' - y',$$

the zeroth-order deformation equation becomes

$$(1-q)\mathcal{L}_2[y] = h\, q\mathcal{N}[y],\ y(0) = 1,\ y'(0) = 0. \tag{4.17}$$

Assuming a three-term solution

$$F_2(x) = y_0(x) + y_1(x)q + y_2(x)q^2,$$

and then balancing powers of q in (4.17), we obtain

$$\begin{aligned}
&\mathcal{L}_2[y_0] = 0,\ \ y_0(0) = 1,\ \ y_0'(0) = 0\,,\\
&\mathcal{L}_2[y_1] = \mathcal{L}_2[y_0] + h\left(y_0'' - y_0^2 - x\right), y_1(0) = 0,\ \ y_1'(0) = 0\,,\\
&\mathcal{L}_2[y_2] = \mathcal{L}_2[y_1] + h\left(y_1'' - 2y_0y_1\right), y_2(0) = 0,\ \ y_2'(0) = 0,\\
&\quad\vdots
\end{aligned}$$

We find successively that

$$y_0(x) = 1\,,$$

$$y_1(x) = 1 + h + hx - \frac{1}{2}e^{-x} - \left(\frac{1}{2} + h\right)e^{x}\,,$$

and so on.

In order to determine the accuracy of this three-term approximation, we compute the accumulated L^2 norm of the residual error of $F_2(x)$ over $x \in [0, 1]$:

$$E_2(h) = \int_0^1 (\mathcal{N}[F_2(x;h)])^2 dx = \mu_0^2 + \mu_1^2 h + \cdots + \mu_8^2 h^8\,,$$

where μ_0^2 through μ_8^2 are constants. $E_2(h)$ is positive definite, so a global minimum exists. We find that

$$h_2^* = \underset{h\in\mathbb{R}}{\operatorname{argmin}}\, E(h) = -1.0867\,,$$

which gives minimal error of $E_2(h_2^*) = 1.32057$. The residual error is very bad. So, the use of L_2 has resulted in much worse error.

4.2.1.3. $\mathcal{L}_3[y] = y''$

For the final choice of linear operator, we select

$$\mathcal{L}_3[y] = y''.$$

The zeroth-order deformation equation then becomes

$$(1-q)\mathcal{L}_3[y] = h\, q\mathcal{N}[y],\ y(0) = 1,\ y'(0) = 0. \tag{4.18}$$

Assuming a three-term solution

$$F_3(x) = y_0(x) + y_1(x)q + y_2(x)q^2,$$

and then balancing powers of q in (4.18), we obtain

$$\begin{aligned}
\mathcal{L}_3[y_0] &= 0,\ \ y_0(0) = 1,\ \ y_0'(0) = 0,\\
\mathcal{L}_3[y_1] &= \mathcal{L}_3[y_0] + h\left(y_0'' - y_0^2 - x\right),\ \ y_1(0) = 0,\ \ y_1'(0) = 0,\\
\mathcal{L}_3[y_2] &= \mathcal{L}_3[y_1] + h\left(y_1'' - 2y_0y_1\right),\ y_2(0) = 0,\ \ y_2'(0) = 0,\\
&\ \ \vdots
\end{aligned}$$

We find successively that

$$y_0(x) = 1,$$

$$y_1(x) = -\frac{h}{6}x^3 - \frac{h}{2}x^2,$$

and so on. Note that these expressions are more simple than those obtained previously, so the present choice of auxiliary linear operator seems to maintain computational efficiency.

In order to determine the accuracy of the three-term approximation $F_3(x)$, we compute the accumulated L^2 norm of the residual error of $F_3(x)$ over $x \in [0,1]$:

$$E_3(h) = \int_0^1 (\mathcal{N}[F_3(x;h)])^2 dx = \mu_0^3 + \mu_1^3 h + \cdots + \mu_8^3 h^8,$$

where μ_0^3 through μ_8^3 are constants. $E_3(h)$ is positive definite, so a global minimum exists. We find that

$$h_3^* = \operatorname*{argmin}_{h\in\mathbb{R}} E(h) = -1.24296,$$

which gives minimal error of $E(h_3^*) = 4.71645 \times 10^{-3}$. This residual error is good, though not nearly as good as we found for the operator $\mathcal{L}_1$.

In summary, the operator $\mathcal{L}_1$ was the best choice, in terms of both control of residual errors and computational efficiency, with the other auxiliary linear operators chosen displaying drawbacks.

4.2.2. *Lane–Emden equation*

The Lane–Emden equation of the first kind (with power index three) reads

$$y'' + \frac{2}{x}y' + y^3 = 0\,, \tag{4.19}$$

with associated initial conditions

$$y(0) = 1 \qquad \text{and} \qquad y'(0) = 0\,. \tag{4.20}$$

Lane–Emden problems are of great interest in the recent literature [32] (and see also [56–73]), due both to their application in stellar physics and because of their nonlinearity and singularity at $x = 0$.

4.2.2.1. $\mathcal{L}_1[y] = y''$

Using the auxiliary linear operator

$$\mathcal{L}_1[y] = y''\,,$$

the zeroth-order deformation equation reads

$$(1-q)\mathcal{L}_1[y] = h\, q\mathcal{N}[y],\ y(0) = 1,\ y'(0) = 0. \tag{4.21}$$

Assuming a four-term solution

$$F_1(x) = y_0(x) + y_1(x)q + y_2(x)q^2 + y_3(x)q^3\,,$$

and then balancing powers of q in (4.21) (after which we set $q = 1$, to recover the three-term approximate solution to the nonlinear problem), we obtain

$$\begin{aligned}
&\mathcal{L}_1[y_0] = 0\,, y_0(0) = 1,\ \ y_0'(0) = 0\,,\\
&\mathcal{L}_1[y_1] = \mathcal{L}_1[y_0] + h\left(y_0'' + \frac{2}{x}y_0' + y_0^3\right), y_1(0) = 0,\ \ y_1'(0) = 0,\\
&\mathcal{L}_1[y_2] = \mathcal{L}_1[y_1] + h\left(y_1'' + \frac{2}{x}y_1' + 3y_0^2 y_1\right), y_2(0) = 0,\ \ y_2'(0) = 0,\\
&\mathcal{L}_1[y_3] = \mathcal{L}_1[y_2] + h\left[y_2'' + \frac{2}{x}y_2' + 3y_0\left(y_0y_2 + y_1^2\right)\right], y_3(0) = 0,\ y_3'(0) = 0.
\end{aligned}$$

We find successively that

$$
\begin{aligned}
y_0(x) &= 1\,, \\
y_1(x) &= \frac{h}{2}x^2\,, \\
y_2(x) &= \frac{h^2}{8}x^4 + \frac{h}{2}\left(1+3h\right)x^2\,, \\
y_3(x) &= \frac{3}{80}h^3x^6 + \frac{h^2}{12}\left(7h+3\right)x^4 + \frac{h}{2}\left(9h^2+6h+1\right)x^2\,.
\end{aligned}
$$

In order to determine the accuracy of this four-term approximation $F_1(x;h)$, we compute the accumulated L^2 norm of the residual error of $F_1(x;h)$ over $x \in [0,1]$:

$$E_1(h) = \int_0^1 \left(\mathcal{N}[F_1(x;h)]\right)^2 dx = \mu_0^1 + \mu_1^1 h + \cdots + \mu_{18}^1 h^{18}\,,$$

where μ_0^1 through μ_{18}^1 are constants. $E_1(h)$ is positive definite, so a global minimum exists. We find that

$$h_1^* = \underset{h\in\mathbb{R}}{\operatorname{argmin}}\, E_1(h) = -0.373817\,,$$

which gives minimal error of $E_1(h_1^*) = 2.364958 \times 10^{-4}$. This error will improve with the addition of higher order terms.

4.2.2.2. $\mathcal{L}_2[y] = y'' + y'$

Using the linear operator

$$\mathcal{L}_2[y] = y'' + y'\,,$$

the zeroth-order deformation equation reads

$$(1-q)\mathcal{L}_2[y] = h\,q\,\mathcal{N}[y],\; y(0) = 1,\; y'(0) = 0\,. \tag{4.22}$$

Assuming a four-term solution

$$F_2(x) = y_0(x) + y_1(x)q + y_2(x)q^2 + y_3(x)q^3\,,$$

and then balancing powers of q in (4.22) (after which we set $q = 1$, to recover the three-term approximate solution to the nonlinear problem), we obtain

$$\mathcal{L}_2[y_0] = 0, \;\; y_0(0) = 1, \;\; y_0'(0) = 0,$$
$$\mathcal{L}_2[y_1] = \mathcal{L}_2[y_0] + h\left(y_0'' + \frac{2}{x}y_0' + y_0^3\right), \; y_1(0) = 0, \;\; y_1'(0) = 0,$$
$$\mathcal{L}_2[y_2] = \mathcal{L}_2[y_1] + h\left(y_1'' + \frac{2}{x}y_1' + 3y_0^2 y_1\right), \; y_2(0) = 0, \;\; y_2'(0) = 0,$$
$$\mathcal{L}_2[y_3] = \mathcal{L}_2[y_2] + h\left[y_2'' + \frac{2}{x}y_2' + 3y_0\left(y_0 y_2 + y_1^2\right)\right], \; y_3(0) = 0, \; y_3'(0) = 0.$$

We find successively that

$$y_0(x) = 1, \;\; y_1(x) = h(x - 1 + e^{-x}),$$

and so on. However, at the evaluation of the term $y_2(x)$, we obtain an integral involving the Ei function (an exponential integral) under the integral:

$$y_2(x) = -h\int_0^x \left\{2e^{-\xi}\mathrm{Ei}(1,-\xi) + (1 - 4\xi h)e^{-\xi} - 1 + (6 - 3\xi)h\right\} d\xi\,.$$

Upon successive iterations, this expression would only become more complicated. As a consequence, the residual error function cannot be tabulated in a useful way, hence we have no good way to control the error through the convergence-control parameter.

In summary, the operator $\mathcal{L}_2 = y'' + y'$ is not effective for use as an auxiliary linear operator for the present problem.

4.2.2.3. $\mathcal{L}_3[y] = y'' + \frac{2}{x}y'$

For our final choice, we use the linear operator

$$\mathcal{L}_3[y] = y'' + \frac{2}{x}y'\,.$$

The zeroth-order deformation equation reads

$$(1 - q)\mathcal{L}_3[y] = h\,q\,\mathcal{N}[y], \;\; y(0) = 1, \; y'(0) = 0\,. \tag{4.23}$$

Assuming a four-term solution

$$F_3(x) = y_0(x) + y_1(x)q + y_2(x)q^2 + y_3(x)q^3\,,$$

and then balancing powers of q in (4.23) (after which we set $q = 1$, to recover the three-term approximate solution to the nonlinear problem), we obtain

$$\mathcal{L}_3[y_0] = 0, \;\; y_0(0) = 1, \;\; y_0'(0) = 0\,,$$
$$\mathcal{L}_3[y_1] = \mathcal{L}_3[y_0] + h\left(y_0'' + \frac{2}{x}y_0' + y_0^3\right), \; y_1(0) = 0, \;\; y_1'(0) = 0\,,$$
$$\mathcal{L}_3[y_2] = \mathcal{L}_3[y_1] + h\left(y_1'' + \frac{2}{x}y_1' + 3y_0^2 y_1\right), \; y_2(0) = 0, \; y_2'(0) = 0,$$
$$\mathcal{L}_3[y_3] = \mathcal{L}_3[y_2] + h\left(y_2'' + \frac{2}{x}y_2' + 3y_0\left(y_0 y_2 + y_1^2\right)\right), \; y_3(0) = 0, \; y_3'(0) = 0\,.$$

We find successively that

$$y_0(x) = 1, \; y_1(x) = \frac{h}{6}x^2, \; y_2(x) = \frac{h^2}{40}x^4 + \frac{h}{6}(1+h)\,x^2,$$

and

$$y_3(x) = \frac{19}{5040}h^3 x^6 + \frac{h^2}{20}(1+h)\,x^4 + \frac{h}{6}\left(2h^2 + h + 1\right)x^2\,.$$

In order to determine the accuracy of this four-term approximation $F_3(x;h)$, we compute the accumulated L^2 norm of the residual error of $F_3(x;h)$ over $x \in [0,1]$:

$$E_3(h) = \int_0^1 (\mathcal{N}[F_3(x;h)])^2 dx = \mu_0^3 + \mu_1^3 h + \cdots + \mu_{18}^3 h^{18}\,,$$

where μ_0^3 through μ_{18}^3 are constants. $E_3(h)$ is positive definite, so a global minimum exists. We find that

$$h_3^* = \underset{h\in\mathbb{R}}{\operatorname{argmin}}\, E_3(h) = -0.892269\,,$$

which gives minimal error of $E_3(h_3^*) = 4.35017 \times 10^{-7}$. This error will improve with the addition of higher order terms.

Note that the obtained error for the four-term homotopy analysis method solution obtained using $\mathcal{L}_3$ is drastically better than that obtained using $\mathcal{L}_1$ (the order 10^{-7} error of the solution corresponding to $\mathcal{L}_3$ is much better than the order 10^{-4} error of the solution corresponding to $\mathcal{L}_1$). Hence, selecting an operator which is a better representation of the original problem is helpful in order to ensure rapid convergence of the approximate solutions obtained via the homotopy analysis method.

4.2.3. *Flow over a nonlinearly stretching sheet*

The boundary value problem governing the flow of a fluid over a nonlinearly stretching sheet reads

$$f''' + ff'' - \rho f'^2 = 0\,, \tag{4.24}$$

$$f(0) = 0\,, \qquad f'(0) = 1 \qquad \text{and} \qquad \lim_{\eta\to\infty} f'(\eta) = 0\,. \tag{4.25}$$

This and a number of related problems appear as a major area of work in fluid mechanics; indeed, there have been innumerable references in this area in recent years. For some of the author's references, see [74–92], and for more detail see the references therein.

4.2.3.1. $\mathcal{L}_1[f] = f''' + f''$

We select the auxiliary linear operator

$$\mathcal{L}_1[u] = u''' + u''\,,$$

with general solution

$$u(\eta) = \alpha_0 + \alpha_1\eta + \alpha_2 e^{-\eta}\,.$$

Let $\mathcal{N}$ represent the nonlinear differential in question, i.e.

$$\mathcal{N}[f] = f''' + ff'' - \rho f'^2\,.$$

Assuming a homotopy of the form

$$H(q)[f] = (1-q)\mathcal{L}_1[f] - hq\mathcal{N}[f]\,,$$

and assuming a four-term approximate solution of the form

$$F(\eta;h) = f_0(\eta) + f_1(\eta;h)q + f_2(\eta;h)q^2 + f_3(\eta;h)q^3\,, \tag{4.26}$$

enforcing $H(q)[F] \equiv 0$ implies

$$\mathcal{L}_1[f_0] = 0\,, f_0(0) = 0\,,\ f_0'(0) = 1\,,\ \lim_{\eta\to\infty} f_0' = 0\,,$$

$$\mathcal{L}_1[f_1] = h(f_0''' + f_0f_0'' - \rho f_0'^2)\,,\ f_1(0) = 0\,,\ f_1'(0) = 0\,,\ \lim_{\eta\to\infty} f_1' = 0\,,$$

$$\mathcal{L}_1[f_2] = \mathcal{L}_1[f_1] + h(f_1''' + f_1f_0'' + f_0f_1'' - 2\rho f_0'f_1')\,,$$

$$f_2(0) = 0\,, \qquad f_2'(0) = 0\,, \qquad \lim_{\eta\to\infty} f_2' = 0\,,$$

and

$$\mathcal{L}_1[f_3] = \mathcal{L}_1[f_2] + h(f_2''' + f_2 f_0'' + f_1 f_1'' + f_0 f_2'' - 2\rho f_0' f_2' - \rho {f_1'}^2),$$

$$f_3(0) = 0, \qquad f_3'(0) = 0, \qquad \lim_{\eta\to\infty} f_3' = 0.$$

We may then solve this system, successively obtaining the higher order terms. Observe that

$$f_0(\eta) = 1 - e^{-\eta},$$

which happens to be the exact solution to the $\rho = 1$ problem. The other terms are given by

$$f_1(\eta) = \frac{h}{4}(\rho - 1)\left(1 - e^{-\eta}\right)^2, \tag{4.27}$$

$$f_2(\eta) = h(\rho - 1)\left\{\frac{h(5\rho + 14) + 9}{36} - \frac{h(8\rho + 11 + 6\eta) + 12}{24}e^{-\eta}\right.$$
$$\left. + \frac{h\rho + 1}{4}e^{-2\eta} - \frac{h(4\rho - 5)}{72}e^{-3\eta}\right\}, \tag{4.28}$$

and so forth. In order to determine the accuracy of this four-term approximation, we compute the accumulated L^2 norm of the residual error of $F(\eta)$,

$$E(h) = \int_0^\infty \left(\mathcal{N}[F(\eta; h)]\right)^2 d\eta = \mu_0 + \mu_1 h + \cdots + \mu_{12} h^{12},$$

where μ_0 through μ_{12} depend on ρ. $E(h)$ is positive definite, so a global minimum exists. In general, this shall depend on the value of ρ. In the case where $\rho = 0.5$, we find that

$$h^* = \underset{h\in\mathbb{R}}{\operatorname{argmin}}\, E(h) = -0.72731,$$

which gives minimal error of $E(h^*) = 1.9528 \times 10^{-6}$.

4.2.3.2. $\mathcal{L}_2[f] = f''' - f'$

We now select the different auxiliary linear operator

$$\mathcal{L}_2[u] = u''' - u',$$

with general solution

$$u(\eta) = \alpha_0 + \alpha_1 e^{\eta} + \alpha_2 e^{-\eta}.$$

Assuming a homotopy of the form

$$H(q)[f] = (1-q)\mathcal{L}_2[f] - hq\mathcal{N}[f]\,,$$

and assuming a four-term approximate solution of the form

$$G(\eta;h) = f_0(\eta) + f_1(\eta;h)q + f_2(\eta;h)q^2 + f_3(\eta;h)q^3\,,$$

enforcing $H(q)[F] \equiv 0$ implies

$$\mathcal{L}_2[f_0] = 0\,,\ f_0(0) = 0\,,\ f_0'(0) = 1\,,\ \lim_{\eta\to\infty} f_0' = 0\,,$$

$$\mathcal{L}_2[f_1] = h(f_0''' + f_0 f_0'' - \rho f_0'^2)\,,\ f_1(0) = 0\,,\ f_1'(0) = 0\,,\ \lim_{\eta\to\infty} f_1' = 0\,,$$

$$\mathcal{L}_2[f_2] = L_2[f_1] + h(f_1''' + f_1 f_0'' + f_0 f_1'' - 2\rho f_0' f_1')\,,$$

$$f_2(0) = 0\,, \qquad f_2'(0) = 0\,, \qquad \lim_{\eta\to\infty} f_2' = 0\,,$$

and

$$\mathcal{L}_2[f_3] = L_2[f_2] + h(f_2''' + f_2 f_0'' + f_1 f_1'' + f_0 f_2'' - 2\rho f_0' f_2' - \rho f_1'^2)\,,$$

$$f_3(0) = 0\,, \qquad f_3'(0) = 0\,, \qquad \lim_{\eta\to\infty} f_3' = 0\,.$$

We may then solve this system, successively obtaining the higher order terms. Observe that the order zero term is identical to the previous result, as it satisfies both linear operators. So,

$$f_0(\eta) = 1 - e^{-\eta}\,.$$

The other terms will differ, and are given by

$$f_1(\eta) = \frac{h}{6}(\rho - 1)\left(1 - e^{-\eta}\right)^2\,,$$

and so forth. Note the difference in the order one term, with the factor 1/6 as opposed to 1/4 which was true for the previous linear operator. So, differences appear at the first order when we selected different linear operators.

The accumulated L^2 norm of the residual error of $G(\eta)$,

$$E_2(h) = \int_0^\infty (\mathcal{N}[G(\eta;h)])^2 d\eta = \nu_0 + \nu_1 h + \cdots + \nu_{12} h^{12}\,,$$

where ν_0 through ν_{12} depend on ρ. Of course, $\nu_j \neq \mu_j$ in general, where μ_j is from the previous example. $E_2(h)$ is positive definite, so a global minimum exists. In general, this shall depend on the value of ρ. Considering again the case where $\rho = 1/2$, we find that

$$h^* = \underset{k_1 \in \mathbb{R}}{\operatorname{argmin}}\, E(k_1) = -1.1661\,,$$

which gives minimal error of $E(h^*) = 2.0593 \times 10^{-6}$.

A few things are clear. First, the value of the residual error given a four-term approximation is not drastically different due to the choice of either $\mathcal{L}_1$ or $\mathcal{L}_2$ as our auxiliary linear operator. However, the value of the convergence-control parameter is quite different in each case. What this implies is that we can have reasonable freedom to select the auxiliary linear operator, provided that we are able to choose the convergence-control parameter, h, in an optimal way. In other words, it appears that the proper selection of the convergence-control parameter is more important than the selection of the auxiliary linear operator. However, when selecting the auxiliary linear operator, we still need to ensure that the zeroth-order approximation is in the kernel.

4.3. Time evolution PDEs and auxiliary linear operators

Consider the time evolution equation

$$u_t = F(u, u_x, u_{xx})\,, \tag{4.29}$$

subject to the initial data

$$u(x, 0) = f(x)\,. \tag{4.30}$$

The nonlinear operator is then

$$\mathcal{N}[u] = u_t - F(u, u_x, u_{xx})\,. \tag{4.31}$$

We shall discuss some ways of choosing the auxiliary linear operator and the convergence-control parameter for such nonlinear PDEs and even systems of nonlinear PDEs.

4.3.1. *Simple polynomial evolution*

Here we consider the basic auxiliary linear operator

$$\mathcal{L}[u] = u_t\,. \tag{4.32}$$

Then $\mathcal{N}[u] = \mathcal{L}[u] - F(u, u_x, u_{xx})$. With this form of $\mathcal{L}$, we will construct a solution in terms of polynomials in t. Such a solution method is logically equivalent to Taylor series solutions in t: on each iteration, we gain a power in t. However, through proper use of the convergence-control parameter, we can obtain solutions more effectively than the standard method of obtaining a Taylor series solution. That said, the convergence can only be ensured on a finite domain for most nonlinear PDEs, since the true solution is not a power series in t.

For instance, assume we have initial data $u(x,0) = f(x)$. Then the order zero solution satisfies

$$u_{0,t} = 0\,,$$

which implies that u_0 is constant in t, i.e.

$$u_0(x,t) = f(x)\,.$$

The next term is governed by

$$u_{1,t} = hF(u_0, u_{0,x}, u_{0,xx}) = hF(f(x), f'(x), f''(x))\,.$$

Solving subject to $u_1(x,0) = 0$, we will have

$$u_1(x,t) = htF(f(x), f'(x), f''(x))\,,$$

and in general higher order terms will take the form of polynomials in t with coefficients in x. Such solutions can be useful locally for small enough t, but seldom can be useful for large t. Therefore, in order to get global solutions for all $t > 0$, this method is not useful.

4.3.2. *Evolution and exponential temporal decay*

In order to take advantage of exponential decay, we construct the auxiliary linear operator

$$\mathcal{L}[u] = u_t + u\,. \tag{4.33}$$

Then $\mathcal{N}[u] = \mathcal{L}[u] - u - F(u, u_x, u_{xx})$. This form of $\mathcal{L}$ will allow decaying exponential base functions. The resulting homotopy solutions will then allow the possibility of convergence as $t \to \infty$.

For example, assume $u(x,0) = f(x)$. Then the order zero solution satisfies

$$u_{0,t} + u_0 = 0\,,$$

which gives

$$u_0(x,t) = e^{-t} f(x)\,.$$

The next order term is governed by

$$u_{1,t} + u_1 = hF(e^{-t}f(x), e^{-t}f'(x), e^{-t}f''(x)),$$

which admits a solution

$$u_1(x,t) = h\int_0^t e^{-(t-s)}F(e^{-s}f(x), e^{-s}f'(x), e^{-s}f''(x))ds,$$

which often will be much more well-behaved than a simple polynomial in t.

In the following sections, we shall demonstrate how to apply this form of evolution operator $u_t + u$ in order to obtain accurate approximations for some nonlinear PDEs. First, we consider a nonlinear Klein–Gordon equation that was recently solved in Russo and Van Gorder [93] using optimal homotopy analysis. Then, we shall turn our attention to a system of PDEs, the Zakharov system with dissipation, which was recently solved by Mallory and Van Gorder [43]. In each of these examples, optimal convergence-control parameters are selected in order to ensure convergence of the solutions after a relatively small number of terms are computed. Such an approach is computationally efficient. In the latter example, three convergence-control parameters are used, since there is a system of three nonlinear PDEs. It is found that these parameters may be jointly optimized, in order to obtain solutions with minimal residual error. A similar approach was also recently applied by Van Gorder [37] for the construction of solutions to the Föppl–von Kármán equations governing deflections of a thin flat plate (which is a system of two nonlinear PDEs).

4.3.3. *Nonlinear Klein–Gordon equation*

One may apply the homotopy analysis method to study solutions of nonlinear evolution equations. In the recent paper of Russo and Van Gorder [93], we considered the nonlinear Klein–Gordon equation

$$u_{tt} - u_{xx} = F(u), \tag{4.34}$$

$$u(x,0) = f(x), \quad u_t(x,0) = g(x). \tag{4.35}$$

Here, $F \in C^\infty(\mathbb{R})$ is an arbitrary analytic function which will serve as the nonlinear term, while $f, g \in C^\infty(\mathbb{R})$ are the analytic initial data. We shall highlight the significance of the results obtained in [93] for nonlinear evolution PDEs.

Additionally, some authors have considered the homotopy analysis of some special cases for $F(u)$. Sun [94] considered the quasilinear cubic Klein–Gordon equation, and used homotopy analysis method to obtain a solution form in terms of a trigonometric basis. However, such solutions were for the restriction to the travelling wave case, which reduces (4.34) from a partial differential equation to an ordinary differential equation. More recently, Iqbal *et al.* [95] considered the same equation from the standpoint of the optimal homotopy analysis method, and similar results were obtained. None of these studies considered arbitrary initial data. Approximate homotopy analysis solutions for the sin-Gordon equation were recently discussed by Yücel [96], however no discussion of error was provided (the author simply applied the so-called h-curve in order to deduce possible regions of convergence of the solutions).

Particularly interesting examples are the quasilinear Klein–Gordon [97–99] equation ($F(u) = u^3 - \alpha u$), the modified Liouville [100, 101] equation ($F(u) = e^{\beta u}$), the sinh-Gordon [102–108] equation ($F(u) = \sinh(u)$) and the tanh-Gordon [109] equation ($F(u) = \tanh(u)$).

4.3.3.1. *HAM for the quasilinear Klein–Gordon equation*

The quasilinear Klein–Gordon equation corresponds to $F(u) = u^3 - \alpha u$, where α is a real-valued parameter. Let

$$\psi(x,t;h) = u_0(x,t) + u_1(x,t;h) + u_2(x,t;h) + \cdots$$

be the homotopy solution evaluated at $q = 1$. We then find that the first several terms of $F(\psi(x,t;h))$ are given by

$$\begin{aligned} F(\psi(x,t;h)) = & \left(u_0^3 - \alpha u_0\right) + \left(3u_0^2 u_1 - \alpha u_1\right) q \\ & + \left(3(u_0^2 u_2 + u_0 u_1^2) - \alpha u_2\right) q^2 + \cdots . \end{aligned}$$

For the present problems, we shall consider a linear operator of the type

$$\mathcal{L}[\psi] = \psi_t + \psi\,,$$

which permits a decaying exponential basis in t.

The zeroth-order approximation remains u_0, while the higher order approximations are governed by

$$\mathcal{L}[u_1] = h\left((u_0)_{tt} - (u_0)_{xx} - u_0^3 + \alpha u_0\right)\,,$$

$$\mathcal{L}[u_2] = (u_1)_{tt} + 2(u_1)_t + u_1 + h\left((u_1)_{tt} - (u_1)_{xx} - 3u_0^2 u_1 + \alpha u_1\right)\,,$$

$$\mathcal{L}[u_3] = (u_2)_{tt} + 2(u_2)_t + u_2 + h\left((u_2)_{tt} - (u_2)_{xx} - 3(u_0^2 u_2 + u_0 u_1^2) + \alpha u_2\right)\,,$$

etc. We find that u_1 is given by

$$\begin{aligned}
u_1(x,t;h) &= \frac{h}{6}\{(f(x)+g(x)-f''(x)-g''(x))\,t \\
&\quad -3\,(f(x)+2g(x)+f''(x))\}\,t^2e^{-t} \\
&\quad -h\int_0^t (t-s)e^{-(t-s)}\Big\{(f(x)+(f(x)+g(x))\,s)^3\,e^{-3s} \\
&\quad -\alpha\,(f(x)+(f(x)+g(x))\,s)\,e^{-s}\Big\}\,ds \\
&= hP_{1,1}(x,t)e^{-t}+hP_{1,3}(x,t)e^{-3t}\,, \qquad (4.36)
\end{aligned}$$

where $P_{1,1}$ and $P_{1,3}$ are polynomials in t of degree no greater than three. Higher order inversion formulas may be obtained in a similar manner. In general, we find that for $n \geq 2$

$$\begin{aligned}
u_n(x,t;h) &= P_{n,1}(x,t;h)e^{-t}+P_{n,3}(x,t;h)e^{-3t} \\
&\quad +\cdots+P_{n,2n+1}(x,t;h)e^{-(2n+1)t}\,,
\end{aligned}$$

where the $P_{n,k}$'s are polynomials in t with coefficients involving $f(x)$, $g(x)$ and their derivatives. These coefficients also depend on h, which we shall use to our advantage.

4.3.3.2. *Initial data $f(x) = sech(x)$, $g(x) = 0$*

Consider the initial data $f(x) = \text{sech}(x)$, $g(x) = 0$. We find that

$$u_0(x,t) = \text{sech}(x)\,(1+t)\,e^{-t}\,,$$

and, upon placing this expression into (4.36), we have

$$\begin{aligned}
u_1(x,t;h) &= h\text{sech}(x)\left(\frac{\alpha}{6}t^2(3+t)+\left(\tanh^2(x)+1\right)\frac{t^3}{3}\right. \\
&\quad \left.-\tanh^2(x)t^2-\frac{19t-23}{8}\text{sech}^2(x)\right)e^{-t} \\
&\quad -\frac{h}{8}\text{sech}^3(x)\left(2t^3+12t^2+27t+23\right)e^{-3t}\,.
\end{aligned}$$

In order to obtain an accurate approximation, we must still select the convergence-control parameter, h. Due to the complicated expressions obtained for the first and second order terms, we compute discrete residuals. We take a double sum over a finite collection of points in order to approximate the squared residual error over a finite subset of the problem domain.

Taking $T = 5$ and $\ell = 10$, we have

$$E(h,\alpha) = \int_0^5 \int_{-10}^{10} \left(\frac{\mathrm{Res}(x,t;h,\alpha)}{U_2(x,t;h,\alpha)} \right)^2 dxdt$$
$$\approx 0.05 \sum_{i=0}^{50} \sum_{j=-20}^{20} \left(\frac{\mathrm{Res}(0.5j, 0.1i; h,\alpha)}{U_2(0.5j, 0.1t; h,\alpha)} \right)^2, \tag{4.37}$$

where we have sampled $(41)(51) = 2091$ points in the set $(x,t) \in [-10,10] \times [0,5]$, and the normalization factor is computed by the product of the two step sizes.

4.3.3.3. *Initial data* $f(x) = e^{-|x|}$, $g(x) = 0$

Consider next the peaked initial data with exponential decay corresponding to $f(x) = e^{-|x|}$, $g(x) = 0$. Note that $f(x)$ is not analytic; all derivatives fail to exist at $x = 0$. Note that we will be exclusively interested in odd order derivatives of $f(x)$. Then, note that for all $x \neq 0$,

$$\frac{d^2 f}{dx^2} = \frac{d}{dx}\left(\mathrm{sgn}(x) e^{-|x|}\right)$$
$$= \left(\mathrm{sgn}^2(x) - 2\delta(x)\right) e^{-|x|} = e^{-|x|} = f(x), \tag{4.38}$$

hence $f^{(2k)}(x) = f(x)$ for all $x \neq 0$. We then find that

$$u_0(x,t) = (1+t)e^{-t}e^{-|x|},$$

$$u_1(x,t;h) = h\left(\frac{\alpha}{6}(3+t) - 1\right) t^2 e^{-t} e^{-|x|} - \frac{h}{8}(19t - 23)\, e^{-t} e^{-3|x|}$$
$$- \frac{h}{8}\left(2t^3 + 12t^2 + 27t + 23\right) e^{-3t} e^{-3|x|},$$

and so on. In order to analyze error, we shall use the same discrete residual errors as in (8.17), with the exception that we no longer include the points of the form $(0,t)$, as $x = 0$ is undefined.

4.3.4. *Zakharov system with dissipation*

One may also apply the homotopy analysis method to systems of nonlinear PDEs with certain initial data. In the very recent paper [43] of Mallory and Van Gorder, we considered the control of error in the homotopy analysis of solutions to the Zakharov system with dissipation. The relevant real-valued system is

$$\begin{cases} \rho_t = \sigma v - \nabla^2 \sigma\,, \\ \sigma_t = -\rho v + \nabla^2 \rho\,, \\ v_{tt} = \nabla^2 \left(v - \rho^2 - \sigma^2\right) - v_t - v\,. \end{cases} \tag{4.39}$$

Initial data takes the form $\rho(\mathbf{x},0) = R(\mathbf{x})$ and $\sigma(\mathbf{x},0) = S(\mathbf{x})$. In what follows, we shall highlight the significant results of the paper of Mallory and Van Gorder [43] which are relevant for systems of nonlinear evolution PDEs.

We should note that the Zakharov system describes the propagation of Langmuir waves in an ionized plasma [110–112]. Soliton solutions were considered in [113–115]. Homoclinic tube solutions and chaos were studied in [116]. Numerical simulations were recently presented in [117, 118].

Let us define the operators

$$\mathcal{L} = \frac{\partial}{\partial t} + 1,$$

$$\mathcal{N}_1[\rho,\sigma,v] = \rho + \sigma v - \nabla^2\sigma,$$

$$\mathcal{N}_2[\rho,\sigma,v] = \sigma - \rho v + \nabla^2\rho,$$

$$\mathcal{N}_3[\rho,\sigma,v] = v_t + \nabla^2\left(v - \rho^2 - \sigma^2\right).$$

We shall also make use of the square of $\mathcal{L}$, $\mathcal{L}^2$, which is of the form

$$\mathcal{L}^2 = \frac{\partial^2}{\partial t^2} + 2\frac{\partial}{\partial t} + 1\,.$$

Then, Zakharov system with dissipation (4.39) takes the form

$$\begin{cases} \mathcal{L}[\rho] = \mathcal{N}_1[\rho,\sigma,v]\,, \\ \mathcal{L}[\sigma] = \mathcal{N}_2[\rho,\sigma,v]\,, \\ \mathcal{L}^2[v] = \mathcal{N}_3[\rho,\sigma,v]\,. \end{cases} \tag{4.40}$$

Then, we construct the zeroth-order deformation equations

$$\begin{cases} H_1(\hat{\rho}(x,t,q),\hat{\sigma}(x,t,q),\hat{v}(x,t,q);q) = (1-q)\mathcal{L}[\hat{\rho}(x,t,q) - \rho_0(x,t)] \\ \qquad - qh_1(\mathcal{L}[\hat{\rho}(x,t,q)] - \mathcal{N}_1[\hat{\rho}(x,t,q),\hat{\sigma}(x,t,q),\hat{v}(x,t,q)])\,, \\ H_2(\hat{\rho}(x,t,q),\hat{\sigma}(x,t,q),\hat{v}(x,t,q);q) = (1-q)\mathcal{L}[\hat{\sigma}(x,t,q) - \sigma_0(x,t)] \\ \qquad - qh_2(\mathcal{L}[\hat{\sigma}(x,t,q)] - \mathcal{N}_2[\hat{\rho}(x,t,q),\hat{\sigma}(x,t,q),\hat{v}(x,t,q)])\,, \\ H_3(\hat{\rho}(x,t,q),\hat{\sigma}(x,t,q),\hat{v}(x,t,q);q) = (1-q)\mathcal{L}^2[\hat{v}(x,t,q) - v_0(x,t)] \\ \qquad - qh_3(\mathcal{L}^2[\hat{v}(x,t,q)] - \mathcal{N}_3[\hat{\rho}(x,t,q),\hat{\sigma}(x,t,q),\hat{v}(x,t,q)])\,, \end{cases} \tag{4.41}$$

where h_1, h_2, and h_3 denote the unique convergence-control parameters and q is the embedding parameter. We take $\hat{\rho}$, $\hat{\sigma}$, and $\hat{v}$ to be the solutions dependent upon q and ρ_0, σ_0, and v_0 to be the approximate general solutions to the equations constructed from the linear operators. Using the embedding parameter, we may represent the linearities by taking $q = 0$ and the original nonlinear differential equations with $q = 1$. We assume the solutions $\hat{\rho}$, $\hat{\sigma}$, and $\hat{v}$ may be represented by the infinite series

$$\hat{\rho}(x,t,q) = \rho_0(x,t) + \rho_1(x,t)q + \rho_2(x,t)q^2 + \dots, \tag{4.42}$$

$$\hat{\sigma}(x,t,q) = \sigma_0(x,t) + \sigma_1(x,t)q + \sigma_2(x,t)q^2 + \dots, \tag{4.43}$$

$$\hat{v}(x,t,q) = v_0(x,t) + v_1(x,t)q + v_2(x,t)q^2 + \dots, \tag{4.44}$$

which we may substitute into the zeroth-order deformation equation (4.41) and collect common powers of q.

Utilizing the linear operators given in (4.40), we begin by solving for the zeroth-order functions required for our solutions $\hat{\rho}$, $\hat{\sigma}$, and $\hat{v}$. Representing ordinary differential equations in t, these are defined as

$$\frac{\partial}{\partial t}\rho_0 + \rho_0 = 0\,,$$

$$\frac{\partial}{\partial t}\sigma_0 + \sigma_0 = 0\,,$$

$$\frac{\partial^2}{\partial t^2}v_0 + 2\frac{\partial}{\partial t}v_0 + v_0 = 0\,.$$

Solving these linear equations subject to initial data, we obtain the initial approximations

$$\rho_0(x,t) = R(x)\mathrm{e}^{-t}\,,$$

$$\sigma_0(x,t) = S(x)\mathrm{e}^{-t}\,,$$

$$v_0(x,t) = (U(x) + (U(x) + W(x))t)\mathrm{e}^{-t}\,.$$

In accordance with the homotopy analysis method, we note that all subsequent functions occur recursively.

Our system (4.40) may be rewritten as

$$\begin{cases} \mathcal{L}[\rho(x,t)] - \mathcal{N}_1[\rho(x,t),\sigma(x,t),v(x,t)] = 0\,, \\ \mathcal{L}[\sigma(x,t)] - \mathcal{N}_2[\rho(x,t),\sigma(x,t),v(x,t)] = 0\,, \\ \mathcal{L}^2[v(x,t)] - \mathcal{N}_3[\rho(x,t),\sigma(x,t),v(x,t)] = 0\,. \end{cases} \tag{4.45}$$

In order to determine the deviation of our approximations from the exact solutions, we employ residual error techniques. Accordingly, for any $x \in (-\infty, \infty)$ and $t \geq 0$, we may calculate the residual error at some point (x^*, t^*) by

$$\epsilon_1(x,t) = \left|\mathcal{L}[\hat{\rho}(x,t)] - \mathcal{N}_1[\hat{\rho}(x,t), \hat{\sigma}(x,t), \hat{v}(x,t)]\right| \Bigg|_{x=x^*,t=t^*},$$

$$\epsilon_2(x,t) = \left|\mathcal{L}[\hat{\sigma}(x,t)] - \mathcal{N}_2[\hat{\rho}(x,t), \hat{\sigma}(x,t), \hat{v}(x,t)]\right| \Bigg|_{x=x^*,t=t^*},$$

$$\epsilon_3(x,t) = \left|\mathcal{L}^2[\hat{v}(x,t)] - \mathcal{N}_3[\hat{\rho}(x,t), \hat{\sigma}(x,t), \hat{v}(x,t)]\right| \Bigg|_{x=x^*,t=t^*}.$$

Therefore, to sum the residual error over the domain $x \in (-\infty, \infty)$, $t \in (0, T)$, with $T > 0$ as some stopping time, we take

$$R_{\epsilon_1}(x,t) = \int_0^T \int_{-\infty}^{\infty} \left|\mathcal{L}[\hat{\rho}(x,t)] - \mathcal{N}_1[\hat{\rho}(x,t), \hat{\sigma}(x,t), \hat{v}(x,t)]\right| \mathrm{d}x\, \mathrm{d}t\,,$$

$$R_{\epsilon_2}(x,t) = \int_0^T \int_{-\infty}^{\infty} \left|\mathcal{L}[\hat{\sigma}(x,t)] - \mathcal{N}_2[\hat{\rho}(x,t), \hat{\sigma}(x,t), \hat{v}(x,t)]\right| \mathrm{d}x\, \mathrm{d}t\,,$$

$$R_{\epsilon_3}(x,t) = \int_0^T \int_{-\infty}^{\infty} \left|\mathcal{L}^2[\hat{v}(x,t)] - \mathcal{N}_3[\hat{\rho}(x,t), \hat{\sigma}(x,t), \hat{v}(x,t)]\right| \mathrm{d}x\, \mathrm{d}t\,.$$

However, such integration is typically too difficult to perform, so in order to avoid the difficulty posed by the absolute value, we evaluate the sum of the squared residual error and define

$$E_1\,(h_1, h_2, h_3) = \int_0^T \int_0^{\ell} (\mathcal{L}[\hat{\rho}(x,t)] - \mathcal{N}_1[\hat{\rho}(x,t), \hat{\sigma}(x,t), \hat{v}(x,t)])^2 \,\mathrm{d}x\, \mathrm{d}t\,, \tag{4.46}$$

$$E_2\,(h_1, h_2, h_3) = \int_0^T \int_0^{\ell} (\mathcal{L}[\hat{\sigma}(x,t)] - \mathcal{N}_2[\hat{\rho}(x,t), \hat{\sigma}(x,t), \hat{v}(x,t)])^2 \,\mathrm{d}x\, \mathrm{d}t\,, \tag{4.47}$$

$$E_3\,(h_1, h_2, h_3) = \int_0^T \int_0^{\ell} (\mathcal{L}^2[\hat{v}(x,t)] - \mathcal{N}_3[\hat{\rho}(x,t), \hat{\sigma}(x,t), \hat{v}(x,t)])^2 \,\mathrm{d}x\, \mathrm{d}t\,. \tag{4.48}$$

Choosing desired stopping points, denoted ℓ and T for x and t, respectively, we compute (4.46)–(4.48) for given initial data as functions of h_1,

h_2, and h_3. Thus we may utilize these embedded convergence-control parameters to minimize the error of our calculated solutions, thereby finding $({h_1}^*, {h_2}^*, {h_3}^*) \in \mathbb{R}^3$ which offers the best approximation of the true solutions $\hat{\rho}$, $\hat{\sigma}$, and $\hat{v}$.

4.4. The convergence control parameter

In the present section, we revisit the previous examples, in order to determine the behavior of the convergence-control parameter, h, after a finite number of iterations. While the number of iterations were fixed in the previous examples, here we fix a linear operator and determine the optimal value h_k^* of the convergence-control parameter after k iterations. As more iterations of the method are taken, we expect that the residual error will decrease. However, it is natural to wonder if the value of h_k^* will appear to converge to a fixed value once we make k large enough.

4.4.1. *Lane–Emden equation under $\mathcal{L}_1 = y''$*

Let us revisit the Lane–Emden problem (4.19)–(4.20). Consider again the auxiliary linear operator $\mathcal{L}_1 = y''$. The two-term approximation takes the form $F_2(x) = y_0(x) + y_1(x; h)$, and reads

$$F_2(x, h) = 1 + \frac{h}{2} x^2 .$$

The accumulated L^2 norm of the error is given by

$$E_2(h) = \int_0^1 (\mathcal{N}[F_2(x,h)])^2 d\eta = \mu_0^2 + \mu_1^2 h + \cdots + \mu_6^2 h^6 ,$$

where μ_j^2 is a constant for all $j = 0, 1, \ldots, 6$. The function E_2 is, by construction, positive definite, so there exists a global minimum. We define

$$h_2^* = \operatorname*{argmin}_h E_2(h) = -0.28584$$

to be this minimizing value of the convergence-control parameter corresponding to $F_2(x, h)$. The minimal value of the function $E_2(h)$ reads $E_2(-0.28584) = 1.2871 \times 10^{-2}$.

The three-term approximation takes the form $F_3(x; h) = y_0(x) + y_1(x; h) + y_2(x; h)$, so that

$$F_3(x; h) = 1 + \frac{h}{2} x^2 + \frac{h^2}{8} x^4 + \frac{h}{2}(1 + 3h)\, x^2 .$$

The accumulated L^2 norm of the error is given by

$$E_3(h) = \int_0^\infty (\mathcal{N}[F_3(x,h)])^2 d\eta = \mu_0^3 + \mu_1^3 h + \cdots + \mu_{12}^3 h^{12},$$

where μ_j^3 is a constant for all $j = 0, 1, \ldots, 12$. The function E_3 is positive definite, so there exists a global minimum. We define

$$h_3^* = \underset{h}{\operatorname{argmin}}\, E_3(h) = -0.23975$$

to be this minimizing value of the convergence-control parameter corresponding to $F_3(x,h)$. The minimal value of the function $E_3(h)$ reads $E_3(-0.23975) = 5.4550 \times 10^{-3}$.

The four-term approximation takes the form $F_4(x;h) = y_0(x) + y_1(x;h) + y_2(x;h) + y_3(x;h)$. Then

$$F_4(x;h) = 1 + \frac{h}{2}x^2 + \frac{h^2}{8}x^4 + \frac{h}{2}(1+3h)\,x^2 + \frac{3}{80}h^3x^6 + \frac{h^2}{12}(7h+3)\,x^4 + \frac{h}{2}\left(9h^2+6h+1\right)x^2 .$$

The accumulated L^2 norm of the residual error of $F_4(x;h)$ over $x \in [0,1]$ is

$$E_4(h) = \int_0^1 (\mathcal{N}[F_4(x;h)])^2 dx = \mu_0^4 + \mu_1^4 h + \cdots + \mu_{18}^1 h^{18},$$

where μ_0^4 through μ_{18}^4 are constants. $E_4(h)$ is positive definite, so a global minimum exists. We find that

$$h_4^* = \underset{h\in\mathbb{R}}{\operatorname{argmin}}\, E_1(h) = -0.373817,$$

which gives minimal error of $E_1(h_4^*) = 2.364958 \times 10^{-4}$.

What we see is that the optimal value of the convergence-control parameter tends to oscillate near $h = -0.3$. If there is convergence of the sequence of h_k's, it may not be monotone.

4.4.2. *Lane–Emden equation under* $\mathcal{L}_3 = y'' + \frac{2}{x}y'$

Consider again the Lane–Emden problem (4.19)–(4.20), this time with the auxiliary linear operator $\mathcal{L}_3 = y'' + \frac{2}{x}y'$. We shall now compare optimal values for the convergence-control parameter under multiple iterations of the solution process.

The two-term approximation takes the form $F_2(x) = y_0(x) + y_1(x;h)$, and reads

$$F_2(x,h) = 1 + \frac{h}{6}x^2 .$$

The accumulated L^2 norm of the error is given by

$$E_2(h) = \int_0^1 (\mathcal{N}[F_2(x,h)])^2 d\eta = \mu_0^2 + \mu_1^2 h + \cdots + \mu_6^2 h^6 \,,$$

where μ_j^2 is a constant for all $j = 0,1,\ldots,6$. The function E_2 is, by construction, positive definite, so there exists a global minimum. We define

$$h_2^* = \operatorname*{argmin}_h E_2(h) = -0.85752 \,.$$

The minimal value of the function $E_2(h)$ reads $E_2(-0.85752) = 1.28713 \times 10^{-2}$.

The three-term approximation takes the form $F_3(x;h) = y_0(x) + y_1(x;h) + y_2(x;h)$, so that

$$F_3(x;h) = 1 + \frac{h}{6}x^2 + \frac{h^2}{40}x^4 + \frac{h}{6}\,(1+h)\,x^2 \,.$$

The accumulated L^2 norm of the error is given by

$$E_3(h) = \int_0^\infty (\mathcal{N}[F_3(x,h)])^2 d\eta = \mu_0^3 + \mu_1^3 h + \cdots + \mu_{12}^3 h^{12} \,,$$

where μ_j^3 is a constant for all $j = 0,1,\ldots,12$. The function E_3 is positive definite, so there exists a global minimum. We define

$$h_3^* = \operatorname*{argmin}_h E_3(h) = -0.87869$$

to be this minimizing value of the convergence-control parameter corresponding to $F_3(x,h)$. The minimal value of the function $E_3(h)$ reads $E_3(-0.87869) = 8.67763 \times 10^{-5}$.

The four-term approximation takes the form $F_4(x;h) = y_0(x) + y_1(x;h) + y_2(x;h) + y_3(x;h)$. Then

$$\begin{aligned} F_4(x;h) = 1 &+ \frac{h}{6}x^2 + \frac{h^2}{40}x^4 + \frac{h}{6}\,(1+h)\,x^2 \\ &+ \frac{19}{5040}h^3x^6 + \frac{h^2}{20}\,(1+h)\,x^4 + \frac{h}{6}\left(2h^2+h+1\right)x^2 \,. \end{aligned}$$

$$E_4(h) = \int_0^1 (\mathcal{N}[F_4(x;h)])^2 dx = \mu_0^4 + \mu_1^4 h + \cdots + \mu_{18}^4 h^{18} \,,$$

where μ_0^4 through μ_{18}^4 are constants. $E_4(h)$ is positive definite, so a global minimum exists. We find that

$$h_4^* = \operatorname*{argmin}_{h\in\mathbb{R}} E_3(h) = -0.892269 \,,$$

which gives minimal error of $E_4(-0.892269) = 4.35017 \times 10^{-7}$. In this example, we see that the optimal value of the convergence-control parameter appears to converge to near $h^* = -0.9$ as the number of iterations increases. As was shown in a previous section, the choice of $\mathcal{L} = \mathcal{L}_3$ was superior to the choice $\mathcal{L} = \mathcal{L}_1$, in terms of obtaining minimal error approximations. So, it appears that when an auxiliary linear operator is selected properly, we may expect the optimal value of the convergence-control parameter to converge more rapidly to some fixed value. Such behavior can be a sign that we are on the right track to selecting the auxiliary linear operator.

4.4.3. *Flow over a nonlinearly stretching sheet*

Returning to the problem of a flow over a nonlinearly stretching sheet (4.24)–(4.25), and selecting operator $\mathcal{L}_1$ given by $\mathcal{L}_1[f] = f''' + f''$, we may give the first few low-order approximations.

The two-term approximation takes the form $F_2(\eta) = f_0(\eta) + f_1(\eta, h)$, and reads

$$F_2(\eta, h) = 1 - e^{-\eta} + \frac{h}{4}(\rho - 1)\left(1 - e^{-\eta}\right)^2 .$$

The accumulated L^2 norm of the error is given by

$$E_2(h) = \int_0^\infty (\mathcal{N}[F_2(\eta, h)])^2 d\eta = \mu_0 + \mu_1 h + \cdots + \mu_4 h^4 ,$$

where μ_j is in general a function of ρ for all $j = 0, 1, 2, 3, 4$. The function E_2 is, by construction, positive definite, so there exists a global minimum. We define

$$h_2^* = \underset{h}{\operatorname{argmin}}\, E_2(h)$$

to be this minimizing value of the convergence-control parameter corresponding to $F_2(\eta, h)$. For the sake of demonstration, at the value $\rho = 0.5$, we find that $h_2^* = -0.69447$. At this value, we see that the minimal value of the function $E_2(h)$ reads $E_2(-0.69447) = 1.4553 \times 10^{-4}$.

The three-term approximation takes the form $F_3(\eta) = f_0(\eta) + f_1(\eta, h) + f_2(\eta, h)$. We shall omit the lengthy expression, but the functions f_k are listed elsewhere in this chapter. The accumulated L^2 norm of the error is given by

$$E_3(h) = \int_0^\infty (\mathcal{N}[F_3(\eta, h)])^2 d\eta = \mu_0 + \mu_1 h + \cdots + \mu_8 h^8 ,$$

where μ_j is in general, a function of ρ for all $j = 0, 1, \ldots, 8$. The function E_3 is positive definite, so there exists a global minimum. We define

$$h_3^* = \underset{h}{\operatorname{argmin}}\, E_3(h)$$

to be this minimizing value of the convergence-control parameter corresponding to $F_3(\eta, h)$. At the value $\rho = 0.5$, we find that $h_3^* = -0.69116$. At this value, we see that the minimal value of the function $E_3(h)$ reads $E_3(-0.69116) = 4.1367 \times 10^{-5}$.

The four-term approximation takes the form $F_4(\eta) = f_0(\eta) + f_1(\eta, h) + f_2(\eta, h) + f_3(\eta, h)$. We shall omit the lengthy expression, but the functions f_k are listed elsewhere in this chapter. The accumulated L^2 norm of the error is given by

$$E_4(h) = \int_0^\infty (\mathcal{N}[F_4(\eta, h)])^2 d\eta = \mu_0 + \mu_1 h + \cdots + \mu_{12} h^{12},$$

where μ_j is in general, a function of ρ for all $j = 0, 1, \ldots, 12$. The function E_4 is positive definite, so there exists a global minimum. We define

$$h_4^* = \underset{h}{\operatorname{argmin}}\, E_4(h)$$

to be this minimizing value of the convergence-control parameter corresponding to $F_4(\eta, h)$. At the value $\rho = 0.5$, we find that $h_4^* = -0.72732$. At this value, we see that the minimal value of the function $E_3(h)$ reads $E_3(-0.72732) = 1.95074 \times 10^{-6}$.

4.5. Modifying the homotopy

In contrast to the standard linear homotopy connecting $\mathcal{L}$ and $\mathcal{N}$, one may consider more exotic forms of the homotopy. To illustrate this point, we return to the problem of a flow over a nonlinearly stretching sheet.

4.5.1. *The general homotopy*

Consider the homotopy

$$H(q)[f] = A(q)\mathcal{L}[f] - B(q)\mathcal{N}[f] = 0. \tag{4.49}$$

Here $\mathcal{L}$ is a linear differential operator and $\mathcal{N}$ is the nonlinear differential operator of interest. A and B are functions of the homotopy parameter q, and have the property that $A(1) = 0 = B(0)$ and $B(1) \neq 0 \neq A(0)$. Then, when $q = 1$, the above equation reduces to the nonlinear operator of interest, while when $q = 0$, the equation is linear. We wish to approximate solutions to $\mathcal{N}$ as partial sums of power series in q. With this in mind, we attempt to select A and B in such a way to minimize error. In many recent papers, this has been done by letting $A(q) = 1 - q$ and $B(q) = cq$, where c is a convergence-control parameter, usually a constant. Now, we consider a new strategy: that A and B are not linear functions of q. We let

$$A(q) = (1 - q)\, a(q), \quad B(q) = q\, b(q).$$

These functions meet the criteria outlined above, as long as $b(1) \neq 0$. In general, as we wish to expand (4.49) in powers of q, we let $a(q)$ and $b(q)$ be polynomials in q:

$$a(q) = c_0 + c_1 q + c_2 q^2 + \ldots + c_n q^n, \tag{4.50}$$

$$b(q) = k_1 + k_2 q + k_3 q^2 + \ldots + k_{n+1} q^n. \tag{4.51}$$

Here, n is one less than the desired degree of the approximation to be computed via (4.49), and c_i and k_i are convergence-control parameters. In general, these extra convergence-control parameters can only improve the error of the method. However, having so many of them increases computational complexity if truly optimal solutions are sought.

4.5.2. *Standard homotopy analysis method*

We again consider the equation for nonlinear flow over a nonlinearly stretching sheet:

$$f''' + ff'' - \rho f'^2 = 0, \tag{4.52}$$

subject to the conditions $f(0) = 0$, $f'(0) = 1$, and $f'(\infty) = 0$. Here $\rho \in (0, 2)$ describes the power law stretching of the sheet; $\rho = 1$ corresponds to linear stretching.

We seek approximate solutions to (4.52) by the homotopy analysis method. We select the linear operator

$$\mathcal{L}[u] = u''' + u'',$$

with general solution

$$u(\eta) = \alpha_0 + \alpha_1 \eta + \alpha_2 e^{-\eta}.$$

$\mathcal{N}$ shall represent the nonlinear differential in question:

$$\mathcal{N}[f] = f''' + ff'' - \rho f'^2.$$

Assuming a homotopy of the form

$$H(q)[f] = (1-q)\mathcal{L}[f] - k_1 q\mathcal{N}[f]\,,$$

and assuming a four-term approximate solution of the form

$$F(\eta;k_1) = f_0(\eta) + f_1(\eta;k_1)q + f_2(\eta;k_1)q^2 + f_3(\eta;k_1)q^3\,, \tag{4.53}$$

the condition $H(q)[F] \equiv 0$ implies

$$\mathcal{L}[f_0] = 0\,,\quad f_0(0) = 0\,,\quad f_0'(0) = 1\,,\quad \lim_{\eta\to\infty} f_0' = 0\,,$$

$$\begin{cases} \mathcal{L}[f_1] = k_1(f_0''' + f_0 f_0'' - \rho f_0'^2)\,, \\ f_1(0) = 0\,,\quad f_1'(0) = 0\,,\quad \lim_{\eta\to\infty} f_1' = 0\,, \end{cases}$$

$$\begin{cases} \mathcal{L}[f_2] = \mathcal{L}[f_1] + k_1(f_1''' + f_1 f_0'' + f_0 f_1'' - 2\rho f_0' f_1')\,, \\ f_2(0) = 0\,,\qquad f_2'(0) = 0\,,\qquad \lim_{\eta\to\infty} f_2' = 0\,, \end{cases}$$

$$\begin{cases} \mathcal{L}[f_3] = \mathcal{L}[f_2] + k_1(f_2''' + f_2 f_0'' + f_1 f_1'' + f_0 f_2'' - 2\rho f_0' f_2' - \rho f_1'^2)\,, \\ f_3(0) = 0\,,\qquad f_3'(0) = 0\,,\qquad \lim_{\eta\to\infty} f_3' = 0\,. \end{cases}$$

We may then solve this system, successively obtaining the higher order terms. Observe that

$$f_0(\eta) = 1 - e^{-\eta}\,,$$

which happens to be the exact solution to the $\rho = 1$ problem. The other terms are given by

$$\begin{aligned} f_1(\eta) &= \frac{k_1}{4}(\rho-1)\left(1-e^{-\eta}\right)^2\,, \\ f_2(\eta) &= k_1(\rho-1)\left\{\frac{k_1(5\rho+14)+9}{36} - \frac{k_1(8\rho+11+6\eta)+12}{24}e^{-\eta}\right. \\ &\quad \left. + \frac{k_1\rho+1}{4}e^{-2\eta} - \frac{k_1(4\rho-5)}{72}e^{-3\eta}\right\}\,, \end{aligned}$$

and so on.

We then compute the accumulated L^2 norm of the residual error of $F(\eta)$ as given in (4.53),

$$E(k_1) = \int_0^\infty (\mathcal{N}[F(\eta;k_1)])^2 d\eta = \mu_0 + \mu_1 k_1 + \cdots + \mu_{12} k_1^{12}\,,$$

where μ_0 through μ_{12} depend on ρ. $E(k_1)$ is positive definite, so a global minimum exists. In general, this shall depend on the value of ρ. In the case where $\rho = 0.5$, we find that

$$k_1^* = \operatorname*{argmin}_{k_1 \in \mathbb{R}} E(k_1) = -0.72731\,,$$

which gives minimal error of $E(k_1^*) = 1.9528 \times 10^{-6}$.

4.5.3. *A homotopy quadratic in q*

With the same linear and nonlinear operators as those given in the preceding section, define the modified homotopy of the form

$$H(q)[f] = (1-q)\mathcal{L}[f] - (k_1 + k_2 q) q \mathcal{N}[f]\,,$$

so that $B(q)$ is now quadratic, with two free convergence-control parameters. When $k_2 = 0$, we recover the previous example. Hence, we expect that the added freedom of having an additional free parameter will allow us to better control the residual error.

Assuming a four-term approximate solution of the form

$$F(\eta; k_1, k_2) = f_0(\eta) + f_1(\eta; k_1) q + f_2(\eta; k_1, k_2) q^2 + f_3(\eta; k_1, k_2) q^3\,,$$

the condition $H(q)[F] \equiv 0$ implies

$$\mathcal{L}[f_0] = 0\,, \quad f_0(0) = 0\,, \quad f_0'(0) = 1\,, \quad \lim_{\eta\to\infty} f_0' = 0\,,$$

$$\begin{cases} \mathcal{L}[f_1] = k_1 (f_0''' + f_0 f_0'' - \rho f_0'^2)\,, \\ f_1(0) = 0\,, \quad f_1'(0) = 0\,, \quad \lim_{\eta\to\infty} f_1' = 0\,, \end{cases}$$

$$\begin{cases} \mathcal{L}[f_2] = \mathcal{L}[f_1] + k_1 (f_1''' + f_1 f_0'' + f_0 f_1'' - 2\rho f_0' f_1') \\ \qquad\qquad + k_2 (f_0''' + f_0 f_0'' - \rho f_0'^2)\,, \\ f_2(0) = 0\,, \quad f_2'(0) = 0\,, \quad \lim_{\eta\to\infty} f_2' = 0\,, \end{cases}$$

$$\begin{cases} \mathcal{L}[f_3] = \mathcal{L}[f_2] + k_1 (f_2''' + f_2 f_0'' + f_1 f_1'' + f_0 f_2'' - 2\rho f_0' f_2' - \rho f_1'^2) \\ \qquad\qquad + k_2 (f_1''' + f_1 f_0'' + f_0 f_1'' - 2\rho f_0' f_1')\,, \\ f_3(0) = 0\,, \quad f_3'(0) = 0\,, \quad \lim_{\eta\to\infty} f_3' = 0\,. \end{cases}$$

Note that the k_2 parameter begins to appear in the order two equation, with the effect being the addition of the previous order inhomogeneity due to the nonlinearity in lower order terms. For this reason, $f_1(\eta; k_1)$ is exactly the same as was given before (while $f_0(\eta)$ will always be the same, with the present choice of linear operator).

Calculating

$$E(k_1, k_2) = \int_0^\infty (\mathcal{N}[F(\eta; k_1, k_2)])^2 d\eta\,,$$

we obtain a complicated expression involving powers of k_1 and k_2, with coefficients given in terms of ρ. Consider again the case of $\rho = 0.5$. We find that

$$(k_1^*, k_2^*) = \underset{(k_1,k_2)\in\mathbb{R}^2}{\operatorname{argmin}}\, E(k_1, k_2) = (-0.77075, -0.016403)\,,$$

which gives minimal error $E(k_1^*, k_2^*) = 1.7836 \times 10^{-6}$.

This shows that the added complexity has not gained us much in the way of improved error. On the other hand, the number of calculations has almost doubled for those terms of order two or higher. Hence, for the very modest improvement in error seen, the method is not efficient. Meanwhile, adding additional terms appears to improve error by an order of one to two magnitudes each iteration. Therefore, it is likely more efficient to obtain higher order approximations with the standard linear homotopy than it is to modify the structure of the homotopy, at least in this example. Note that these results agree with similar conclusions obtained in [42] for the Blasius problem.

4.6. What about the auxiliary function $H(x)$?

In the present section, we shall revisit the so-called auxiliary function $H(x)$ which sometimes appears in the homotopy linking $\mathcal{L}$ and $\mathcal{N}$. While this function is often taken to one, a recent paper by the present author [38] demonstrated a clear advantage to taking the function $H(x)$ to be more complicated in certain circumstances. We shall highlight the main results of that paper here.

Let us consider the Fitzhugh–Nagumo equation

$$\frac{\partial u}{\partial t} = \frac{\partial^2 u}{\partial x^2} + u(u-\alpha)(1-u)\,. \tag{4.54}$$

This equation, and related variants, have been studied recently through a variety of techniques; sees [119–123] and references therein.

The Fitzhugh–Nagumo equation has various applications in the fields of logistic population growth, flame propagation, neurophysiology, autocatalytic chemical reaction, branching Brownian motion process and nuclear reactor theory; see, e.g., [124–126]. There is also a density dependent diffusion Nagumo equation, where the diffusion coefficient is a simple power function. This equation is used in modeling electrical pulse propagation in nerve axons and in population genetics; see [127–132] and the references therein.

Under the assumption of travelling wave solutions, we set $z = x - ct$ and consider solutions of the type $u(x,t) = w(z)$, where c denotes the wave speed. From (4.54), we have

$$w'' + cw' + w(w-\alpha)(1-w) = 0\,. \tag{4.55}$$

In the case of a solitary wave over the whole real line, natural boundary conditions are

$$w \to 0 \quad \text{as} \quad z \to +\infty \quad \text{and} \quad w \to 0 \quad \text{as} \quad z \to -\infty\,. \tag{4.56}$$

As is standard in the method of homotopy analysis, we construct a homotopy

$$\mathcal{H}[\phi(z,q)] = (1-q)\mathcal{L}[\phi(z,q) - w_0(z)] - qhH(z)\mathcal{N}[\phi(z,q)]\,,$$

where $\mathcal{H}$ denotes the homotopy between a nonlinear operator $\mathcal{N}$ (which is the operator describing the nonlinear differential equation we wish to solve) and an auxiliary linear operator $\mathcal{L}$. Here $q \in [0,1]$ is the embedding parameter, h is the convergence-control parameter, and $H(z)$ is the auxiliary function. It is clear that the nonlinear differential operator should take the form

$$\mathcal{N}[\phi(z,q)] = \frac{\partial^2\phi}{\partial z^2} + c\frac{\partial\phi}{\partial z} + \phi(\phi-\alpha)(1-\phi)\,,$$

while the linear operator should be chosen in order to permit the initial approximation taken.

Note that when $q=0$ we have the initial approximation, i.e. $\phi(z,0) = w_0(z)$, whereas when $q=1$ we see that $\phi(z,1)$ is a solution to the nonlinear differential equation of interest, i.e. $\mathcal{N}[\phi(z,1)] = 0$. Considering a series expansion in q (treating q as a "small parameter"), we have

$$w(z) = w_0(z) + \sum_{n=1}^{\infty} w_n(z) \tag{4.57}$$

where

$$w_n(z) = \frac{1}{n!}\frac{\partial^n\phi(z,q)}{\partial q^n}|_{q=0}$$

is the solution to (4.55) provided that the series for $\phi(z,q)$ converges at $q=1$. A more useful recursive formula for the w_n's is given by the nth-order deformation equations

$$\mathcal{L}[w_n(z) - \chi_n w_{n-1}(z)] = hH(z)R_n(z,h) \tag{4.58}$$

where

$$R_n(z,h) = w''_{n-1} + cw'_{n-1} - \alpha w_{n-1} + (1+\alpha)\sum_{i=0}^{n-1} w_i w_{n-1-i} - \sum_{i=0}^{n-1} w_{n-1-i}\sum_{j=0}^{i} w_j w_{i-j}$$

and

$$\chi_n = \begin{cases} 0, & n = 0,1, \\ 1, & n \geq 2. \end{cases}$$

Hence, once we have selected an auxiliary linear operator $\mathcal{L}$, we can recover the terms in the expansion (4.57). The terms in the expansion (4.57) should be solved from (4.58) subject to boundary conditions

$$w_n \to 0 \quad \text{as} \quad z \to +\infty \quad \text{and} \quad w_n \to 0 \quad \text{as} \quad z \to -\infty\,. \tag{4.59}$$

In practice, we will truncate the series (4.57) to some desired number of terms. Thus, we shall be concerned with an approximate solution $\tilde{w}(z)$ with n^*+1 terms:

$$\tilde{w}(z) = w_0(z) + \sum_{n=1}^{n^*} w_n(z)\,. \tag{4.60}$$

Up to this point the choice of the convergence-control parameter h, the auxiliary function $H(z)$ and the linear operator $\mathcal{L}$ have all been kept arbitrary. Once $H(z)$ and $\mathcal{L}$ are selected, we can compute approximations of the form (4.60) for fixed n^*. We will then attempt to minimize the error in such an approximation by way of choosing the convergence-control parameter, h, in an appropriate manner.

We seek a solution over the whole real line, $(-\infty,\infty)$, hence we shall assume that the initial approximation to the solution profile takes the form of a Gaussian, $e^{-\alpha z^2}$. Note that such an initial approximation satisfies both $w \to 0$ as $z \to +\infty$ and $w \to 0$ as $z \to -\infty$, whereas the decaying exponential will only satisfy the first boundary condition.

Let us assume an initial approximation of the form

$$w_0(z) = \exp\left(-(g(z))^2\right) . \tag{4.61}$$

When $g(z) = az$ we obtain the Gaussian approximation, while when $g(z) = az^m$ we obtain a generalized Gaussian approximation. Here a effects the width and m effects the shape of the wave. For greatest generality, we shall consider the latter case in our computations, and take $m = 1$ for the special case of a Gaussian approximation.

In order to construct an appropriate linear operator, we should select an operator $\mathcal{L}$ satisfying the properties

$$\begin{aligned} &(1): \quad \mathcal{L}[C_1 w_0(z) + C_2] = 0 \quad \text{for constants } C_1 \text{ and } C_2 \,, \\ &(2): \quad \deg(\mathcal{L}) = \deg(\mathcal{N}) \,. \end{aligned}$$

The first condition permits the initial approximation to satisfy the linear operator, whereas the second condition mandates that the order of the operators $\mathcal{N}$ and $\mathcal{L}$ must match (which is reasonable, as it corresponds to a non-degenerate linearization of $\mathcal{N}$). Let us search for an operator

$$\mathcal{L} = \frac{d^2}{dz^2} + \zeta(z)\frac{d}{dz} \tag{4.62}$$

as such an operator satisfies the second condition, as well as part of the first condition ($\mathcal{L}[C_2] = 0$). Placing the initial approximation (4.61) into (4.62) we find that

$$\zeta(z) = \frac{2g^2g'^2 - g'^2 - gg''}{gg'}$$

gives a linear operator

$$\mathcal{L} = \frac{d^2}{dz^2} + \frac{2g^2g'^2 - g'^2 - gg''}{gg'}\frac{d}{dz}$$

which satisfies the first condition. Given the choice $g(z) = az^m$, this reduces to

$$\mathcal{L} = \frac{d^2}{dz^2} + \frac{2m(a^2z^{2m} - 1) + 1}{z}\frac{d}{dz} . \tag{4.63}$$

Now that we have found an appropriate linear operator, we may return to the higher order deformation equations (4.58). In order to solve these for the w_n's, we must invert $\mathcal{L}$. Note that a solution to the inhomogeneous differential equation $\mathcal{L}[U(z)] = F(z)$ is given by

$$U(z) = \int^z \left\{ \int^s F(t)e^{a^2t^{2m}}t^{-2m}|t|dt + A \right\} e^{-a^2s^{2m}}s^{2m}|s|^{-1}ds + B$$

where A and B are constants of integration. Applying this general formula to the higher order deformation equations (4.58), we find that

$$\begin{aligned} w_n(z) &= \chi_n w_{n-1}(z) + B_n \\ &+ \int_0^z \left\{ \int_0^s hH(t) R_n(t,h) e^{a^2 t^{2m}} t^{-2m} |t| dt + A_n \right\} \\ &\times e^{-a^2 s^{2m}} s^{2m} |s|^{-1} ds \,. \end{aligned} \tag{4.64}$$

Enforcing the boundary conditions (4.59), we obtain expressions for the A_n's and B_n's:

$$A_n = ma^2(I_n(-\infty) - I_n(\infty)) \,, \quad B_n = -\frac{1}{2}(I_n(-\infty) + I_n(\infty)) \,,$$

where

$$I_n(z) = h \int_0^z \int_0^s H(t)\; R_n(t,h) e^{a^2 t^{2m}} |t| t^{-2m} dt \;\; e^{-a^2 s^{2m}} s^{2m} |s|^{-1} ds \,.$$

From here, it is clear that the amplitude satisfies

$$w(0) = 1 + \sum_{n=1}^{\infty} B_n \,.$$

Physically, we expect a unique value of the amplitude. Hence, when plotting the h-curve for the problem at hand, it makes sense to use the amplitude $w(0)$ as an invariant in order to determine a proper domain for the convergence-control parameter, h.

Due to the singular nature of the integrand in (4.64) at $t = 0$, we need to exercise care when selecting the auxiliary function $H(z)$. In many cases, this function is set to unity, while the convergence-control parameter is chosen so that convergent solutions are obtained. In the present case, we see that taking $H(z) = 1$ would lead to a singular integrand at $t = 0$ for all $m \geq 1$. In addition to selecting $H(z)$ to permit regularity of solutions, we can select $H(z)$ such that the integration in (4.64) is simplified. Note that if a factor of t^{2m-1} is present inside the double integral (4.64) then the substitution $\mu = t^{2m}$ and $d\mu = 2mt^{2m-1}$ can be taken in order to speed the integration process when computing the approximate solutions.

We find that we may pick a functional form of $H(z)$ which takes both concerns, namely regularity of solutions and efficiency of calculation, into account. To do this, we set $H(z) = z^{2m-1}|z|^{2m-1}$. Note that $H(0) = 0$ whereas $H(z) \neq 0$ for all $z \neq 0$. Hence, the structure of the nonlinear differential equation is not lost over any open interval in $\mathbb{R}$. With this choice of $H(z)$, (4.64) is reduced to

$$\begin{aligned} w_n(z) &= \chi_n w_{n-1}(z) + B_n \\ &\quad + \int_0^z \left\{ \int_0^s hR_n(t,h) e^{a^2 t^{2m}} t^{2m-1} dt + A_n \right\} \\ &\quad \times e^{-a^2 s^{2m}} s^{2m} |s|^{-1} ds\,, \end{aligned}$$

while

$$I_n(z) = \int_0^z \int_0^s hR_n(t,h) e^{a^2 t^{2m}} t^{2m-1} dt \;\; e^{-a^2 s^{2m}} s^{2m} |s|^{-1} ds\,.$$

In the case of a Gaussian initial approximation to the profile, these quantities reduce to

$$\begin{aligned} w_n(z) &= \chi_n w_{n-1}(z) + B_n \\ &\quad + \int_0^z \left\{ \int_0^s hR_n(t,h) e^{a^2 t^2} t dt + A_n \right\} e^{-a^2 s^2} |s| ds \end{aligned} \tag{4.65}$$

and

$$I_n(z) = \int_0^z \int_0^s hR_n(t,h) e^{a^2 t^2} t dt \;\; e^{-a^2 s^2} |s| ds\,.$$

With the formulation given in the previous sections, we are now in a position to compute approximate analytical solutions for the Fitzhugh-Nagumo equation by use of the homotopy analysis method. While the parameter a can be used to influence the basis functions selected, we shall set $a = 1$ as this is sufficient to study the behavior of the solutions we seek. Starting with $w_0 = e^{-z^2}$, we iteratively compute the w_k's from the relation given in (4.65). We obtain lengthy algebraic expressions depending on x, c, α and the convergence-control parameter, h. We omit the lengthy expressions, and focus on the solutions for specific values of the physical parameters. Hence, our solutions are numerical in nature. However, in order to demonstrate the form of the solutions given, we give the first and second order solutions, which were computed recursively in the computer algebra program Maple 13. We find that

$$\begin{aligned} w_1(z) &= \frac{hc\sqrt{\pi}}{4}(1 - \operatorname{erf}(z)) - \frac{h}{24} e^{-3z^2} + \frac{h(1+\alpha)}{8} e^{-2z^2} \\ &\quad + \left[\frac{h(\alpha-2)}{4} + \frac{hc}{2} z + \frac{h(\alpha-2)}{4} z^2 + \frac{hc}{3} z^3 - \frac{h}{2} z^4 \right] e^{-z^2} \end{aligned}$$

and so on. Here

$$\mathrm{erf}(z) = \frac{2}{\sqrt{\pi}} \int_0^z e^{-t^2} dt\,.$$

We take the residual error

$$\epsilon(x, c, \alpha, h) = \mathcal{N}[\tilde{w}]$$

for the $n^* + 1$ term approximation. Integrating the residual errors over the domain (the entire real line) is computationally difficult, so we consider a discrete form (see Liao [10])

$$\hat{\epsilon}(c, \alpha, h) = \frac{1}{21} \sum_{k=-10}^{10} (\epsilon(k, c, \alpha, h))^2.$$

The solutions always decay rapidly, so the profiles become negligible past $x = \pm 10$. For fixed c and α, we are able to find h which minimize the square of these errors. We find that there are typically two critical values of h, one positive and one negative. The positive value of h leads to solutions which blow-up, and hence is not reasonable. The negative value of h leads to the physically meaningful solutions.

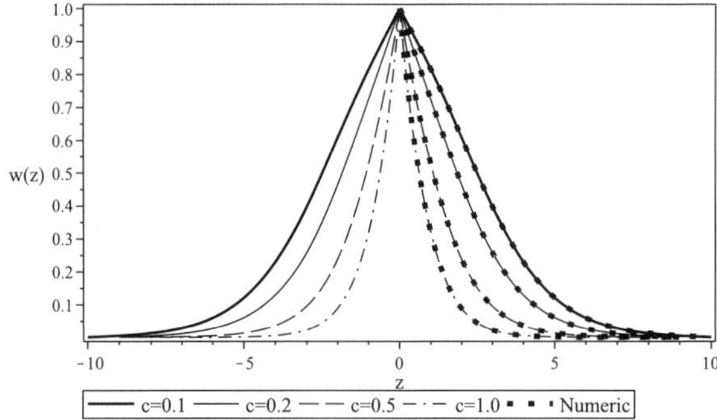

Fig. 4.1. Profiles of the Gaussian wave profiles, $w(z)$, for various values of the wave speed, c. Here we have taken $\alpha = 0.5$. We plot the HAM solutions of order 6 with various types of lines, whereas the numerical solution to the problem (the numerical solution is set up as an initial value problem, and hence is valid over the positive region) is given by dots. The error minimizing values of the convergence control parameter, h, are given by $h = -0.90026$ ($c = 0.1$), $h = -0.88407$ ($c = 0.2$), $h = -0.86139$ ($c = 0.5$), $h = -0.79109$ ($c = 1.0$). With these values, error in the approximate solutions was controlled to within 10^{-4}. The numerical solutions were controlled to an error tolerance of 10^{-5}.

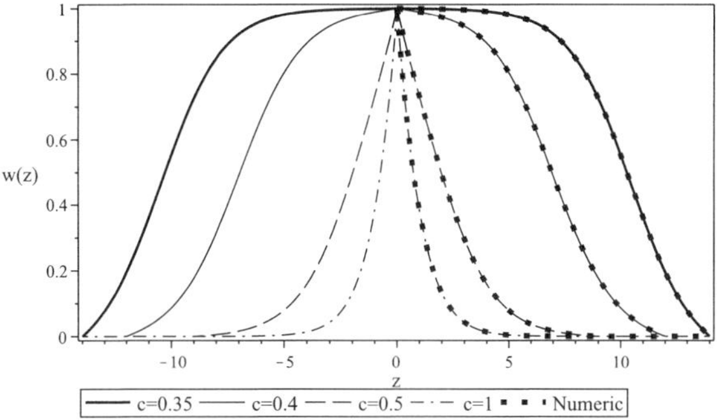

Fig. 4.2. Profiles of the Gaussian wave profiles, $w(z)$, for various values of the wave speed, c. Here we have taken $\alpha = 0.2$. We plot the HAM solutions of order 6 with various lines, whereas the numerical solution to the problem (the numerical solution is set up as an initial value problem, and hence is valid over the positive region) is given by dots. The error minimizing values of the convergence control parameter, h, are given by $h = -0.98612$ ($c = 0.35$), $h = -0.96980$ ($c = 0.4$), $h = -0.94635$ ($c = 0.5$), $h = -0.88060$ ($c = 1.0$). With these values, error in the approximate solutions was controlled to within 10^{-4}. The numerical solutions were controlled to an error tolerance of 10^{-5}.

In Fig. 4.1, for the sake of demonstration we fix $w(0) = 1$, $\alpha = 0.5$, and allow the wave speed c to change. We see that the derivative of the solution as one approaches from either side of the origin strongly depends on the choice of c. By symmetry, $\lim_{z\to 0^-} w'(z) = \lim_{z\to 0^+} |w'(z)|$. We are able to compare the solution branch on the positive half-line to a relevant initial value problem solved numerically by RKF-45 method, and the agreement is very good.

In Fig. 4.2, we provide a similar plot, only taking $\alpha = 0.2$. Here we observe that, for smaller values of the wave speed, the solutions appear to have continuous derivatives at $z = 0$. This makes sense because, as we have discussed above, the condition $w'(0) = 0$ implies that $w(0)$ is a nonlinear function of the wave speed, the parameter α, and the physical invariant $\int_0^\infty (w'(z))^2\, dz$. So, for appropriate values of the parameters, we can expect to find solutions permitting $w'(0) = 0$, whereas for other values, solutions may be continuous but not differentiable at $z = 0$.

4.7. Avoiding singularities

Often, singularities arise in mathematical problems. Either they are removable singularities, like in the case of the Lane–Emden problem, or they may be essential. In the latter case, they can seriously impede the solution process for such problems. However, the homotopy analysis method gives us a way to deal with such problems, given that we are able to adequately choose the auxiliary linear operator.

The strongly singular problem for the Ernst equation governing axially symmetric stationary vacuum gravitational fields was recently considered by Baxter and Van Gorder [44]. We shall highlight the findings of this study relevant to recovering solutions near singularities.

Physically, the Ernst equation serves as a model of axially symmetric stationary vacuum gravitational fields [133–136]. Harrison [137] shows that the Ernst equation admits a Bächlund transform. Perturbation theory has been considered [138]. In [139], the Ernst equation is used to completely separate the vacuum Einstein equations for an arbitrary stationary axisymmetric space-time. In [140], the Virasoro algebra is shown to exist in the solution space of the Ernst equation. Rational approximations of the flip angle dependence of an MRI signal are derived using half-angle trigonometric substitutions in the Ernst equation in [141]. In [142], the class of hyperelliptic solutions to the Ernst equation are derived using Riemann–Hilbert techniques. The inverse scattering method [143] and the algebra-geometric ideology [144] have also been discussed.

The transformed equation reads

$$u'' + \frac{2s}{s^2-1}u' - \frac{2}{s^2-1}uu' = 0\,, \tag{4.66}$$

where prime denotes differentiation with respect to s. We may renormalize the domain so that $s \in [1+\epsilon, \infty)$, where $0 < \epsilon$ is a parameter. Meaningful boundary conditions are

$$u(1+\epsilon) = A \qquad \text{and} \qquad \lim_{s\to\infty} u(s) = 0\,, \tag{4.67}$$

where A is a constant.

We may apply the homotopy analysis method to the reduced Ernst equation (4.66) with initial data (4.67). We shall select the linear operator

$$\mathcal{L}[u] = u'' + \frac{2s}{s^2-1}u'\,. \tag{4.68}$$

The zeroth-order approximation is then governed by the boundary value problem

$$\mathcal{L}[u_0] = 0\,, \qquad u_0(1+\epsilon) = A\,, \qquad \lim_{s\to\infty} u(s) = 0\,.$$

The zeroth-order approximation is then given by

$$u_0(s) = A\frac{\ln(s+1) - \ln(s-1)}{\ln(2+\epsilon) - \ln(\epsilon)}. \tag{4.69}$$

The following will be useful in computing the higher order terms. Consider the initial value problem

$$\mathcal{L}[U(s)] = Y(s), \qquad U(1+\epsilon) = 0, \qquad \lim_{s\to\infty} U(s) = 0.$$

Let us define the function

$$I(s;\epsilon) = \int_{1+\epsilon}^{s} \frac{1}{\zeta^2 - 1} \int_{1+\epsilon}^{\zeta} (\xi^2 - 1) Y(\xi) d\xi d\zeta.$$

The solution obeying the condition at $s = 1+\epsilon$ takes the form

$$\begin{aligned} U(s) &= I(s;\epsilon) + C\int_{1+\epsilon}^{s} \frac{d\xi}{\xi^2 - 1} \\ &= I(s;\epsilon) + \frac{C}{2}\left(\ln\left(\frac{s-1}{s+1}\right) - \ln\left(\frac{\epsilon}{2+\epsilon}\right)\right), \end{aligned} \tag{4.70}$$

where C is a constant to be determined. Evaluating this as $s \to \infty$ and using the remaining boundary condition, we have

$$0 = \lim_{s\to\infty} I(s;\epsilon) - \frac{C}{2}\ln\left(\frac{\epsilon}{2+\epsilon}\right),$$

which gives

$$C = \frac{2I(\infty;\epsilon)}{\ln\left(\frac{\epsilon}{2+\epsilon}\right)},$$

where the numerator is defined in the limiting sense. With this, we have the solution

$$\begin{aligned} U(s) &= I(s;\epsilon) + \frac{I(\infty;\epsilon)}{\ln\left(\frac{\epsilon}{2+\epsilon}\right)}\left(\ln\left(\frac{s-1}{s+1}\right) - \ln\left(\frac{\epsilon}{2+\epsilon}\right)\right) \\ &= I(s;\epsilon) - I(\infty;\epsilon) + \frac{I(\infty;\epsilon)}{\ln\left(\frac{\epsilon}{2+\epsilon}\right)}\ln\left(\frac{s-1}{s+1}\right). \end{aligned} \tag{4.71}$$

The Homotopy between the reduced Ernst equation and the auxiliary linear operator $\mathcal{L}$, given in (4.68), is

$$0 \equiv \mathcal{H}[u, u_0; q] = (1-q)\mathcal{L}[U - u_0] - hq\mathcal{N}[U], \tag{4.72}$$

where $u_0(s)$ is an initial approximate solution, h is the so-called convergence-control parameter, and

$$\mathcal{N}[U] = \mathcal{L}[U] - \frac{2}{s^2-1}U\frac{dU}{ds}. \tag{4.73}$$

The initial approximation is (4.69). The first-order deformation equation reads

$$\mathcal{L}[u_1] = -\frac{2h}{s^2-1}u_0(s)u_0'(s),$$

which simplifies to

$$\mathcal{L}[u_1] = \frac{4hA^2}{\ln(\frac{2+\epsilon}{\epsilon})^2}\ln\left(\frac{s+1}{s-1}\right)\frac{1}{(s^2-1)^2}$$

upon using (4.69). The solution, according to (4.71), is

$$u_1(s) = -\frac{hA^2}{6\ln(\frac{\epsilon}{2+\epsilon})^2}\ln\left(\frac{s-1}{s+1}\right)\left\{\ln\left(\frac{s-1}{s+1}\right)^2 - \ln\left(\frac{\epsilon}{2+\epsilon}\right)^2\right\}.$$

The second order deformation equation is

$$\mathcal{L}[u_2] = (1+h)\mathcal{L}[u_1] - \frac{2h}{s^2-1}\left(u_0(s)u_1'(s) + u_1(s)u_0'(s)\right).$$

Using (4.69), the solution is

$$\begin{aligned} u_2(s) = &-\frac{hA^2}{90\ln(\frac{2+\epsilon}{\epsilon})^3}\left\{\ln\left(\frac{2+\epsilon}{\epsilon}\right)\left(15(1+h) - 2Ah\ln\left(\frac{2+\epsilon}{\epsilon}\right)\right)\right. \\ &\left.+ 3Ah\ln\left(\frac{s-1}{s+1}\right)^2\right\}\ln\left(\frac{s-1}{s+1}\right) \\ &\times\left\{\ln\left(\frac{s-1}{s+1}\right)^2 - \ln\left(\frac{\epsilon}{2+\epsilon}\right)^2\right\}. \end{aligned} \tag{4.74}$$

The sum of the first three terms gives us the approximate solution

$$\widehat{U}(s,h,A,\epsilon) = u_0(s) + u_1(s) + u_2(s).$$

We run $\widehat{U}$ through the original nonlinear operator corresponding to (4.66), then we set $\epsilon = 1$ and define the residual error $\mathcal{N}[\widehat{U}(s,h,A,1)]$. To get the error over the interval $[2,\infty)$, we compute the sum of squared residual errors

$$E(h,A) = \int_2^\infty (\mathcal{N}[\widehat{U}(s,h,A,1)])^2 ds. \tag{4.75}$$

What we get is that $E(h, A)$ is an eighth-degree polynomial in h. To get numerical estimates of the error, care must be taken in using approximations in the coefficients of $E(h, A)$ when actually plotting the error numerically. When using 5 or 10 digits of accuracy for the coefficients of h, rounding errors can be very large. Upon explicit evaluation, (4.75) becomes

$$E(h, A) = \mu_8(A)h^8 + \mu_7(A)h^7 + \cdots + \mu_0(A)\,. \tag{4.76}$$

When $A = 1$ we have a minimum at $h = -0.982$. This gives the approximate solution of $\widehat{U}(s, -0.982, 1, 1)$ with a sum of squares residual error of $E(-0.982, 1) = 3.19 \times 10^{-5}$. The error increases with the value in A, however. When $A = 0.1$, the error is 3.46×10^{-13}, but when $A = 5$, the error is 0.567. This makes sense: given a larger value of A, the adjustable parameter ϵ is closer to the singularity, hence the singularity's effect is stronger.

To see the role increasing A has on making the solution more singular, note that the solution to our choice of linear operator $\mathcal{L}$ has Laurent expansion near $s = 1$ of the form

$$U(s) = \frac{C_\gamma}{(s-1)^\gamma} + \text{higher order terms}\,,$$

for arbitrary γ where C_γ is the leading order coefficient. Then, applying $A = u(1+\epsilon)$, we have

$$A = \frac{C_\gamma}{\epsilon^\gamma}$$

(neglecting higher order terms, since the singularity dominates near $s = 1$). If ϵ is small yet fixed, the parameter γ scales as $\gamma \sim \ln(A)$. So, as A becomes large, the strength of the singularity increases. For such strongly singular cases, the approximation method breaks down. However, for small or moderate values of A, the three-term expansions approximate the solutions remarkably well. The connection with the singularity and the value A was explored in greater detail in the paper; see [44] for more details.

4.8. Conclusions

Using optimal homotopy analysis method through adequate selection of the auxiliary linear operator and convergence-control parameter, we have been able to study a number of features of homotopy analysis. When choosing different linear operators, we find that the rate of convergence is strongly tied to our choice. Indeed, for some auxiliary linear operators, the rate of convergence may be quite slow, where for others it can be rather fast. So, it is important to properly select the linear operator. Similar comments hold in the case of nonlinear partial differential equations. The rate of convergence is important if we wish only to compute few terms. For complicated equations, particularly PDEs, each iteration of the method can become very computationally demanding. Hence, perhaps only a few terms can even be calculated in minimal time. Therefore, the choice of the auxiliary linear operator is vital for computational efficiency.

When applying an optimal form of the homotopy analysis, we treat the convergence-control parameter as an optimization parameter, and attempt to minimize some function of the error or residual error. Doing so, we arrive at an error minimizing value of the convergence-control parameter. It is natural to ask what happens to this parameter as we increase the number of terms in our homotopy approximations. What we find from solving some nonlinear ordinary differential equations is that when the auxiliary linear operator is properly chosen, the values of the optimal convergence-control parameter appear to stabilize around some fixed value as we increase the number of iterations. On the other hand, for poor choices of operators, it seems that the optimal value of the convergence-control parameter is more variable on each iteration, owing to the fact that a poor choice of linear operator results in slower convergence.

There are a number of other ways one can think of modifying the homotopy analysis method. For instance, instead of the standard homotopy

$$H(q)[U] = (1-q)\mathcal{L}[U] - hq\mathcal{N}[U]\,,$$

it is possible to consider more general forms of the homotopy

$$H(q)[U] = (1-q)a(q)\mathcal{L}[U] - hqb(q)\mathcal{N}[U]\,.$$

However, such a generalization is not always useful. We show that when we have a quadratic homotopy in the embedding parameter q, there is no significant improvement in the control of error of the solution (even through we introduce two, as opposed to one, convergence-control parameters), yet the computations are far more complicated. Again, whether such an approach is useful will usually depend on the specific problem. So, for some problems, this approach might be worthwhile and useful.

Another form of the homotopy is

$$H(q)[U] = (1-q)\mathcal{L}[U] - hH(x)q\mathcal{N}[U]\,.$$

This was the traditional type introduced by Liao, however many authors omit the general function $H(x)$. From what we now know of optimal homotopy, we can view $hH(x)$ as a type of variable convergence-control term. Using this term in it's most general form, it is possible to greatly simplify the solution process involved in obtaining higher order approximations. We illustrate this by constructing solutions to a nonlinear model over the real line, with base functions of the type $e^{-\alpha x^2}$. Which such base functions are usually too hard and computationally demanding to work with, we show that it is possible to pick $H(x)$ in such a way as to simplify the process. Hence, the choice of $H(x)$ can be useful for certain problems.

Finally, we present a highly singular problem, and we demonstrate that the optimal selection of the convergence-control parameter still can be useful for such situations. The method also gives an indication of the strength of the singularity.

In summary, the homotopy analysis method gives us great freedom in solving nonlinear ordinary and partial differential equations. By appropriately selecting the auxiliary linear operator, the convergence-control parameter, and some other features when needed, we can obtain computationally efficient approximate solutions with low residual errors for a number of nonlinear problems.

References

[1] S.J. Liao, *On the proposed homotopy analysis techniques for nonlinear problems and its application, Ph.D. dissertation.* (Shanghai Jiao Tong University, 1992).

[2] S.J. Liao, *Beyond Perturbation: Introduction to the Homotopy Analysis Method.* (Chapman & Hall/CRC Press, Boca Raton, 2003).

[3] S.J. Liao, An explicit, totally analytic approximation of Blasius viscous flow problems, *International Journal of Non-Linear Mechanics.* **34**, 759–778, (1999).

[4] S.J. Liao, *On the homotopy analysis method for nonlinear problems, Applied Mathematics and Computation.* **147**, 499–513, (2004).

[5] S.J. Liao and Y. Tan, A general approach to obtain series solutions of nonlinear differential equations, *Studies in Applied Mathematics.* **119**, 297–354, (2007).

[6] S.J. Liao, Notes on the homotopy analysis method: some definitions and theorems, *Communications in Nonlinear Science and Numerical Simulation.* **14**, 983–997, (2009).

[7] S.J. Liao, *Homotopy Analysis Method in Nonlinear Differential Equations.* (Springer & Higher Education Press, Heidelberg, 2012).

[8] R.A. Van Gorder and K. Vajravelu, On the selection of auxiliary functions, operators, and convergence-control parameters in the application of the Homotopy Analysis Method to nonlinear differential equations: A general

approach, *Communications in Nonlinear Science and Numerical Simulation.* **14**, 4078–4089, (2009).

[9] K. Vajravel and R.A. Van Gorder, *Nonlinear Flow Phenomena and Homotopy Analysis: Fluid Flow and Heat Transfer.* (Springer & Higher Education Press, Heidelberg, 2013).

[10] S. Liao, An optimal homotopy-analysis approach for strongly nonlinear differential equations, *Communications in Nonlinear Science and Numerical Simulation.* **15**, 2315–2332, (2010).

[11] S. Abbasbandy, The application of homotopy analysis method to nonlinear equations arising in heat transfer, *Physics Letters A.* **360**, 109–113, (2006).

[12] S. Abbasbandy, Homotopy analysis method for heat radiation equations, *Int. J. Heat and Mass Transfer.* **34**, 380–387, (2007).

[13] S.J. Liao, J. Su and A.T. Chwang, Series solutions for a nonlinear model of combined convective and radiative cooling of a spherical body, *Int. J. Heat and Mass Transfer.* **49**, 2437–2445, (2006).

[14] S.J. Liao and A. Campo, Analytic solutions of the temperature distribution in Blasius viscous flow problems, *Journal of Fluid Mechanics.* **453**, 411–425, (2002).

[15] S.J. Liao, An explicit, totally analytic approximation of Blasius viscous flow problems, *International Journal of Non-Linear Mechanics.* **34**, 759–778, (1999).

[16] S.J. Liao, A uniformly valid analytic solution of 2D viscous flow past a semi-infinite flat plate, *Journal of Fluid Mechanics.* **385**, 101–128, (1999).

[17] S. J. Liao, On the analytic solution of magnetohydrodynamic flows of non-Newtonian fluids over a stretching sheet, *Journal of Fluid Mechanics.* **488**, 189–212, (2003).

[18] F.T. Akyildiz, K. Vajravelu, R.N. Mohapatra, E. Sweet, and R.A. Van Gorder, Implicit Differential Equation Arising in the Steady Flow of a Sisko Fluid, *Applied Mathematics and Computation.* **210**, 189–196, (2009).

[19] X. Hang, Z.L. Lin, S.J. Liao, J.Z. Wu and J. Majdalani, Homotopy based solutions of the Navier-Stokes equations for a porous channel with orthogonally moving walls, *Physics of Fluids.* **22**, 053601, (2010).

[20] M. Sajid, T. Hayat and S. Asghar, Comparison between the HAM and HPM solutions of thin film flows of non-Newtonian fluids on a moving belt, *Nonlinear Dynamics.* **50**, 27–35, (2007).

[21] T. Hayat and M. Sajid, On analytic solution for thin film flow of a fourth grade fluid down a vertical cylinder, *Physics Letters A.* **361**, 316–322, (2007).

[22] M. Turkyilmazoglu, Purely analytic solutions of the compressible boundary layer flow due to a porous rotating disk with heat transfer, *Physics of Fluids.* **21**, 106104, (2009).

[23] S. Abbasbandy and F.S. Zakaria, Soliton solutions for the fifth-order KdV equation with the homotopy analysis method, *Nonlinear Dynamics.* **51**, 83–87 (2008).

[24] W. Wu and S.J. Liao, Solving solitary waves with discontinuity by means of the homotopy analysis method, *Chaos, Solitons & Fractals.* **26**, 177–185,

(2005).

[25] E. Sweet and R.A. Van Gorder, Analytical solutions to a generalized Drinfel'd - Sokolov equation related to DSSH and KdV6, *Applied Mathematics and Computation.* **216**, 2783–2791, (2010).

[26] E. Sweet and R.A. Van Gorder, Exponential type solutions to a generalized Drinfel'd - Sokolov equation, *Physica Scripta.* **82**, 035006, (2010).

[27] Y. Wu, C. Wang and S.J. Liao, Solving the one-loop soliton solution of the Vakhnenko equation by means of the homotopy analysis method, *Chaos, Solitons & Fractals.* **23**, 1733–1740, (2005).

[28] J. Cheng, S.J. Liao, R.N. Mohapatra and K. Vajravelu, Series solutions of Nano-boundary-layer flows by means of the homotopy analysis method, *Journal of Mathematical Analysis and Applications.* **343**, 233–245, (2008).

[29] R.A. Van Gorder, E. Sweet and K. Vajravelu, Nano boundary layers over stretching surfaces, *Communications in Nonlinear Science and Numerical Simulation.* **15**, 1494–1500, (2010).

[30] A.S. Bataineh, M.S.M. Noorani and I. Hashim, Solutions of time–dependent Emden–Fowler type equations by homotopy analysis method, *Physics Letters A.* **371**, 72–82, (2007).

[31] A.S. Bataineh, M.S.M. Noorani and I. Hashim, Homotopy analysis method for singular IVPs of Emden-Fowler type, *Communications in Nonlinear Science and Numerical Simulation.* **14**, 1121–1131, (2009).

[32] R.A. Van Gorder and K. Vajravelu, Analytic and numerical solutions to the Lane–Emden equation, *Physics Letters A.* **372**, 6060–6065, (2008).

[33] S. Liao, A new analytic algorithm of Lane–Emden type equations, *Applied Mathematics and Computation.* **142**, 1–16, (2003).

[34] M. Turkyilmazoglu, The Airy equation and its alternative analytic solution, *Phys. Scr.* **86**, 055004, (2012).

[35] A.M. Wazwaz, Solitary waves solutions for extended forms of quantum Zakharov–Kuznetsov equations, *Phys. Scr.* **85**, 025006, (2012).

[36] M. Turkyilmazoglu, An effective approach for approximate analytical solutions of the damped Duffing equation, *Phys. Scr.* **86**, 015301, (2012).

[37] R.A. Van Gorder, Analytical method for the construction of solutions to the Föppl – von Kármán equations governing deflections of a thin flat plate, *International Journal of Non-Linear Mechanics.* **47**, 1–6, (2012).

[38] R.A. Van Gorder, Gaussian waves in the Fitzhugh-Nagumo equation demonstrate one role of the auxiliary function $H(x)$ in the homotopy analysis method, *Communications in Nonlinear Science and Numerical Simulation.* **17**, 1233–1240, (2012).

[39] R.A. Van Gorder, Control of error in the homotopy analysis of semilinear elliptic boundary value problems, *Numerical Algorithms.* **61**, 613–629, (2012).

[40] M. Ghoreishi, A.I.B. Ismail, A.K. Alomari, A. Sami Bataineh, The comparison between Homotopy Analysis Method and Optimal Homotopy Asymptotic Method for nonlinear age structured population models, *Communications in Nonlinear Science and Numerical Simulation.* **17**, 1163–1177, (2012).

[41] S. Abbasbandy, E. Shivanian, K. Vajravelu, Mathematical properties of h-curve in the frame work of the homotopy analysis method, *Communications in Nonlinear Science and Numerical Simulation.* **16**, 4268–4275, (2011).

[42] S. Liao, An optimal homotopy-analysis approach for strongly nonlinear differential equations, *Communications in Nonlinear Science and Numerical Simulation.* **15**, 2315–2332, (2010).

[43] K. Mallory and R.A. Van Gorder, Control of error in the homotopy analysis of solutions to the Zakharov system with dissipation, *Numerical Algorithms.* in press (2013) DOI: 10.1007/s11075-012-9683-6.

[44] M. Baxter and R.A. Van Gorder, Exact and analytic solutions of the Ernst equation governing axially symmetric stationary vacuum gravitational fields, *Physica Scripta.* **87**, 035005, (2013).

[45] R.A. Van Gorder, A linearization approach for rational nonlinear models in mathematical physics, *Communications in Theoretical Physics.* **57**, 530–540, (2012).

[46] J. Chazy, Sur les équations différentielles de troisiéme ordre et d'ordre supérieur dont l'intégrale générale a ses points critiques fixés, *Acta Math.* **33**, 317–385, (1911).

[47] R. Fuchs, Über lineare homogene Differentialgleichungen zweiter Ordnung mit drei im Endlichen gelegene wesentlich singulare Stellen, *Math. Ann.* **63**, 301–321, (1907).

[48] B. Gambier, Sur les équations différentielles du second ordre et du premier degré dont l'intégrale générale est á points critique fixés, *Acta. Math.* **33**, 1–55, (1910).

[49] R. Garnier, Sur des équations différentielles du troisiéme ordre dont l'intégrale est uniform et sur une classe d'équations nouvelles d'ordre supérieur dont l'intégrale générale a ses point critiques fixés, *Ann. Sci. de l'ENS.* **29**, 1–126, (1912).

[50] R. Garnier, Étude de l'intégrale générale de l'équation (VI) de M. Painlevé dans le voisinage de ses singularités transcendantes, *Ann. Sci. Ecole Norm. Sup.* **34**, 239–353, (1917).

[51] R. Garnier, Sur une classe de systémes differentiels abéliens deduits de la théorie des équations linéaires, *Rend. Circ. Mat. Palermo.* **43**, 155–191, (1918-19).

[52] P. Painlevé, Memoire sur les équations différentielles dont l'intégrale générale est uniforme, *Bull. Soc. Math. Phys. France.* **28**, 201–261, (1900).

[53] P. Painlevé, Sur les équations différentielles du second ordre et d'ordre supérieur dont l'intégrale générale est uniforme, *Acta Math.*, **21**, 1–85, (1902).

[54] L. Schlesiner, Über eine Klasse von Differentialsystemen beliebiger Ordnung mit festen kritischer Punkten, *J. für Math.* **141**, 96–145,(1912).

[55] E.L. Ince, *Ordinary differential equations.* (Dover, New York, 1956).

[56] R.A. Van Gorder, An elegant perturbation solution for the Lane–Emden equation of the second kind, *New Astronomy.* **16**, 65–67, (2011).

[57] R.A. Van Gorder, Analytical solutions to a quasilinear differential equation related to the Lane–Emden equation of the second kind, *Celestial Mechan-*

ics and Dynamical Astronomy. **109**, 137–145, (2011).
[58] R.A. Van Gorder, Exact first integrals for a Lane–Emden equation of the second kind modelling a thermal explosion in a rectangular slab, *New Astronomy.* **16**, 492–497, (2011).
[59] K. Boubaker and R.A. Van Gorder, Application of the BPES to Lane–Emden equations governing polytropic and isothermal gas spheres, *New Astronomy.* **17**, 565–569, (2012).
[60] C. Harley and E. Momoniat, First integrals and bifurcations of a Lane–Emden equation of the second kind, *J. Math. Anal. Appl.* **344**, 757–764, (2008).
[61] J. Binney and S. Tremaine, *Glactic Dynamics.* (Princeton University Press, Princeton, 1987).
[62] W.B. Bonnor, Boyle's Law and gravitational instability, *Mon. Not. R. Astron. Soc.* **116**, 351, (1956).
[63] S. Chandrasekhar, *An introduction to the study of stellar structure.* (Dover Publications Inc., New York 1939).
[64] R. Ebert, Über die Verdichtung von H I-Gebieten, *Zeitschrift für Astrophysik.* **37**, 217, (1955).
[65] R. Emden, *Gaskugeln Anwendungen der Mechan. Warmtheorie.* (Druck und Verlag Von B. G. Teubner, Leipzig and Berlin, 1907).
[66] H. Goenner and P. Havas, Exact solutions of the generalized Lane–Emden equation, *J. Math. Phys.* **41**, 7029–7042, (2000).
[67] C. Hunter, Series solutions for polytropes and the isothermal sphere, *Mon. Not. R. Astron. Soc.* **328**, 839–847, (2001).
[68] R. Kippenhahn and A. Weigert, *Stellar Structure and Evolution.* (Springer, Berlin, 1990).
[69] J.H. Lane, On the theoretical temperature of the sun under the hypothesis of a gaseous mass maintaining its volume by its internal heat and depending on the laws of gases known to terrestrial experiment, *Am. J. Sci. Arts* **50**, 57, (1870).
[70] E. Momoniat and C. Harley, Approximate implicit solution of a Lane–Emden equation, *New Astronomy.* **11**, 520–526, (2006).
[71] A. M. Wazwaz, A new algorithm for solving differential equations of Lane–Emden type, *Appl. Math. Comp.* **118**, 287–310, (2001).
[72] P. Mach, All solutions of the $n = 5$ Lane–Emden equation, *Journal of Mathematical Physics.* **53**, 062503, (2012).
[73] D. Britz, J. Strutwold and O. Osterby, Digital simulation of thermal reactions, *Applied Mathematics and Computation.* **218**, 1280–1290, (2011).
[74] R.A. Van Gorder and K. Vajravelu, Hydromagnetic stagnation point flow of a second grade fluid over a stretching sheet, *Mechanics Research Communications.* **37**, 113–118, (2010).
[75] R.A. Van Gorder and K. Vajravelu, Existence and uniqueness results for a nonlinear differential equation arising in stagnation point flow in a porous medium, *Acta Mechanica.* **210**, 215–220, (2010).
[76] F.T. Akyildiz, D.A. Siginer , K. Vajravelu, J.R. Cannon and R.A. Van Gorder, Similarity solutions of the boundary layer equation for a nonlin-

early stretching sheet, *Mathematical Methods in the Applied Sciences.* **33**, 601–606, (2010).

[77] R.A. Van Gorder, High-order nonlinear boundary value problems admitting multiple exact solutions with application to the fluid flow over a sheet, *Applied Mathematics and Computation.* **216**, 2177–2182, (2010).

[78] R.A. Van Gorder and K. Vajravelu, A note on flow geometries and the similarity solutions of the boundary layer equations for a nonlinearly stretching sheet, *Archive of Applied Mechanics.* **80**, 1329–1332, (2010).

[79] R.A. Van Gorder and K. Vajravelu, Existence and uniqueness results for a nonlinear differential equation arising in viscous flow over a nonlinearly stretching sheet, *Applied Mathematics Letters.* **24**, 238–242, (2011).

[80] R.A. Van Gorder and K. Vajravelu, Multiple solutions for hydromagnetic flow of a second grade fluid over a stretching or shrinking sheet, *Quarterly of Applied Mathematics.* **69**, 405–424, (2011).

[81] T.R. Mahapatra, S.K. Nandy, K. Vajravelu and R.A. Van Gorder, Stability analysis of fluid flow over a nonlinearly stretching sheet, *Archive of Applied Mechanics.* **81**, 1087–1091, (2011).

[82] R.A. Van Gorder, K. Vajravelu and I. Pop, Hydromagnetic stagnation point flow of a viscous fluid over a stretching or shrinking sheet, *Meccanica.* **47**, 31–50, (2012).

[83] T.R. Mahapatra, S.K. Nandy, K. Vajravelu and R.A. Van Gorder, Dual solutions for the magnetohydrodynamic stagnation-point flow of a power-law fluid over a shrinking sheet, *ASME Journal of Applied Mechanics.* **79**, 024503, (2012).

[84] R.A. Van Gorder and K. Vajravelu, Convective heat transfer in a conducting fluid over a permeable stretching surface with suction and internal heat generation / absorption, *Applied Mathematics and Computation.* **217**, 5810–5821, (2011).

[85] K. Vajravelu, K.V. Prasad, J. Lee, C. Lee, I. Pop and R.A. Van Gorder, Convective heat transfer in the flow of viscous Ag-water and Cu-water nanofluids over a stretching surface, *International Journal of Thermal Sciences.* **50**, 843–851, (2011).

[86] R.A. Van Gorder, Analysis of nonlinear BVPs motivated by fluid film flow over a surface, *Applied Mathematics and Computation.* **217**, 8068–8079, (2011).

[87] K.V. Prasad, K. Vajravelu and R.A. Van Gorder, Non-Darcian flow and heat transfer along a permeable vertical surface with nonlinear density temperature variation, *Acta Mechanica.* **220**, 139–154, (2011).

[88] F.T. Akyildiz, H. Bellout, K. Vajravelu and R.A. Van Gorder, Existence results for third order nonlinear boundary value problems arising in nano boundary layer fluid flows over stretching surfaces, *Nonlinear Analysis Series B: Real World Applications.* **12**, 2919–2930, (2011).

[89] K. Vajravelu, K. V. Prasad, R.A. Van Gorder and J. Lee, Free convection boundary layer flow past a vertical surface in a porous medium with temperature-dependent properties, *Transport in Porous Media.* **90**, 977–992, (2011).

[90] T.R. Mahapatra, S.K. Nandy, K. Vajravelu and R.A. Van Gorder, Stability analysis of the dual solutions for stagnation-point flow over a non-linearly stretching surface, *Meccanica.* **47**, 1623–1632, (2012).

[91] S. Mukhopadhyay, K. Vajravelu and R.A. Van Gorder, Flow and heat transfer in a moving fluid over a moving non-isothermal surface, *International Journal of Heat and Mass Transfer.* **55**, 6632–6637, (2012).

[92] S. Mukhopadhyay, K. Vajravelu and R.A. Van Gorder, Chemically reactive solute transfer in a moving fluid over a moving flat surface, *Acta Mechanica.* **224**, 513–523, (2013).

[93] M. Russo and R.A. Van Gorder, Control of error in the homotopy analysis of nonlinear Klein–Gordon initial value problems, *Applied Mathematics and Computation.* **219**, 6494–6509, (2013).

[94] Q. Sun, Solving the Klein–Gordon equation by means of the homotopy analysis method, *Applied Mathematics and Computation.* **169**, 355–365, (2005).

[95] S. Iqbal, M. Idrees, A.M. Siddiqui and A.R. Ansari, Some solutions of the linear and nonlinear Klein–Gordon equations using the optimal homotopy asymptotic method, *Applied Mathematics and Computation.* **216**, 2898–2909, (2010).

[96] U. Yücel, Homotopy analysis method for the sine-Gordon equation with initial conditions, *Applied Mathematics and Computation.* **203**, 387–395, (2008).

[97] A.H. Nayfeh, *Perturbation Methods.* (Wiley, New York, 1973).

[98] Z. Jiefang and L. Ji, Similarity solutions of the cubic nonlinear Klein–Gordon equation, *International Journal of Theoretical Physics.* **32**, 39–42, (1993).

[99] W. Strauss and L. Vazquez, Numerical solution of a nonlinear Klein–Gordon equation, *Journal of Computational Physics.* **28**, 271–278, (1978).

[100] J. Liouville, Sur l'equation aux diff'erences partielles, d2 logλ/dudv ? λ/2a2 = 0, *J. Math.* **18**, 71–72, (1853).

[101] A.L. Larsen and N. Sanchez, sinh-Gordon, cosh-Gordon, and Liouville equations for strings and multistrings in constant curvature spacetimes, *Physical Review D* **54**, 2801–2807, (1996).

[102] A. Grauel, Sinh-Gordon equation, Painlevé property and Bäcklund transformation, *Physica A.* **132**, 557–568, (1985).

[103] A.M. Grundland and E. Infeld, A family of nonlinear Klein – Gordon equations and their solutions, *Journal of Mathematical Physics.* **33**, 2498–2503, (1992).

[104] R.Z. Zhdanov, Separation of variables in the nonlinear wave equation, *J. Phys. A: Math. Gen.* **27**, L291–L298, (1994).

[105] A.M. Wazwaz, The tanh method: exact solutions of the sine-Gordon and the sinh-Gordon equations, *Applied Mathematics and Computation.* **167**, 1196–1210, (2005).

[106] Y. Xie and J.A. Tang, A unified method for solving sinh-Gordon-type equations, *Nuovo Cimento B.* **121**, 115–120, (2006).

[107] M.J. Ablowitz, D.J. Kaup, A.C. Newell and H. Segur, Nonlinear-evolution

equations of physical significance, *Physical Review Letters.* **31**, 125–127, (1973).

[108] Z. Yan, Jacobi elliptic function solutions of nonlinear wave equations via the new sinh-Gordon equation expansion method, *Journal of Physics A: Mathematical and General.* **36**, 1961–1972, (2003).

[109] S. Bugaychuk, G. Mandula, L. Kovcs and R.A. Rupp: "The sine-Gordon model in optical dynamic holography", Nemzetkzi muhelytallkoz a disszipatv szolitonokrl, Drezda, Nmetorszg, janur 23–29 (2006), P8 (angol nyelvu poszter).

[110] V.E. Zakharov, Collapse of Langmuir waves, *Soviet Journal of Experimental and Theoretical Physics.* **35**, 908–914, (1972).

[111] M. Marklund, Classical and quantum kinetics of the Zakharov system, *Physics of Plasmas.* **12**, 082110, (2005).

[112] R. Fedele, P.K. Shukla, M. Onorato, D. Anderson, and M. Lisak, Landau damping of partially incoherent Langmuir waves, *Physics Letters A.* **303**, 61–66, (2002).

[113] H. Hadouaj, B.A. Malomed, and G.A. Maugin, Dynamics of a soliton in a generalized Zakharov system with dissipation, *Physical Review A.* **44**, 3925, (1991).

[114] B. Malomed, D. Anderson, M. Lisak, M.L. Quiroga-Teixeiro, and L. Stenflo, Dynamics of solitary waves in the Zakharov model equations, *Physical Review E.* **55**, 962, (1997).

[115] A. Borhanifar, M.M. Kabir, and L. Maryam Vahdat, New periodic and soliton wave solutions for the generalized Zakharov system and (2 + 1)-dimensional Nizhnik–Novikov–Veselov system, *Chaos, Solitons & Fractals.* **42**, 1646–1654, (2009).

[116] Z. Dai, J. Huang, M. Jiang, Explicit homoclinic tube solutions and chaos for Zakharov system with periodic boundary, *Physics Letters A.* **352**, 411–415, (2006).

[117] S. Jin, P.A. Markowich, and C. Zheng, Numerical simulation of a generalized Zakharov system, *Journal of Computational Physics.* **201**, 376–395, (2004).

[118] W. Bao, F. Sun, and G.W. Wei, *Journal of Computational Physics.* **190**, 201–228, (2003).

[119] R.A. Van Gorder and K. Vajravelu, Analytical and numerical solutions of the density dependent diffusion Nagumo equation, *Physics Letters A.* **372**, 5152–5158, (2008).

[120] S. Abbasbandy, Soliton solutions for the Fitzhugh-Nagumo equation with the homotopy analysis method, *Applied Mathematical Modelling.* **32**, 2706–2714, (2008).

[121] R.A. Van Gorder and K. Vajravelu, A variational formulation of the Nagumo reaction-diffusion equation and the Nagumo telegraph equation, *Nonlinear Analysis Series B: Real World Applications.* **11**, 2957–2962, (2010).

[122] R.A. Van Gorder and K. Vajravelu, Analytical and Numerical Solutions for a Density Dependent Nagumo Telegraph Equation, *Nonlinear Analysis*

Series B: Real World Applications. **11**, 3923–3929, (2010).

[123] R.A. Van Gorder, Travelling waves for a density dependent diffusion Nagumo equation over the real line, *Communications in Theoretical Physics.* **58**, 5–11, (2012).

[124] R. Fitzhugh, Impulse and physiological states in theoretical models of nerve membrane, *Biophys. J.* **1**, 445–466, (1961).

[125] J.S. Nagumo, S. Arimoto and S. Yoshizawa, An active pulse transmission line simulating nurve axon, *Proc. IRE.* **50**, 2061–2070, (1962).

[126] M. Shih, E. Momoniat and F.M. Mahomed, Approximate conditional symmetries and approximate solutions of the perturbed Fitzhugh–Nagumo equation, *J. Math. Phys.* **46**, 023503, (2005).

[127] D.G. Aronson and H.F. Weinberger, *Nonlinear Diffusion in Population Genetics, Combustion and Nerve Propagation, Partial Differential Equations and Related Topics.* Lecture Notes in Mathematics, vol. **446**. (Springer, Berlin, 1975).

[128] M.E. Gurtin, R.C. MacCamy, On the diffusion of biological populations, *Math. Biosci.* **33**, 35–49, (1977).

[129] Y. Hosono, Travelling wave solutions for some density dependent diffusion equations, *Jpn. J. Appl. Math.* **3**, 163–196, (1986).

[130] J.D. Murray, *Mathematical Biology.* (Springer, Berlin, Heidelberg, New York, 1989).

[131] M.B.A. Mansour, Accurate computation of traveling wave solutions of some nonlinear diffusion equations, *Wave Motion.* **44**, 222–230, (2007).

[132] M.G. Pedersen, Wave speeds of density dependent Nagumo diffusion equations, *J. Math. Biol.* **50**, 683–698, (2005).

[133] A. Tomimatsu and H. Sato, New exact solution for the gravitational field of a spinning mass, *Physical Review Letters.* **29**, 1344, (1972).

[134] A. Tomimatsu and H. Sato, New series of exact solutions for gravitational fields of spinning masses, *Prog. Theor. Phys.* **50**, 95, (1973).

[135] F.J. Ernst, A new family of solutions of the Einstein field equations, *Journal of Mathematical Physics.* **18**, 233, (1977).

[136] C. Reina and A. Treves, Axisymmetric gravitational fields, *J. Gen. Rel. Gravity.* **7**, 817, (1976).

[137] B.K. Harrison, Bäcklund transformation for the Ernst equation of general relativity, *Physical Review Letters.* **41**, 1197, (1978).

[138] F.J. Ernst, New formulation of the axially symmetric gravitational field problem, *Physical Review.* **167**, 1175, (1968).

[139] D. Korotkin and H. Nicolai, Separation of variables and Hamiltonian formulation for the Ernst equation, *Physical Review Letters.* **74**, 1272, (1995).

[140] B.Y. Hou and W. Lee, Virasoro algebra in the solution space of the Ernst equation, *Letters in Mathematical Physics.* **13**, 1, (1987).

[141] G. Helms, H. Dathe, and P. Dechent, Quantitative FLASH MRI at 3 tesla using a rational approximation of the ernst equation, *Magnetic Resonance in Medicine.* **59**, 667, (2008).

[142] C. Klein and O. Richter, Physically realistic solutions to the Ernst equation on hyperelliptic Riemann surfaces, *Physical Review D* **58**, 124018, (1998).

[143] M. Omote and M. Wadati, The Bčklund transformations and the inverse scattering method of the Ernst equation, *Prog. Theor. Phys.* **65**, 1621, (1981).

[144] D.A. Korotkin and V.B. Matveev, Theta function solutions of the Schlesinger system and the Ernst equation, *Functional Analysis and its Applications.* **34**, 252, (2000).

Chapter 5

A Convergence Condition of the Homotopy Analysis Method

Mustafa Turkyilmazoglu

Mathematics Department, University of Hacettepe
06532-Beytepe, Ankara, Turkey
turkyilm@hotmail.com

In this chapter, we present a condition enabling the homotopy analysis method (HAM) to converge to the exact solution of the sought solution of algebraic, highly nonlinear differential-difference, integro-differential, fractional differential and ordinary or partial differential equations or systems. The previous notions of convergence control parameter are carefully reviewed and a novel description is proposed to find out an optimal value for the convergence control parameter, which, although it is completely different from the classical definition by means of the squared residual error as often used in the literature, yields nearly the same interval of convergence and optimal convergence parameters as those found from the squared residual error. When an unknown parameter is embedded into the governing equations, the convergence of the HAM is better pursued by the ratio relevant to this parameter rather than the ratio of other functions involving much harder integrations. An error estimate for the HAM is also provided. Physical and mechanical examples, including the Volterra differential-difference equation, the Fredholm integro-differential equation for the static beam, the Airy equation, the undamped and dumped Duffing oscillators, the Thomas–Fermi equation, the Gelfand problem, the fractional differential equation, the rotating sphere, and more, clearly illustrate the validity of the new approach and further provide knowledge on why the corresponding homotopy series generated by the HAM should converge to the exact solution in the domain of interest.

Contents

5.1. Introduction . . . 182
5.2. A brief description of the homotopy analysis method . . . 184
5.3. A convergence theorem and its error estimate . . . 186
5.3.1. Important consequences and remarks . . . 189
5.4. Convergence control parameter in the homotopy analysis method . . . 190
5.4.1. Interval of convergence . . . 191
5.4.2. Optimum value from an appropriate residual . . . 192
5.4.3. Optimum value from the ratio . . . 194
5.5. Illustrative examples . . . 196
5.5.1. An algebraic equation . . . 196
5.5.2. A nonlinear Volterra differential-difference equation . . . 200
5.5.3. A nonlinear high-order Fredholm integro-differential equation . . . 203
5.5.4. Some ordinary differential equations . . . 206
5.5.5. A nonlinear fractional differential equation . . . 233
5.5.6. Coupled nonlinear differential equations . . . 237
5.5.7. Partial differential equations . . . 244
5.6. Concluding remarks . . . 252
References . . . 254

5.1. Introduction

Since the nonlinear ordinary/partial differential equations or systems for initial and boundary value problems are most challenging in finding their exact solutions, besides the classical perturbation methods, some new perturbation or analytical-nature methods have been introduced and developed by researchers in the literature. Among variety of methods proposed to find analytic approximate solutions of a given nonlinear mathematical model, the most recent popular and powerful technique is the homotopy analysis method (HAM). In this method, which requires neither a small parameter nor a linear term, a homotopy with an embedding parameter $p \in [0, 1]$ is constructed [1]. The solution is considered as the sum of an infinite series converging rapidly and accurately to the exact solutions by means of enjoying the so-called *convergence control parameter*, a missing tool in most of the other techniques. Fundamental characteristics and smart advantages of the HAM over the existing analytical techniques were clearly laid out by Liao in the recent book [2]. In addition to its early success in several nonlinear problems as summarized in the book [3], further numerous nonlinear problems in science, finance and engineering were successfully treated by the method, see among them [4–12]. Particularly, a few new solutions of some nonlinear problems were discovered by means of the method [13], which were unfortunately neglected by other analytic methods and even by numerical techniques. The method was successively applied recently to some series of strongly nonlinear problems, such as the Blasius equation for the flow over a flat plate [14], the system of differential equations concerning the flow over a rotating cone [15], the system of differential equations related to the rotating disk [16–19], the system of differential equations related to the rotating sphere [20], the singularly perturbed boundary layer problems [21, 22], the undamped and dumped Duffing oscillators [23, 24], the limit cycle of Duffing-van der Pol equation [25], the nonlinear pendulum problem [26], the Thomas–Fermi equation [27] and the Airy differential equation [28]. An analytic shooting approach combined with the HAM was also proposed in [29]. After the work of [30], the use of the HAM is more safe now since an optimal parameter controlling the fast convergence can always be picked from the squared residual error ensuring to gain the most accurate results, as also implemented in the above citations. Despite the fact that all these demonstrate the validity and high potential of the homotopy analysis method for strongly nonlinear problems of real life, apart from some general approaches as presented in [2], the question of convergence of the method is yet to be answered.

The present chapter is devoted to the investigation of the homotopy analysis technique from a mathematical point of view to serve to its convergence issue. The aim is thus to analyze the method and to show that under a given constraint the HAM converges to the exact solution desired, with an error estimate, without a prior knowledge of the exact solution. In addition to the well-known interval of convergence control parameter by means of constant h-curves and optimal value for the convergence control parameter via the squared residual error, a new conceptual definition is offered, which makes use of the ratios of the homotopy series based on a properly chosen norm. It is shown through examples that both yield approximately the same values regarding the convergence control parameter, though the newly introduced scheme seems more advantageous in some aspects at least in terms of computational efforts. Embedding an unknown parameter into the studied differential equation from the boundaries is demonstrated to facilitate a great advantage for understanding the convergence through this ratio, that avoids the use of ratios concerning the other physical functions involving heavier integrations. The given convergence criterion is justified exemplifying it by basic commonly-known examples from nonlinear algebraic, differential-difference, integro-differential, the fractional differential, ordinary and partial differential equations and also systems often studied in the literature. The convergence of the HAM for the considered problems is not only guaranteed, but the interval of convergence and further the optimum value for the convergence can also be determined by the presented theory. The new approach also enlightens us about the inevitable failure of the so-called homotopy perturbation method blindly used by some investigators.

5.2. A brief description of the homotopy analysis method

A systematic description of the homotopy analysis method is outlined in this section. Within this purpose, let us consider the following general nonlinear equation

$$\mathcal{N}[u(t)] = 0, \tag{5.1}$$

where $\mathcal{N}$ is either a nonlinear function (concerning the algebraic equations), or a nonlinear operator (concerning the differential or difference equations), t denotes an independent variable, $u(t)$ is an unknown function, respectively. In more general case, equation (8.1) may represent a full set of nonlinear equations. For simplicity, we ignore all boundary or initial conditions in the case of a differential equation (or a system), which can be treated in the similar way. After the early HAM described by Liao in his PhD dissertation [1], the first thing and thus the key point is to construct the so-called *zeroth-order deformation equation* [3, 31]

$$(1-p)\mathcal{L}[\varphi(t,p) - u_0(t)] - p\, h\, H(t)\mathcal{N}[\varphi(t,p)] = 0, \tag{5.2}$$

where $p \in [0,1]$ is called *the homotopy embedding parameter*, h is a nonzero auxiliary parameter which is called *the convergence control parameter*, $\mathcal{L}$ is an auxiliary linear operator, $u_0(t)$ is an initial guess for $u(t)$, $H(t)$ is an auxiliary function to adjust the sought solution, and $\varphi(t,p)$ is an unknown function, respectively. In fact, the success of the HAM substantially relies upon the zeroth-order deformation equation (8.2), which can be further generalized [2]. It is important, that one has great freedom to choose auxiliary parameters in the HAM. Obviously, when $p = 0$ and $p = 1$, it holds

$$\varphi(t,0) = u_0(t), \quad \varphi(t,1) = u(t). \tag{5.3}$$

Thus, as p increases from 0 to 1, the solution $\varphi(t,p)$ varies from the initial guess $u_0(t)$ to the final solution $u(t)$ of the original nonlinear equation (8.1).

Expanding $\varphi(t,p)$ in Maclaurin series with respect to p at $p=0$, it reads

$$\varphi(t,p) = u_0(t) + \sum_{k=1}^{\infty} u_k(t)\, p^k, \tag{5.4}$$

where the series coefficients u_k are defined by

$$u_k(t) = \left. \frac{1}{k!} \frac{\partial^k \varphi(t,p)}{\partial p^k} \right|_{p=0}. \tag{5.5}$$

Here, the series (8.4) is called *the homotopy series* and the expression (8.5) is called *the kth-order homotopy-derivative* of φ, see [31]. If the auxiliary linear operator $\mathcal{L}$, the initial guess $u_0(t)$, the convergence control parameter h and the auxiliary function $H(t)$ are so properly chosen, the homotopy-series (8.4) converges at $p=1$, then using the relationship $\varphi(t,1) = u(t)$, one has the so-called *homotopy series solution*

$$u(t) = u_0(t) + \sum_{k=1}^{\infty} u_k(t), \tag{5.6}$$

which must be one of the solutions of original nonlinear equation (8.1), as proved by Liao [3].

Based on the definition (8.5), the governing equation for the homotopy series (5.6) can be deducted from the zeroth-order deformation equation (8.2). Differentiating the zeroth-order deformation equation (8.2) k times with respect to the homotopy parameter p, setting $p=0$ and finally dividing them by $k!$, we have the so-called *kth-order deformation equation*

$$\mathcal{L}[u_k(t) - \chi_k u_{k-1}(t)] = h\; H(t)\; \mathcal{D}_{k-1}[\varphi(t,p)], \tag{5.7}$$

where $\mathcal{D}_n$ is the so-called *nth-order homotopy derivative operator* given by

$$\mathcal{D}_n[\varphi(t,p)] = \left. \frac{1}{n!} \frac{\partial^n \mathcal{N}[\varphi(t,p)]}{\partial p^n} \right|_{p=0},$$

and

$$\chi_k = 0 \quad \text{for } k \le 1, \quad \chi_k = 1 \quad \text{for } \; k > 1. \tag{5.8}$$

Note that the right-hand side term $\mathcal{D}_{k-1}[\varphi(t,p)]$ of (5.7) is dependent only upon

$$u_0(t), u_1(t), u_2(t), \cdots, u_{k-1}(t),$$

which are known for the kth-order deformation equation described above.

Because of the fact that the higher-order deformation equation (5.7) is linear in nature, an appropriate linear operator $\mathcal{L}$ will easily generate the homotopy terms u_k in homotopy series (5.6) by means of computer algebra systems such as Mathematica, Maple and so on. Finally, *an Mth-order approximate analytic solution* of practical interest is given by truncating the homotopy series (5.6)

$$u_M(t) = u_0(t) + \sum_{k=1}^{M} u_k(t), \tag{5.9}$$

and *the exact solution* is given by the limit

$$u(t) = \lim_{M \to \infty} u_M(t).$$

It should be reminded that the homotopy terms

$$u_1(t), u_2(t), \cdots, u_k(t)$$

in (8.27) strongly depend on both the physical variable t and the convergence control parameter h. In essence, h is an artificial parameter without physical meanings but it can adjust and control the convergence region of the homotopy series solution (8.27). In fact, the use of such an auxiliary parameter distinguishes the HAM from other perturbation-like analytical techniques.

5.3. A convergence theorem and its error estimate

Performing the methodology underlined above in § 5.2, the number of problems treated by the homotopy analysis method approaches a couple of thousands now. Existing theorems about the convergence of the resulting homotopy series of a given nonlinear problem unfortunately have general meanings [2, 3] or the convergence of solution series is believed to take place since the HAM logically contains the famous Euler Theorem as proved by Liao [2]. However, a rigorous and firm answer to the very basic question of why the series (5.6) obtained by setting $p = 1$ in (8.4) should be convergent remains unanswered till today. To make sure of the convergence, the analyticity of solutions is generally presumed, otherwise, a Maclaurin series of a function may not necessarily converge to that function, see pages 23 and 24 in [2]. This may limit the homotopy method leading to divergent homotopy series solutions especially for nonlinear problems with strong nonlinearity. Moreover, although it is fortunate to know that the convergence control parameter h can greatly modify the convergence of the homotopy series solution, the guarantee of convergence still needs a mathematical explanation. To remedy this issue up to a point, we provide the subsequent theorems and the resulting corollaries here. It should be noted that even though the proofs require prescription of convergence control parameter h, how to find a proper value, or even better, to get a fastest convergent one, will be discussed later.

Since the homotopy analysis methodology as described in § 5.2 is a very user-friendly tool among the scientists frequently used for solving complicated highly nonlinear problems, the convergence criterion to be given should also be easy-to-use beyond the generality and in the absence of exact solution to the nonlinear equation under consideration. This is essential in deeper understanding of whether the HAM performed for a specific problem will converge to the true exact solution or not. Such a convergence criterion was made use of in several physical problems, for example the reader may refer to [20, 32] and [24, 27, 28]. In what follows we state the criterion, that is based on the fixed point theorem well known in the functional analysis.

Theorem 5.1. *Suppose that $A \subset R$ be a Banach space donated with a suitable norm $\|\|$ (depending on the physical problem under consideration), over which the functional sequence $u_k(t)$ of (8.4) is defined for a prescribed value of h. Assume also that the initial approximation $u_0(t)$ remains inside the ball of the solution $u(t)$ of (8.1). Taking $r \in R^+$ be a constant, the following statements hold true:*

(i) *For a prescribed convergence control parameter h, if $\|v_{k+1}(t)\| \leq r\|v_k(t)\|$ for all k, provided that $0 < r < 1$, then the series solution $\varphi(t,p)$ defined in (8.4) converges absolutely at $p = 1$ to $u(t)$ given by (5.6) over the domain of definition of t.*

(ii) *For a prescribed convergence control parameter h, if $\|v_{k+1}(t)\| \geq r\|v_k(t)\|$ for all k, given some $r > 1$, then the series solution $\varphi(t,p)$ defined in (8.4) diverges at $p = 1$ over the domain of definition of t.*

Proof. **(i)** If $S_n(t)$ denote the sequence of partial sum of the series (5.6), it is demanded that $S_n(t)$ be a Cauchy sequence in A. For this purpose, the subsequent inequalities are constructed

$$\begin{aligned}\|S_{n+1}(t) - S_n(t)\| &= \|u_{n+1}(t)\| \leq r\|u_n(t)\| \\ &\leq r^2\|u_{n-1}(t)\| \leq \cdots \leq r^{n+1}\|u_0(t)\|. \end{aligned} \tag{5.10}$$

It is remarked that owing to (5.10), all the approximations produced by the homotopy analysis method by (8.2) in § 5.2 will lie within the ball of the solution $u(t)$. For every $m, n \in N$, such that $n \geq m$, the following results in making use of (5.10) and the triangle inequality successively,

$$\begin{aligned}\|S_n(t) - S_m(t)\| &= \|(S_n(t) - S_{n-1}(t)) + \cdots + (S_{m+1}(t) - S_m(t))\| \\ &\le \left(\frac{1 - r^{n-m}}{1 - r}\right) r^{m+1} \|u_0(t)\|. \end{aligned} \tag{5.11}$$

Since by the hypothesis $0 < r < 1$, we get from (5.11)

$$\lim_{n,m\to\infty} \|S_n(t) - S_m(t)\| = 0. \tag{5.12}$$

Therefore, $S_n(t)$ is a Cauchy sequence in the Banach space A, which implies that the series solution (5.6) is indeed convergent. This completes the proof **(i)**.

The proof of **(ii)** follows from the fact that under the hypothesis supplied in **(ii)**, there exist a number l, $l > r > 1$, so that the interval of convergence of the power series (8.4) is $|p| < 1/l < 1$, which obviously excludes the case of $p = 1$. □

Theorem 5.2. *If the series solution defined in (8.4) is convergent at $p = 1$, then the resulting series (5.6) converges to an exact solution of the nonlinear problem given in (8.1).*

Proof. The formal proof can be found in the books by Liao [2, 3]. □

Theorem 5.3. *Assume that the series solution defined in (5.6) is convergent to the solution $u(t)$ for a prescribed value of h. If the truncated series $u_M(t)$ expressed in equation (8.27) is used as an approximation to the solution $u(t)$ of problem (8.1), then an upper bound for the error, that is, $E_M(t)$, is estimated as*

$$E_M(t) \le \frac{r^{M+1}}{1-r} \|u_0(t)\|. \tag{5.13}$$

Proof. Making use of the inequality (5.10) of Theorem 5.1, we immediately obtain

$$\|u(t) - S_M(t)\| \le \left(\frac{1 - r^{n-M}}{1 - r}\right) r^{M+1} \|u_0(t)\|, \tag{5.14}$$

and taking into account the constraint $(1 - r^{n-M}) < 1$, (5.14) leads to the desired formula (5.13). This completes the proof. □

5.3.1. *Important consequences and remarks*

A few important consequences of the aforementioned theorems will be presented now together with some relevant remarks.

Corollary 5.1. *Since the finite number of terms does not affect the convergence, Theorem 5.1 is equally valid if the inequalities stated in* **(i-ii)** *are true for sufficiently large $k's$. Thus, for a preassigned value of h, it is sufficient to keep track of magnitudes of the ratio β defined by*

$$\beta = \frac{\|v_{k+1}(t)\|}{\|v_k(t)\|}, \tag{5.15}$$

and whether it remains less than unity for increasing values of k. An optimal value for the convergence control parameter h could also be determined from (5.15) by requiring the ratio β to be as close to zero as possible, so that for such a value the rate of convergence of homotopy series (8.27) will be the fastest, since then the remainder of the series will most rapidly decay.

Corollary 5.2. *On the condition that the norm is taken in the sense of absolute value, by enforcing the ratio in* **(i)** *to hold true in the infinite limit or, at least for large k, the validity domain for t of the sought solution can also be constructed having provided a value for the convergence control value h, that is,*

$$\lim_{k\to\infty} \frac{|v_{k+1}(t)|}{|v_k(t)|} < 1. \tag{5.16}$$

Corollary 5.3. *The graphical constant h-curves idea of Liao [3] can also be approved by the ratio given in (5.15), in such a way that an analytical interval of convergence for h can be determined by application of Theorem 5.1 to some certain physical quantities, say for instance $u^{(m)}(t_0)$, $m \in N$, $t_0 \in R$ and by solving the inequality*

$$\frac{|u_{k+1}^{(m)}(t_0)|}{|u_k^{(m)}(t_0)|} < 1, \tag{5.17}$$

since $u_k^{(m)}(t)$ is a function of h.

Corollary 5.4. *In the case that the nonlinear equation (8.1) comprises of finding the root of an equation $f(x) = 0$, solution of the inequality for large k*

$$\frac{|x_{k+1}|}{|x_k|} < 1, \tag{5.18}$$

directly yields the interval of convergence control parameter h.

Remark 5.1. If exact or numerical solution $u_e(t)$ is available, then one can always use the absolute error defined by

$$err = \int_{\Gamma} |u_e(t) - u(t)| dt, \tag{5.19}$$

where $u(t)$ is the homotopy solution of (8.1) defined over the domain Γ.

Remark 5.2. Readers should be warned that the condition given in Theorem 5.1 and in the subsequent corollaries is only a sufficient condition for the convergence of the homotopy analysis method. This strongly implies that in the cases where the limit for the ratio in (5.15) cannot be reached or tends to unity, the method may still converge or fail to do so.

Remark 5.3. In the case that there may exist an unknown parameter to be solved together with the differential equation, such as the parameterized differential equations, it would suffice to keep track of the ratio regarding this parameter from the ratio (5.18), that will luckily avoid the use of residuals and ratios which involve complicated integrations. Such examples will be provided later on. Note that every considered system always virtually possesses this sort of unknown parameters, named hereafter as *the pseudo parameters*, at least inserted from the physical boundary conditions which are themselves unknown.

Remark 5.4. The homotopy Padé technique introduced in [3] can always be applied to enlarge the interval of convergence and also to accelerate the rate of convergence for a given nonlinear problem. In addition to this, by means of the freedom on the selection of initial guess, either an optimal initial approximation is sought, or a homotopy iteration approach can be devised in the frame of the HAM, which can greatly speed up the convergence of the homotopy series, see [2] for illustrations.

5.4. Convergence control parameter in the homotopy analysis method

It is widely known that in the general homotopy method improved by Liao [2, 3], the convergence can be controlled by the convergence control parameter h, a concept that plays a key role in the HAM. The prime role of h in a homotopy series is to adjust the convergence at a reasonable rate, fortunately at the quickest rate. Therefore, in recent applications optimal convergence control parameter is generally made use to gain sufficiently accurate approximations with the smallest number of homotopy terms in the homotopy series (5.6). In this section, we first recall the previously used schemes for finding an appropriate value of the convergence control parameter, later on, a new and novel approach for finding an optimum value of h is presented, based on the ratio presented in equation (5.15). It should be stated that if a single parameter h is involved in (8.2), the below procedure of finding h is recently termed as *basic optimal HAM*, see [2]. Besides, although two-parameter, three-parameter and infinite-parameter optimal HAMs, see for instance the so-called *optimal homotopy asymptotic method* in [33], can also be treated without any difficulty, for the sake of brevity and conciseness, we only prefer the basic optimal HAM in what follows. This is satisfactory enough, since the optimal HAMs with more convergence control parameters do not always give better homotopy approximations than the basic optimal HAM in general, thus the basic optimal HAM is strongly suggested to use in practice [2].

How can one identify an appropriate convergence control parameter h that results in a convergent homotopy series solution? There are in fact two different up-to-date ways of determining the convergence control parameter h most frequently used among the HAM society; either by plotting its interval of convergence or so-called effective region via a physical quantity depending on h or by searching for a global optimal value by means of a residual error, as formulated below, which is implemented by a straightforward substitution of homotopy series solution into the original governing equation.

5.4.1. *Interval of convergence*

One of the remarkable property of the HAM is that the value of the auxiliary parameter h can be freely chosen to ensure the convergence, and even more, to increase the convergence rate of the solution series (8.27). However, the freedom of selecting h is subject to the so-called valid regions of h. This notion is first introduced by Liao [3], the creator of the HAM, and has been commonly used in many HAM applications, even today. A physical variable from the problem under consideration, which is unknown, but evaluated analytically at the number of truncated homotopy series M from the homotopy series (8.27); say $u^{(m)}(t_0)$, $m \in N$ (like $u'(0)$ or $u''(0)$) of the nonlinear differential equation (8.1), is usually plotted versus h (in the case of an algebraic equation $f(x) = 0$, the homotopy solutions of x are drawn against h). These curves are so-called as *the constant h-curves* or *curves for the convergence control parameter*, which have been successfully handled in many nonlinear problems, see at least the cited references herein.

It has been found by the HAM researchers that there often exists such an effective region that certain values of h obtained from some physical quantities result in a convergent homotopy series solution. Even though such a region can always be found, with less computational effort as compared to the squared residual, by plotting the curves of these unknown quantities versus h, the information about the interval of convergence can be gained only approximately from the plot. Besides, no optimum value can be in general realizable from the approach, that corresponds to the quickest convergent homotopy series. Just random values from this interval are picked over and over again to carry out the HAM solutions in nonlinear problems. In principle, the valid region of h will expand without bound as the order of homotopy approximation approaches infinity. However, in finite order occasions, it is the case that the solution series corresponding to a fixed value of h may not give a good approximation even when h is chosen within the valid region. This is due to the fact that the h-curve is just a plot of a physical quantity versus h at a specific $t = t_0$. Therefore, one should expect the solution series to converge to the exact solution only locally about $t = t_0$ when one selects a h value from such a plotted region. Nevertheless by chance, there are cases where a valid h value chosen this way does give good approximation for a large range of the independent variable t. In spite of the fact that some investigators still insist to use this idea in their recent HAM publications, it is now abolished and hence out of date.

It is worthy of emphasizing that the inequalities involving the ratios presented in Corollaries 5.3 and 5.4 effectively cover the above idea, being capable of reproducing the constant h-curve intervals. Besides, the most crucial advantage using such ratios is that exact interval of convergence for convergence control parameter h can be obtained by solving these inequalities, as clearly demonstrated in the below examples.

5.4.2. *Optimum value from an appropriate residual*

Instead of approximately locating the interval of convergence as above, it is better and hopefully more economical to try to get a best value of h, from which the homotopy solutions from the series (8.27) are to be generated. In general speaking, an optimal value for the convergence control parameter h

at the Mth-order homotopy approximation may be found from the following norm representing the residual of the governing equation (8.1)

$$Res(h) = \|\mathcal{N}[\sum_{k=0}^{M} u_k(t)]\|,$$

with the norm being understood as L^p. Theoretically speaking, whenever this residual goes to zero, then the obtained homotopy series is a solution of the original governing equation (8.1). Researchers, in time, were adapted themselves to L^2 only, and further, since being more concise after the work of [34], the subsequent squared residual error is often employed to determine an optimal value for the convergence control parameter h

$$Res(h) = \int_{\Gamma} \left\{ \mathcal{N} \left[\sum_{k=0}^{M} u_k(r) \right] \right\}^2 dr, \tag{5.20}$$

where Γ is the region of interest for the problem under consideration. On the other hand, if it is known that the integrand $\mathcal{N}[u(t)]$ is positive, then it is better to use the subsequent residual error based on L^1

$$Res(h) = \int_{\Gamma} \mathcal{N} \left[\sum_{k=0}^{M} u_k(r) \right] dr, \tag{5.21}$$

in order to greatly save from the computation time. It should be reminded that via the residuals (5.20) and (5.21), the optimal auxiliary operator $\mathcal{L}$, the optimal initial approximation u_0 and the optimal auxiliary function H can also be determined, see [2].

Obviously, the more quickly $Res(h)$ in (5.20) or (5.21) decreases to zero, the faster the corresponding homotopy series solution (8.27) converges and the accuracy of the homotopy approximations increases. So, at the truncated order M of approximation, the corresponding so-called *optimal value of the convergence control parameter* h is given by the minimum of $Res(h)$, corresponding generally to a nonlinear algebraic equation to be solved from

$$\frac{dRes(h)}{dh} = 0. \tag{5.22}$$

From this approach, the interval of convergence for h will be more evident than the constant h-curve analysis, besides it yields an optimal value. As a consequence, the convergence control parameter obtained via (5.20–5.22) can be safely supplied into the Theorem 5.1 to compute the ratio β in the convergence analysis of the HAM.

However, since the exact integration in (5.20) or (5.21) is hardly possible to perform especially for strong nonlinear problems, or it is time consuming particularly for large M, as mentioned on page 105 in [30], a discrete form of averaged value of (5.20) can be substituted via the numerical integration accounting for the simple rectangle rule, as again first suggested by Liao [30]. Hence, the discrete squared residual error is expressed by

$$Res(h) \approx \frac{1}{N+1} \sum_{j=0}^{N} \left\{ \mathcal{N} \left[\sum_{k=0}^{M} u_k(t_j) \right] \right\}^2, \tag{5.23}$$

see also equation (3.29) on page 105 and definition (3.44) on page 119 in [30], where N is the number of discrete points with $t_j = j\Delta t$ and

$$\Delta t = \frac{\text{length of } \Gamma}{N}.$$

Alternatively, provided that the integrand is positive, the residual given by (5.21) can be similarly discretized in the form

$$Res(h) \approx \frac{1}{N+1} \sum_{j=0}^{N} \mathcal{N} \left[\sum_{k=0}^{M} u_k(t_j) \right]. \tag{5.24}$$

The above residuals may also be adjusted to the system of differential equations as demonstrated later in the examples.

A few shortcomings of the above residual approach should be mentioned here. Analytical integrations may not always be possible if transcendental functions are involved. Even discretization may not help, in terms of CPU time, due to the requirement of evaluation of strongly nonlinear operator in (5.23) and (5.24). Moreover, if the physical problem is defined over a semi infinite domain, either the computational domain is cut to a finite interval or a plenty of grid points may be required to resolve the infinite region, which becomes a tedious computational task from the above residuals.

5.4.3. *Optimum value from the ratio*

The aforementioned approach of squared residual error is theoretically rigorous, but its minimization task in practice may not always be efficient in computational terms. Taking this into account, better and more effective ways should always be targeted. To serve to this purpose, in the present section, by means of the ratio given in equation (5.15), a novel and easy way of identifying the optimum value of convergence control parameter h will be introduced. As underlined by the above theorems and by its corollaries, for a prescribed h, if the ratio is less than unity, then the convergence of HAM is guaranteed. In addition to this, to get a faster rate of convergence towards the exact solution of (8.1), it is sensible to require that the ratio β is as small as desired, possibly getting close to zero, so that the convergence happens at the fastest rate. Keeping this in mind, it is rational to search for a value of h that gives rise to as small ratio as possible from equation (5.15). This can be simultaneously achieved via practically plotting β versus h in (5.15), as well as theoretically checking whether

$$\frac{d\beta}{dh} = 0,$$

if exists, as a result, it produces *the optimum value for the convergence parameter h*. In fact, the following examples clearly demonstrate that this procedure results in very close values of h to those obtained from the residuals via (5.22). Considering for instance L^p ($p = 1$ or $p = 2$), since integrals in the ratio

$$\beta = \frac{\int_\Gamma u_{k+1}^p(r)dr}{\int_\Gamma u_k^p(r)dr}, \tag{5.25}$$

demand less laborious work as compared to (5.20) and (5.21) (notice that exact integration is always possible from (5.25) unlike the residuals from (5.20) and (5.21)), as also revealed in below examples, equation (5.25) or its discrete counterpart (5.26) given as

$$\beta \approx \frac{\sum_{j=0}^{N}[u_{k+1}(t_j)]^p}{\sum_{j=0}^{N}[u_k(t_j)]^p}, \tag{5.26}$$

brings a more convenient way of evaluating the convergence control parameter h. This is apparent, since β in (5.25) involves only the homotopy terms of the homotopy series (8.27) that are usually integrable, whereas equations (5.20) and (5.21) incorporate the nonlinear operator from equation (8.1). Similar to the squared residual idea, at the given order of approximation, the curves of ratio β versus h indicate not only the effective region for the convergence control parameter h, but also the optimal value of h that corresponds to the minimum of β. An emphasize is deserved such that the present approach seems more promising in terms of computational efficiency and hence can be used in place of the two existing aforementioned methods. We should further remind that the above definitions for ratios may also be adjusted to the system of differential equations as revealed later. Finally, an emphasis should be made that for the convergence of the HAM it is not necessary for the ratio to reach a minimum as described above. In the circumstances that no minimum is attainable for the ratio, then either a value that makes the ratio less than unity is chosen, or it is enough to supply into the ratio the optimum h value worked out from the minimum residual error to check out the convergence of the HAM.

5.5. Illustrative examples

To illustrate the validity and accuracy of the new approach outlined in Section 5.4 through the utilities (5.20–5.26), we take into account the following examples taken from the homotopy analysis studies in the literature, which are algebraic, nonlinear differential-difference equation, linear or strongly nonlinear ordinary and partial differential equations and systems. All the norms are based on L^1, unless otherwise is mentioned in the below examples. We should note that the computational efficiency of the optimal HAM depends strongly on the method of searching for the minimum of either residual or ratio. As mentioned in Liao's book [2], the command **NMinimize** (or sometimes **FindRoot**) with reasonable **WorkingPrecision** is used in the computer algebra system **Mathematica** to get the results provided here.

5.5.1. *An algebraic equation*

Let us first consider the quadratic algebraic equation

$$f(u) = u^2 - 2 = 0, \tag{5.27}$$

whose numerical solution up to the nine significant digit is simply $u = \sqrt{2} = 1.414213562$. In accordance with the homotopy concept given in § 5.2, the initial guess and the auxiliary operator are chosen in the from

$$u_0 = 1, \qquad \mathcal{L}(u) = f(u) - f(u_0)$$

(see for instance pages 19 and 20 in [2]), so that the homotopy series solution via the homotopy approach (8.2) can be straightforwardly constructed. We should remind that any real number could be selected as the initial approximation, that will certainly have impact on the region of convergency of the HAM. In fact, for the algebraic problems like (5.27), the residual and absolute errors can be immediately computed from

$$Res(h) = u^2 - 2, \tag{5.28}$$

$$err = u - \sqrt{2}, \tag{5.29}$$

for which, u denotes the homotopy series solution (8.27) depending upon only the convergence control parameter h. Hence, making use of (5.28), the minimum residual occurs at $h = -0.69$ that corresponds to the optimal value of convergence control parameter at the $22th$-order approximation.

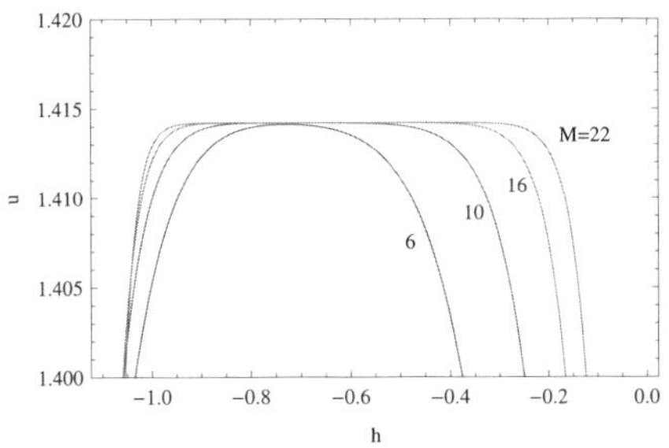

Fig. 5.1. Constant h-curves for equation (5.27).

At the orders $M = 6$, 10, 16 and 22 of approximation, the constant h-curves are displayed in Figure 5.1, which shows that as the order of approximation increases the convergence is guaranteed in the interval $h \in [-1, 0)$, which is also well estimated analytically by solving the inequality $|u_{22}/u_{21}| < 1$ from (5.18) that exactly yields $-1.0366 < h < 0$. Having fixed h as -0.69, Table 5.1 presents the values of the root and the absolute errors (5.28) for a variety of the approximation order M of the algebraic equation (5.27). Table 5.1 also contains the results obtained from the classical modified Newton iteration algorithm (see [2])

$$u_k = u_{k-1} + h\frac{f(u_{k-1})}{f'(u_{k-1})}. \tag{5.30}$$

Tabulated data in Table 5.1 clearly explains why the homotopy analysis method generates completely convergent series solution to the problem (5.27) for the chosen parameter h. It is interesting to observe that the HAM produces even better approximations for the roots as compared to the Newton iteration method. For instance, in order to get accurate solutions of order 10^{-6} from the HAM, it seems enough to use only 8–10 terms of the homotopy approximation, which is still advantageous over the Newton iteration results.

Actually, it was found that when the order of approximation M is an even number, then both the ratio (5.18) and residual (5.28) practically vanish at the same optimum value of h, as revealed on Table 5.2. Now,

Table 5.1. Values of the root of (5.27) and absolute errors at some selected orders of iteration M generated with $h = -0.69$.

M	5	10	15	20
u^{a}	1.413817050	1.414212541	1.414213561	1.414213562
err^{a}	3.9651×10^{-4}	1.0218×10^{-6}	1.7317×10^{-9}	7.9015×10^{-12}
u^{b}	1.413624474	1.414211877	1.414213558	1.414213562
err^{b}	5.8909×10^{-4}	1.6854×10^{-6}	4.8251×10^{-9}	1.3814×10^{-11}

[a]Solutions from homotopy (8.27)
[b]Solutions from Newton iteration (5.30)

Table 5.2. Optimum values for h from both the ratio (5.18) and residual (5.28).

M	10	20	30	50	100	200	300	350
h	−0.7100	−0.6919	−0.6849	−0.6786	−0.6734	−0.6704	−0.6693	−0.6686

let us find some optimum values for the convergence control parameter h, both from (5.28) and from (5.18) using as high odd order homotopy terms as possible. Within this respect, Table 5.3 shows the optimums and corresponding values of β at sufficiently high level of iterations. It is indicated that both the residual using (5.28) and the ratio with (5.18) yield nearly the same values for the optimum of h, as also justified from Figures 5.2 (a–b), evaluated taking $M = 301$. Thus, there is no doubt that in the infinite limit of M, both optimums will collide onto a single value. This proves the validity of the ratio approach as proposed in § 5.4. We should remark here that the limiting value of β towards zero for even number of approximations M and the limiting value of β towards the value as presented in Table 5.3 for odd number of approximations M do not contradict with the Theorem 5.1, since in total the sequence of partial sum for the homotopy series (5.6) will eventually converge to a finite limit.

Table 5.3. Optimum values for h and the resulting ratio β.

M	21	51	101	151	201	251	301	351
h^{a}	−0.6936	−0.6798	−0.6740	−0.6719	−0.6702	−0.6700	−0.6695	−0.6686
h^{b}	−0.6644	−0.6658	−0.6662	−0.6664	−0.6664	−0.6665	−0.6665	−0.6665
β	0.33281	0.33312	0.33323	0.33326	0.33328	0.33329	0.33330	0.33330

[a]Equation (5.28)
[b]Equation (5.18)

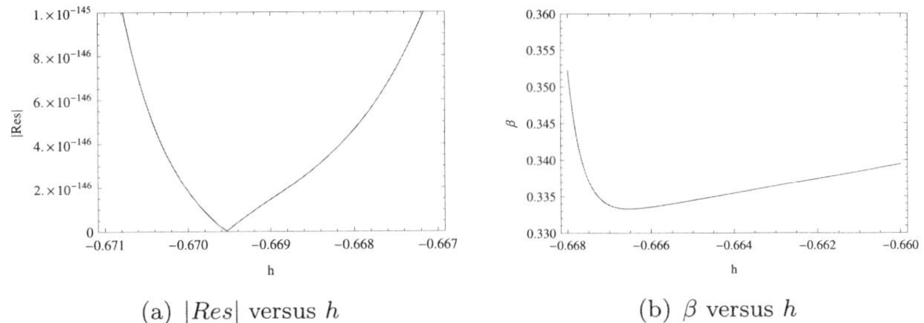

(a) $|Res|$ versus h (b) β versus h

Fig. 5.2. Residual error and ratio for equation (5.27) at the order of approximation $M = 301$.

Even though the convergence of the homotopy series solution is apparent using the optimal values of h shown in Tables 5.2, 5.3 and Figures 5.2 (a–b), the convergence of the homotopy series with the optimum convergence control parameter $h = -0.666$ and having 300th terms can be further assessed by checking the ratio (5.18) from Figure 5.3 and evidently observing the development of the ratio β from Table 5.4. It appears that, after an initial oscillatory character, the ratio β is rapidly settling down to a limiting value 0.33353, remaining less than unity, that further ensures the convergence of the produced homotopy series for the algebraic equation (8.27). It further deserves to mention that while finding the optimums, the CPU time to evaluate the minimum of β from (5.18) is much less than the CPU time to evaluate the minimum of Res from (5.28). For example, it needs 9.938 seconds, 11.466 seconds, 21.7 seconds and 39.234 seconds to calculate the minimums of (5.28) for $M = 151$, 201, 251 and 351, respectively. On the other hand, it takes only 1.279 seconds, 2.075 seconds, 3.697 seconds and 5.367 seconds for the minimums of (5.18) at the same orders of approximations.

Table 5.4. The ratio β evaluated with 300 homotopy terms and $h = -0.666$.

M	100	150	200	250	280	290	296	300
β	0.33325	0.33334	0.33342	0.33348	0.33352	0.33353	0.33353	0.33353

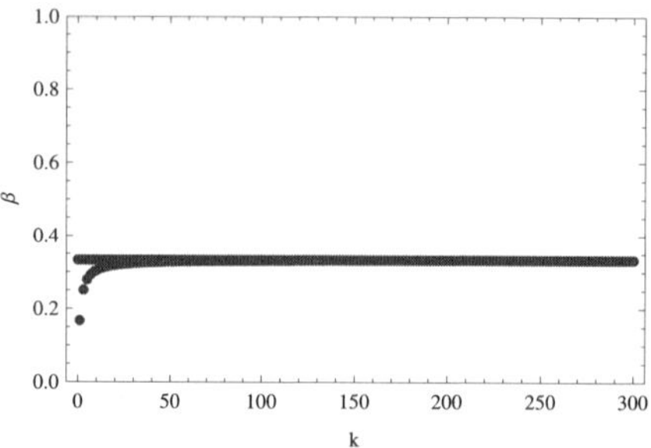

Fig. 5.3. A list plot of the ratio β to reveal the convergence of the HAM solutions for equation (5.27).

5.5.2. *A nonlinear Volterra differential-difference equation*

The differential-difference equations play a crucial role in modeling complicated physical phenomena such as particle vibrations in lattices, current flow in electrical networks and pulses in biological chains. One such equation is the famous nonlinear Volterra differential-difference initial value problem

$$u_n'(t) = u_n(t)(u_{n-1}(t) - u_{n+1}(t)), \quad u_n(0) = n, \tag{5.31}$$

which possesses the exact solution $u_n(t) = \frac{n}{1+2t}$, as mentioned by Wang *et al.* [35]. Without loss of generality, the above system is confined to the time domain [0,1].

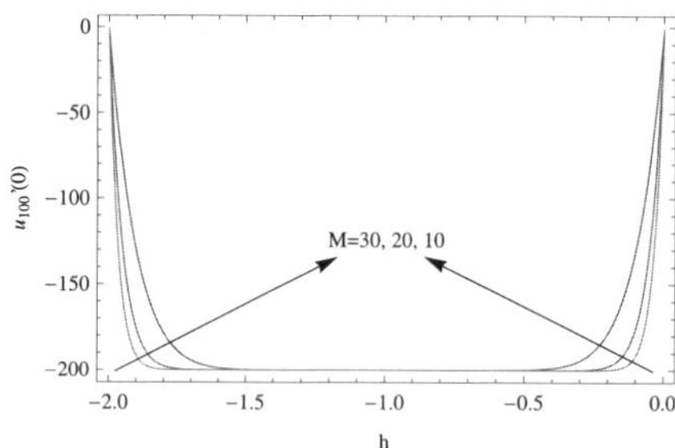

Fig. 5.4. Constant h-curves for equation (5.31).

The approximate solution of the Volterra equation (5.31) is obtained by means of the HAM using the following auxiliary linear operator, the initial guess and the auxiliary function

$$\mathcal{L} = \frac{d}{dt}, \quad u_{n,0}(t) = n - t, \quad H(t) = 1,$$

respectively. Using them, a kind of homotopy is constructed as described in § 5.2, but more details of the HAM regarding such difference equations can be found in [35]. By means of the exact solution of the inequality (5.18) for $u'_{100}(0)$, it is derived that the interval of convergence control parameter h is exactly [−2,0] for all n at all orders of approximation. The constant h-curve plots shown in Figure 5.4 also verify this outcome.

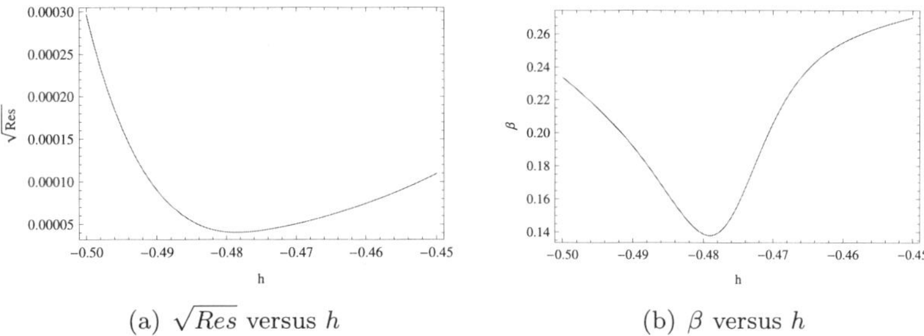

(a) $\sqrt{Res}$ versus h (b) β versus h

Fig. 5.5. Residual error and ratio for equation (5.31) at the 20th-order of HAM approximation.

Carrying out exact integration over $t \in [0, 1]$ for the fixed $n = 100$, the residual (5.20) and the ratio (5.25) (with $p = 2$) are depicted in Figures 5.5 (a–b) at the 20th-order of HAM approximation. It is fascinating to observe that the region of convergence is almost the same from the two methods, and more importantly the optimum value for the convergence control parameter h approaches the same limiting optimal value $h = -0.479$, as also verified in Table 5.5. It also indicates how fast the minimization is performed from the ratio as compared to the squared residual, especially for increasing order of approximations. Thus, the present ratio approach indeed constitutes a strong alternative to the classical squared residual method. Although, Figure 5.4 itself suggests the convergence of the homotopy method for $h = -1$, the residual $\sqrt{Res}$ and ratio β receive the values 1.0691×10^{13} and 3.8139, respectively. This obviously means that the so-called homotopy perturbation method will absolutely fail for the Volterra differential-difference equation (5.31).

Even though the accuracy of the HAM used for the present problem is clearly undoubted from Table 5.5 and Figures 5.4 and 5.5, list plots for the ratio β are further given in Figures 5.6 (a–c) at some different convergence control parameters h. It is clear that the initial behavior ends up with the settlement as the number of homotopy terms is increased. Hence, the accuracy is supported by the convergence of the HAM which can be safely used to gain higher order approximations with better performance.

Table 5.5. The optimum values of h and β evaluated from the minimum of residual and ratio for equation (5.31) (with CPU times in parenthesis).

M	2	6	10	14	20
h^a	−0.4131(2.77)	−0.4522(6.56)	−0.4654(19.41)	−0.4724(26.91)	−0.4785(100)
h^b	−0.4221(0.70)	−0.4567(2.14)	−0.4674(4.20)	−0.4733(6.34)	−0.4791(8.97)
β	0.06119	0.12078	0.13599	0.13904	0.13797

[a] From equation (5.20)
[b] From equation (5.25)

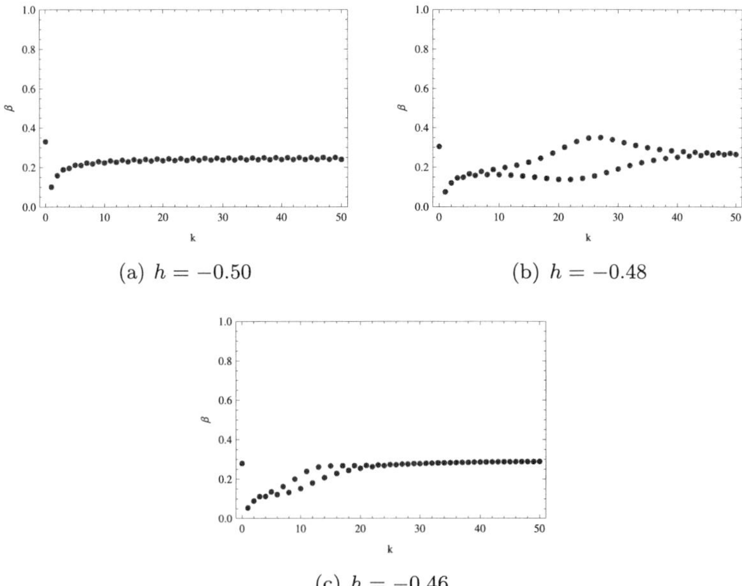

(a) $h = -0.50$ (b) $h = -0.48$

(c) $h = -0.46$

Fig. 5.6. List plots of the ratio β to reveal the convergence of the HAM solutions for equation (5.31).

5.5.3. *A nonlinear high-order Fredholm integro-differential equation*

The Kirchhoff type equation mathematically modeling the deflection of an extensible beam with hinged ends is given by the strongly nonlinear non-dimensional Fredholm integro-differential equation

$$u^{(4)}(t) - u''(t) - u''(t)\int_0^1 (u'(x))^2 dx = 1, \quad 0 < t < 1, \tag{5.32}$$

of fourth-order, accompanied with the boundary conditions

$$u(0) = u(1) = 0, \quad u''(0) = u''(1) = 0, \tag{5.33}$$

where $u(t)$ represents the static deflection of the beam, see [36].

For this highly nonlinear static beam problem we consider two different HAM approaches, with the corresponding auxiliary parameters;
Case 1:

$$\mathcal{L} = \frac{d^4}{dt^4}, \quad u_0(t) = 2\frac{t - 2t^3 + t^4}{53} \quad \text{and}$$

Case 2:

$$\mathcal{L} = \frac{d^4}{dt^4} - \frac{d^2}{dt^2}, \quad u_0(t) = \cosh(t) + \frac{1 - \cosh(1)}{\sinh(1)}\sinh(t) - (1/2)t^2 + (1/2)t - 1.$$

Regarding both cases, the homotopies are formed as described in § 5.2, with the difference here from the previous examples is that the nonlinear operator involves an integration as seen in (5.32).

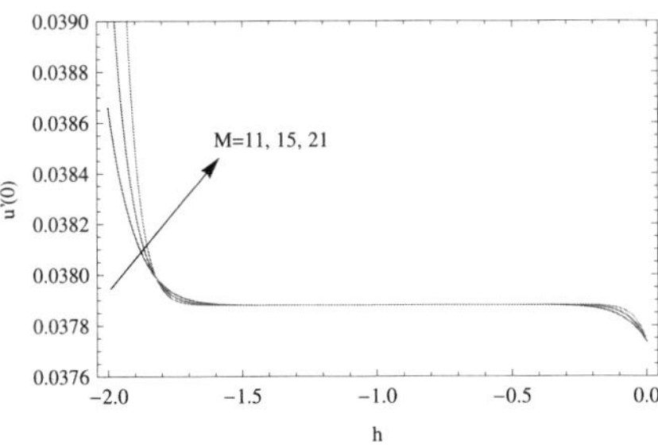

Fig. 5.7. Constant h-curves for equation (5.32).

For the first case, we first draw in Figure 5.7 the constant h-curves concerning the physical unknown $u'(0)$ at three different orders of approximations, $M = 11$, 15 and 21, respectively. Again the figure reflects that the convergence interval is $[\alpha,0]$, where α is close to -2, but it is uncertain how close to or far from -2. To completely sort out this matter, it is necessary to solve the inequality (5.17) for $u'(0)$ analytically, that yields the convergence intervals as

$$[-1.8293, 0], \quad [-1.8214, 0], \quad [-1.8180, 0],$$

respectively for the approximation orders as shown in Figure 5.7.

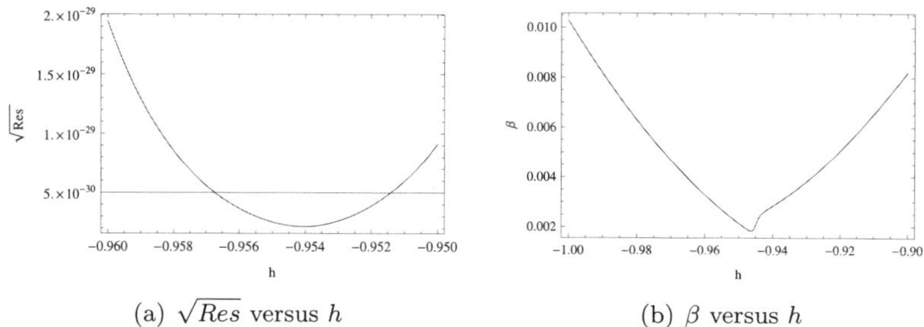

(a) $\sqrt{Res}$ versus h (b) β versus h

Fig. 5.8. Residual error and ratio for equation (5.32) at the 21st-order of HAM approximation.

Exact integration of the squared residual (5.20) is only practical up to $M = 12$ with 68.17 seconds spent, after which the CPU time increases enormously. Therefore, we prefer minimization through the discrete version (5.23) for the residual using 500 equally spaced points. However, no discrete integration is needed for the ratio, thus the exact integration from (5.25) (with $p = 2$) is implemented. At the 21st-order of HAM approximation, the best interval of convergence and also the optimum values are revealed in Figures 5.8 (a–b). It is seen that the optimal values are getting closer to $h = -0.95$ from both the classical squared residual and the proposed ratio methods, which are also summarized in Table 5.6 together with the consumed CPU times. Again for this nonlinear problem the newly-introduced ratio approach even with the exact integration is superior to the classical squared residual method.

The convergence of the HAM in Case 1 can also be conceived from the list plot for the ratio β exhibited in Figure 5.9 computed by means of the convergence control parameter $h = -0.95$. The fast limiting value of β, towards 0.00213, is the best indicator for the rapid convergence of the HAM, too.

As for Case 2, the computational cost to evaluate the homotopy terms

Table 5.6. The optimum values of h and β evaluated from the minimum of residual and ratio for equation (5.32) (with CPU times in parenthesis).

M	1	5	11	15	21
h^{a}	−0.9938(1.75)	−0.9636(11.98)	−0.9566(27.20)	−0.9551(39.42)	−0.9541(72.64)
h^{b}	−0.9091(0.85)	−0.9394(5.19)	−0.9442(9.64)	−0.9452(16.61)	−0.9464(35.25)
β	0.00012	0.00140	0.00167	0.00174	0.00183

[a] From equation (5.23)
[b] From equation (5.25)

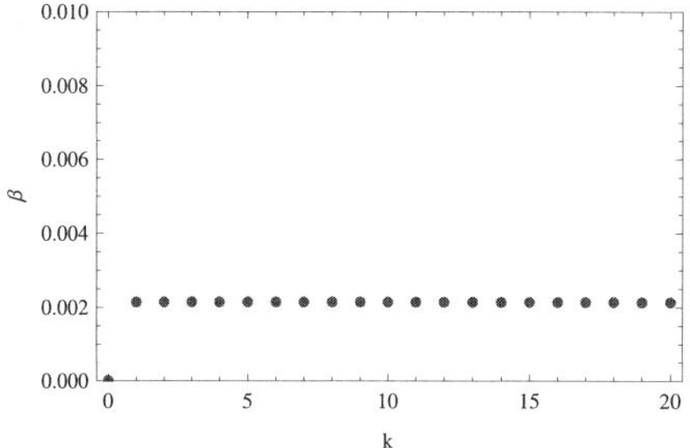

Fig. 5.9. A list plot of the ratio β to reveal the convergence of the HAM solution for equation (5.32).

is too expensive. To illustrate, although approximately 20 minutes are sufficient for evaluation of the whole terms up to $M = 21$ in Case 1, only 4 homotopy series terms are calculated over hours in Case 2. Nevertheless, it is very accurate as also concluded in [36].

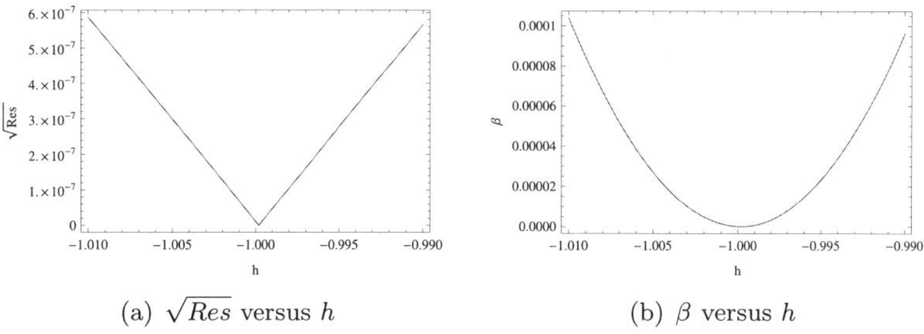

(a) $\sqrt{Res}$ versus h (b) β versus h

Fig. 5.10. Residual error and ratio for equation (5.32) at the 1st-order of HAM approximation.

Only two homotopy terms, i.e. the first-order homotopy approximation, give an exact residual $\sqrt{Res} = 1.34707 \times 10^{-10}$ and an exact optimal convergence control parameter $h = -0.999808$, see Figure 5.10 (a), which are the same values as those found in [36]. The minimization of the residual takes 47.69 seconds. On the other hand, the minimization of the ratio requires just 11.27 seconds to gain the same h, but with a practically vanishing β, see Figure 5.10 (b). This, of course, implies that the convergence of the HAM via the Case 2 is much accelerated as compared to the Case 1, which is also verified from the first four consecutive ratios of β tabulated in Table 5.7.

Table 5.7. First four ratios when $h = -1$.

k	0	1	2	3
β	4.09119×10^{-9}	3.68135×10^{-8}	6.54474×10^{-8}	8.59284×10^{-8}

Finally, the constant h-curve corresponding to the Case 2 is further shown in Figure 5.11. The interval of convergence as depicted in this figure is resolved exactly by means of the ratio (5.17), which leads to the exact interval $[-1.9996, 0]$.

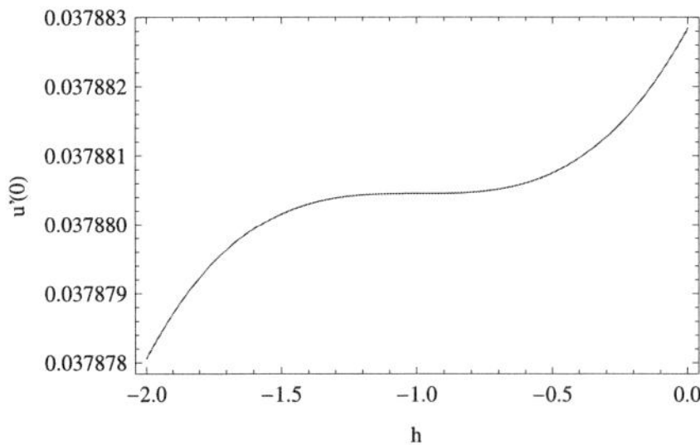

Fig. 5.11. Constant h-curve for equation (5.32).

5.5.4. *Some ordinary differential equations*

5.5.4.1. *Airy differential equation*

Consider now the second-order linear Airy differential equation recently treated with the HAM in [28]

$$u'' - tu = 0, \quad u(0) = Ai, \quad u(\infty) = 0, \tag{5.34}$$

where Ai is the value of Airy function at $t = 0$. For further information about the Airy equation (5.34) and its usage in science, please refer to [28] and references therein. Also, we are interested in the decaying solution, although the blowing Airy solution was also found using HAM in [28].

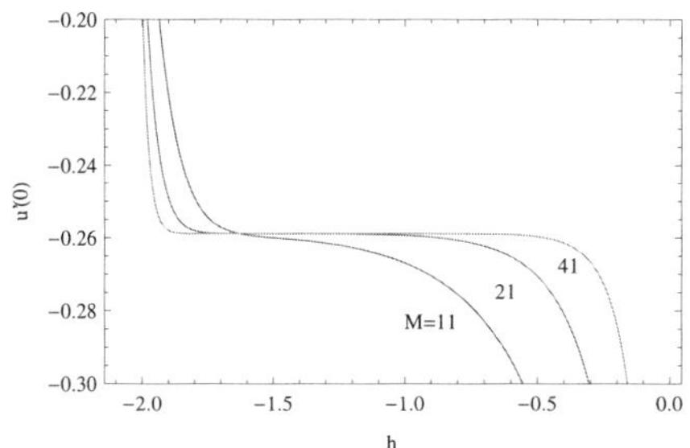

Fig. 5.12. Constant h-curves for equation (5.34).

The corresponding auxiliary linear operator, the initial guess and the auxiliary function in the frame of the HAM are chosen as follows:

$$\mathcal{L} = \frac{d^2}{dt^2} - c^2, \quad u_0(t) = Ai\, e^{-ct}, \quad H(t) = 1,$$

where $c = 3$ was found to be optimum by minimizing the residual within a few order of approximations. Together with these employing the HAM in § 5.2, the valid interval for the convergence control parameter h using $u'(0)$ is shown approximately as [−2,0] in Figure 5.12 at different orders of the HAM approximations. This interval is better predicted using the inequality (5.17) corresponding to the exact intervals [−2.048,0], [−2.025,0] and [−2.0124,0], respectively, from the 11th-order, 21st-order and 41st-order homotopy series approximations.

Based on the norm L^2, the interval of integration is divided into 100 equal discrete points in the interval [0,10]. Afterwards, the curves of the residual (5.23) and the ratio (5.26) versus h are plotted at different order of approximations $M = 6$, 8 and 10, respectively, to perceive the region of convergence and also the optimum values of h in Figure 5.13. It is found that the minimums of discrete squared residuals decrease in the region $-1.65 \leq h \leq -1.3$ shifting towards $h = -1.6$ as the order of approximation increases, which indicates that the homotopy series most quickly converges for such value of h. The strong indication is that the residual attains its minimum at $h = -1.6$. Therefore, the curves of the discrete residuals versus

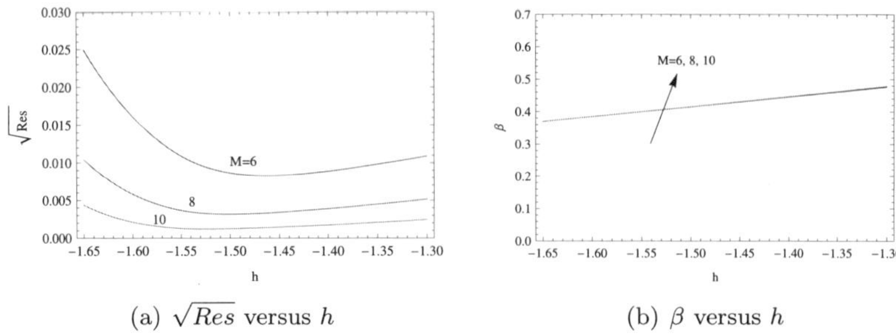

(a) $\sqrt{Res}$ versus h (b) β versus h

Fig. 5.13. Residual error and ratio for equation (5.34) at different order M of the HAM approximation.

h provide both the effective region of the convergence control parameter as well as the optimal value of h that gives the optimal homotopy series that converges fastest, as also emphasized by Liao [2]. For those values of h leading to optimum values from the residuals, the ratios remain almost constant signifying to the convergence for the h interval shown. In addition to this,

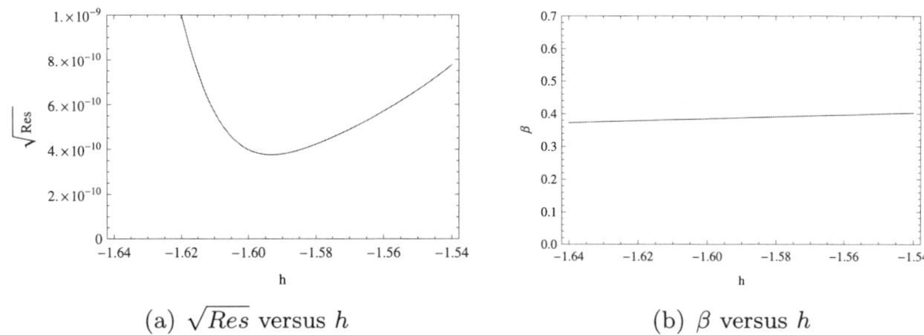

(a) $\sqrt{Res}$ versus h (b) β versus h

Fig. 5.14. Residual error and ratio for equation (5.34) at the 41st-order of the HAM approximation.

at the 41st-order of approximation, as supported by Figures 5.14 (a–b), the values of the convergence control parameter calculated from equations (5.23) and (5.26) almost coincidentally reside within the interval shown, so that the optimal h could be chosen as $h = -1.6$, as also indicated by Figure 5.12.

Hence, with this value in mind, the homotopy (8.2) generates a homotopy series (8.27) whose $u'(0)$, absolute errors and ratio β are tabulated in Table 5.8 at several order of approximations. As observed from the Ta-

Table 5.8. Values of $u'(0)$, absolute error err and ratio β calculated from the homotopy series corresponding to equation (5.34).

M	1	10	20	30	40
$u'(0)$	−0.228795857	−0.260777885	−0.258833058	−0.258819509	−0.258819405
err^{a}	1.309×10^{-1}	5.084×10^{-3}	4.014×10^{-5}	3.279×10^{-7}	2.715×10^{-9}
β	0.19631	0.38481	0.38466	0.38467	0.38467

[a]Equation (5.19)

ble 5.8, the convergence of the homotopy series to the exact solution takes place at a reasonably fast rate. Notice that the exact value of $u'(0)$ is −0.2588194038.

Just for the purpose of comparison, Table 5.9 shows the minimum of the residuals obtained from the exact integration (5.20) and the discrete one (5.23), the corresponding optimal values of h and also the used CPU time for these evaluations. According to the Table 5.9, both exact and discrete residuals result in very close results, but much less CPU time is needed by means of the discrete integration, especially for high-order approximations. Therefore, the discrete squared residual (5.23) can be reliably used to obtain the optimum values and the interval of convergence for h.

Table 5.9. Minimum of the exact and discrete residual errors and the corresponding optimal values of h with the used CPU time (in seconds) for (5.34).

M	$\sqrt{Res}^{a}$	h	CPU time	$\sqrt{Res}^{b}$	h	CPU time
2	0.18604	−1.3212	2.73	0.06209	−1.2549	2.32
5	0.03915	−1.4835	4.49	0.01344	−1.4344	5.32
10	0.00341	−1.5558	10.62	0.00121	−1.5230	11.55
13	0.00079	−1.5722	30.34	0.00028	−1.5441	15.99
15	0.00030	−1.5792	44.07	0.00011	−1.5537	19.18

[a]Equation (5.20)
[a]Equation (5.23)

The convergence of the homotopy series in case of $h = -1.6$ can be further confirmed by checking the ratio β, as shown in Figure 5.15 and Table 5.8. It is most likely that the ratio tends to 0.38467, remaining less than unity. This ensures the convergence of the HAM approximation for the current physical problem. Figure 5.16 clearly demonstrates how the convergence takes place towards the exact solution of Airy function by means of $h = -1.6$ (see also Figure 1 in [28], which was obtained using $h = -1.1$).

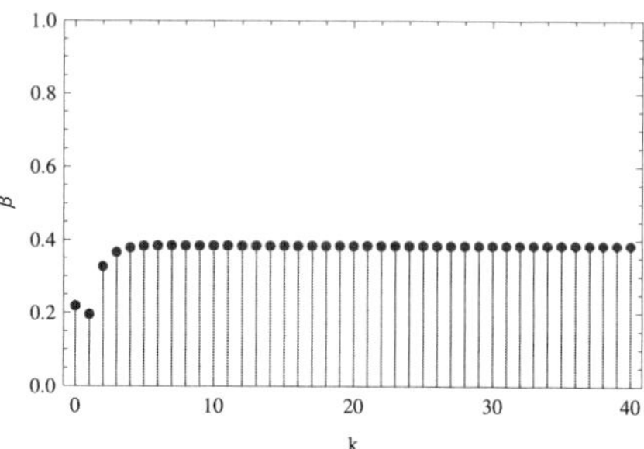

Fig. 5.15. A list plot of the ratio β to reveal the convergence of the HAM solutions for equation (5.34).

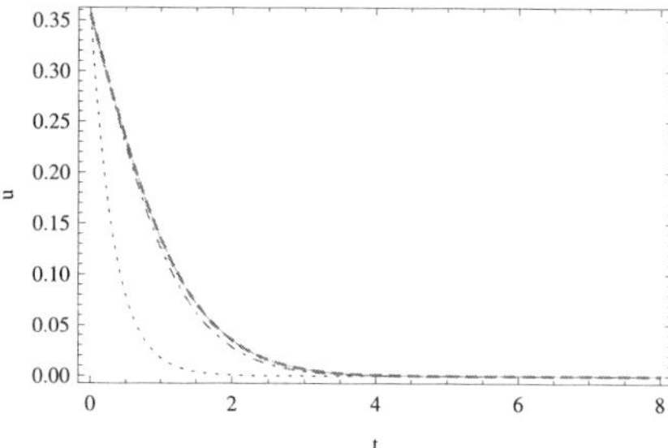

Fig. 5.16. Solution of Airy equation (5.34). Solid line: the exact solution; Dashed-line: the 20th-order HAM approximation; Dashed-dotted line: the 5th-order HAM approximation; Dotted line: the initial approximation.

Finally, Figure 5.17 demonstrates that the divergence of the HAM approximation becomes inevitable if the value of $h = -2.1$ is assigned wrongly *outside* the interval of convergence, as worked out for this example. When such a choice is made, in consistent with the Theorem 5.1, a limiting value of ratio greater than one causes the divergence of the series. This once again indicates the importance of the so-called convergence control parameter, which differs the HAM from all other analytic approximation methods.

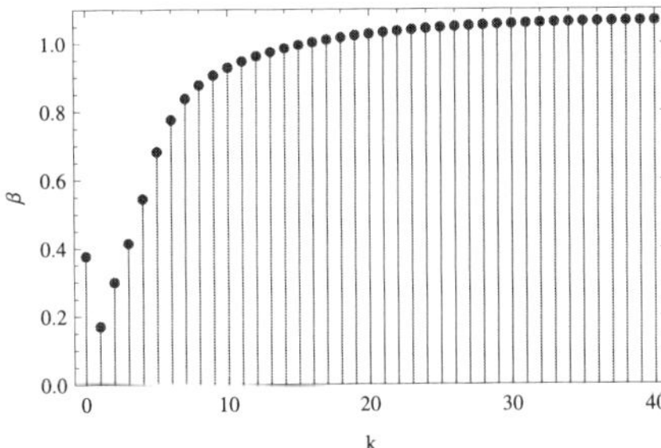

Fig. 5.17. A list plot of the ratio β to reveal the divergence of the HAM approximations in case of $h = -2.1$ for equation (5.34).

5.5.4.2. *A first order nonlinear differential equation*

Let us consider the first-order nonlinear differential equation

$$u' + u^2 = 1, \quad u(0) = 0, \tag{5.35}$$

which was used by Liao [3] to describe the basic ideas of the HAM. Equation (5.35) is known to govern the free-falling body problems. Its exact solution $u_e(t) = \tanh t$ will be used for comparison purposes with the solutions obtained by means of the HAM. We actually take into account the two different cases. In the first case, equation (5.35) is solved over the finite domain $t \in [0, 1]$. In the second case, we consider the physical problem (5.35) over the semi-infinite region $t \in [0, \infty)$. In both situations, we employ the following initial guess, auxiliary linear operator and auxiliary function auxiliary parameters

$$u_0(t) = 1 - e^{-2t}, \quad \mathcal{L} = \frac{d}{dt} + 2, \quad H(t) = 1,$$

respectively. We initially use the quantity $u'(0)$ to evaluate the constant h-curves at the orders of approximations $M = 11$, 21 and 41, respectively. According to Figure 5.18, the region of convergence is best suited to $h \in [-2, 0]$. This is indeed the case from the relation (5.17), which fortunately leads to the exact formula $\beta = |h + 1|$. Table 5.10 presents evolution of the ratio (5.15), using the exact formula (5.25) with L^1 norm and physical domain as $[0,\infty)$ for a variety of the preassigned values of the convergence control parameters h for the problem (5.35). Additionally, the last column of Table 5.10 gives the error defined by (5.19), taking into account the exact solution $u_e(t) = \tanh t$ for the problem (5.35). Data displayed in Table 5.10 clearly explains why the HAM generates completely convergent

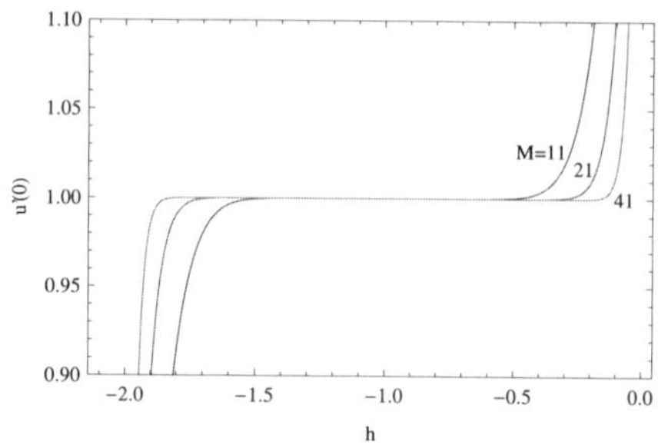

Fig. 5.18. Constant h-curves for equation (5.35).

Table 5.10. The evolution of the ratio β (5.15) for equation (5.35).

h	$M = 1$	$M = 10$	$M = 20$	$M = 30$	$M = 40$	err^{a}
−1.5	0.00001	0.42690	0.45457	0.46875	0.47619	2.8411×10^{-17}
−1.4	0.06667	14.795	0.44736	0.38142	0.38143	2.2129×10^{-21}
−1.3	0.13333	0.32498	0.33436	0.33899	0.34159	1.2674×10^{-21}
−1.2	0.20000	0.36561	0.38152	0.38736	0.39040	3.7223×10^{-19}
−1.0	0.33333	0.45833	0.47727	0.48438	0.48810	5.1726×10^{-15}
−0.5	0.66667	0.70189	0.71969	0.72824	0.73308	2.3609×10^{-7}

[a]Equation (5.19)

homotopy series solution to the problem (5.35) for the chosen parameters h. According to the Table 5.10, the homotopy series given by different values of h has different rate of convergence. It is found that the homotopy series converges fastest for the values of h between -1.4 and -1.3, and thus an optimum convergence control parameter exists within this interval. Note that, in practice, it is unnecessary to locate the exact value of the optimal convergence control parameter, instead it is enough to use a value close to it.

The optimal values of convergence control parameter h are further calculated from the exact formulae, respectively for residual (5.21) and for ratio (5.25) (with $p = 1$), taking $M = 22$ as $h = -1.3567$, see Figure 5.19 (a), and $h = -1.2678$, see Figure 5.19 (b) for the finite domain case. Moreover, taking $M = 30$, the infinite domain case results in the optimums which are evaluated as $h = -1.3947$, see Figure 5.20 (a) and from the ratio as $h = -1.3253$, see Figure 5.20 (b). When the number of homotopy terms increases, it is time consuming to evaluate the integrals for both finite and infinite domain cases from (5.21), but a straightforward evalua-

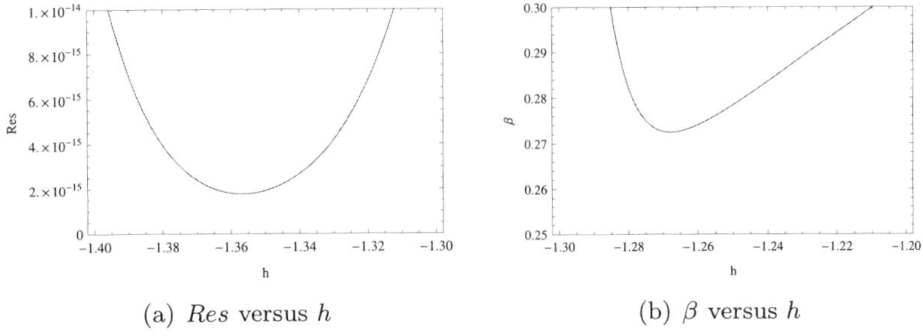

(a) *Res* versus h (b) β versus h

Fig. 5.19. Residual error and ratio for equation (5.35) with finite domain at the 22nd-order of HAM approximation.

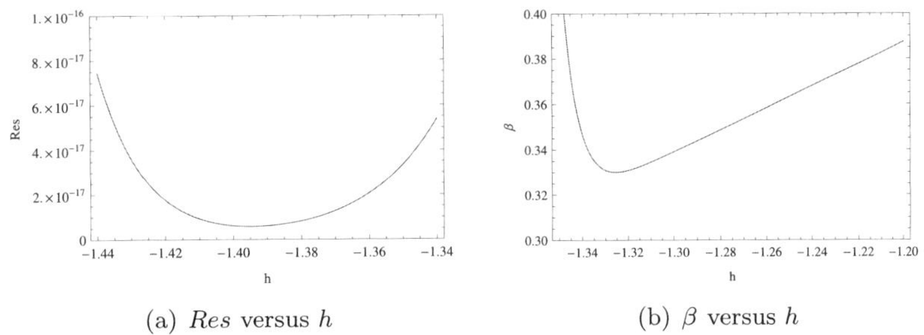

(a) *Res* versus h (b) β versus h

Fig. 5.20. Residual error and ratio for equation (5.35) with infinite domain at the 30th-order of HAM approximation.

tion from (5.25) still yielded the optimums $h = -1.2714$ for the finite and $h = -1.3273$ for the infinite domain cases, even for $M = 40$, which strongly indicates the reason of the faster converge of the homotopy series near this value, see Table 5.10. Table 5.11 further reveals the optimum values of h evaluated at different orders of approximation M. Note that, due to above reason, numerical integration is preferred with the residual (5.24), whereas the ratio β is exactly computed from (5.25). During the numerical scheme, the infinity is truncated at 5 and the region is divided into 400 equally-spaced points. It is anticipated that as the order of approximation is increased, both the residual and ratio converge to the same optimum value for h.

To better admire the advantage of ratio β in terms of computational efficiency, Table 5.12 demonstrates the CPU times to evaluate the minimums of residuals and ratios computed exactly from equations (5.21) and

Table 5.11. The optimum values (together with CPU times) of h from the residual and from the ratio for equation (5.35) at different orders of approximation M.

M	10	20	30	40
h^a	−1.3598(57.45)	−1.3373(150.01)	−1.3323(411.70)	−1.3305(1069.34)
h^b	−1.3101(12.98)	−1.3214(8.721)	−1.3253(31.761)	−1.3273(97.766)

[a]Equation (5.21)
[b]Equation (5.25)

(5.25), respectively, for the infinite domain case. Special attention should be given to CPU times, which clearly shows that the approach introduced in this chapter is essential in terms of computational time, if exact integration is demanded. Even the CPU time from the discrete residual takes longer than the exact evaluation of the integrals from the ratio, as clearly shown in Table 5.11. So, it is more economical to find the optimum h from the ratio (5.25) than from the residual (5.21).

Table 5.12. CPU times to obtain optimum values of h from the residual and from the ratio for equation (5.35) at different orders of approximation M.

M	2	6	10	20	30
CPU time [a]	0.624	10.20	4.867	227.04	724.56
CPU time [b]	0.624	3.915	12.98	8.721	31.761

[a] Equation (5.21)
[b] Equation (5.25)

Moreover, the uniform validity region of the homotopy series solution (8.27) can also be analytically resolved for the special value of $h = -1$. In this particular case, considering the absolute value norm, the homotopy terms in the homotopy series (8.27) satisfy the ratio

$$\left|\frac{u_{n+1}(t)}{u_n(t)}\right| = \left|\frac{1}{2}(1-e^{-2t})\right|. \tag{5.36}$$

Hence, concerning the equation (5.35), Theorem 1 assures the convergence of the corresponding homotopy series (8.27) for all values of t valid in the interval $t > -\frac{\ln 3}{2}$, that covers the entire physical domain. With the fixed value $h = -1.27$ for the finite and $h = -1.3$ for the infinite domain cases, list plots of β for the convergence history are demonstrated in Figures 5.21 (a–b). It seems that the even and odd homotopy series terms approach the same limit at different rates, which causes the oscillations in the ratios in the figures. The limiting values turn out to be 0.28 and 0.34, respectively which are sufficient to explain the convergency of the HAM approximations applied to the physical problem defined in (5.35).

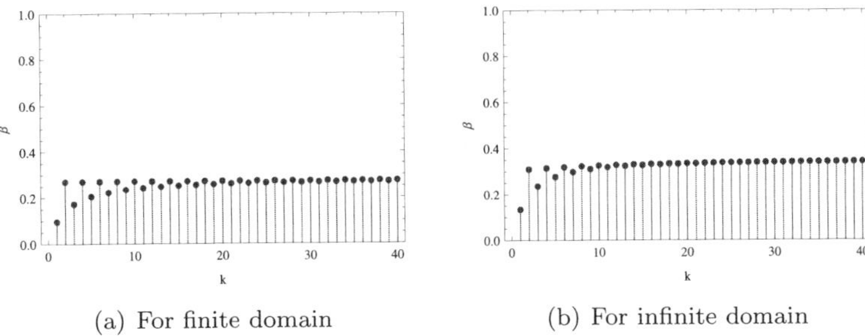

(a) For finite domain (b) For infinite domain

Fig. 5.21. List plots of the ratio β to reveal the convergence of the HAM solutions for equation (5.35).

5.5.4.3. *A parameterized nonlinear differential equation*

Let us reconsider now the first-order nonlinear differential equation mentioned in the above example

$$u' + u^2 = 1, \quad u(0) = 0, \tag{5.37}$$

to be treated over the finite interval $[0,\alpha]$, where $\alpha \in R$ with $\alpha = 1$, without loss of generality. Although such a finite interval problem can be solved as in the previous example, we differ it from that by assuming that, at the end point $\alpha = 1$, the solution takes the value λ, i.e. $u(1) = \lambda$, where λ is unknown *in priori*, called the *pseudo parameter*. Using the transformation $u(t) = \lambda v(t)$, equation (5.37) is transformed into a parameterized differential equation

$$\lambda v' + \lambda^2 v^2 = 1, \quad v(0) = 0, \quad v(1) = 1. \tag{5.38}$$

In the system described by (5.38), both v and λ are expanded into homotopy series (8.27) via the homotopy built in (8.2). Each homotopy term for the pseudo parameter λ is then evaluated using either of the boundary conditions in (5.37). It is remarked that only the zeroth-order term λ_0 for λ satisfies a quadratic algebraic equation, otherwise the rest of the terms satisfy linear equations. Thus, the homotopy solution for (5.38) is obtained by virtue of the following initial guess, auxiliary linear operator and auxiliary function

$$v_0(t) = t, \quad \mathcal{L} = \frac{d}{dt}, \quad H(t) = 1.$$

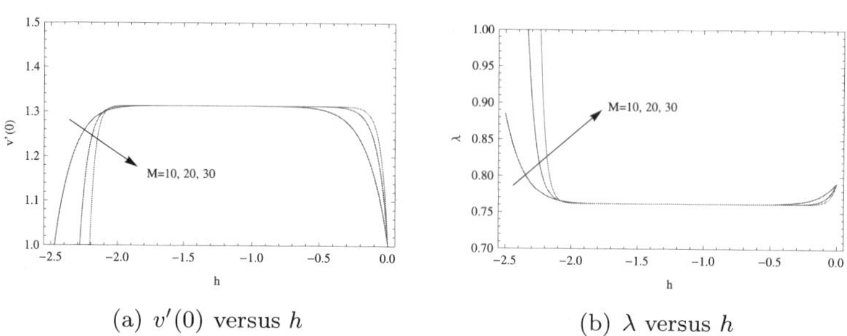

(a) $v'(0)$ versus h (b) λ versus h

Fig. 5.22. Constant h-curves for equation (5.38).

The constant h curves for the unknown variable $v'(0)$ and for the unknown parameter λ at the orders of approximations $M = 10$, 20 and 30, respectively, are displayed in Figures 5.22 (a–b). Note that the exact value of λ is $\tanh(1) = 0.761594$, which is truly estimated by the HAM approximations, as shown in Figure 5.22 (b). The intervals of h for convergence shown by the graphs extend to the point $h = -2$, suggesting that the convergence interval is close to $h \in [-2, 0]$. On the other hand, the exact interval can only be figured out by solving analytically the inequalities in (5.17) and (5.18), that yield $h \in [-2.024, 0]$ at 30th-order of approximation.

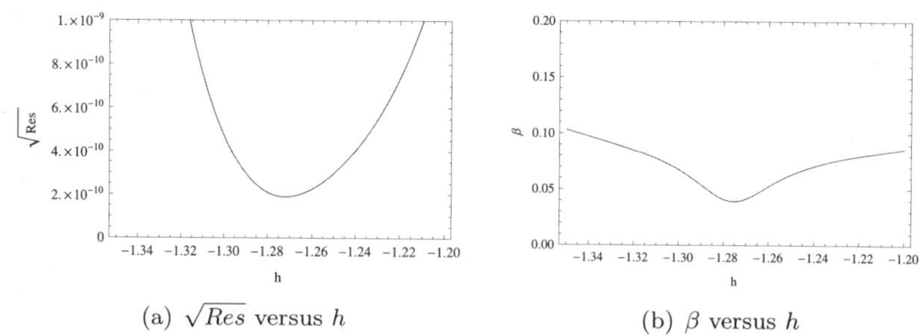

(a) $\sqrt{Res}$ versus h (b) β versus h

Fig. 5.23. Residual error and ratio for equation (5.38) at the 41st-order of HAM approximation.

Employing the exact residual (5.20) and ratio (5.25) (with $p = 2$), the minimization of them results in the optimums demonstrated in Figures 5.23

(a–b) at the 41st-order of approximation. Moreover, several optimum values are tabulated in Table 5.13 at various orders of approximations. It is apparent again that both the formerly-used squared residual approach in the literature and the present ratio approach give nearly the same optimum values of h even at small orders of approximation, unlike the HAM technique in the previous example. The optimums seem to converge to $h = -1.27$ in the large truncation limit.

Table 5.13. The optimum values of h and the ratio β evaluated from the minimum of residual and ratio for equation (5.38).

M	2	6	10	12	14
h^{a}	−1.2955	−1.2818	−1.2753	−1.2733	−1.2720
h^{b}	−1.1607	−1.2817	−1.2769	−1.2764	−1.2754
β	0.06096	0.04154	0.02705	0.03097	0.03912

[a]Equation (5.20)
[b]Equation (5.25)

The convergence of the HAM is ensured by Theorem 5.1 following the ratios related to the variables (5.15) or parameters (5.18) existing in the considered problem. For this purpose, list plots of β when $h = -1.31$ are demonstrated in Figures 5.24 (a–b) for both v and λ. It is clear that both ratios tend to finite limits, less than unity. Although both are good in explaining the convergency of the HAM approximations of the physical problem defined by (5.37) and (5.38), it needs much more CPU time to gain the ratio about v. For example, 2400 seconds (using exact integration) are needed to calculate the ratios in Figure 5.24 (a), whereas only 0.031 seconds for Figure 5.24 (b). Thus, consistent with the statements made in Remark 3, the proposed ratio approach together with introducing an unknown parameter into the considered physical problem works very efficiently for the present problem. Finally, it is found that keeping the original equation (5.37) over the interval [0,1] with the same auxiliary parameters does not give converging ratios as in Figures 5.24 (a–b), but instead, it yields wildly oscillatory ratios for the whole convergence control parameters h. This illustrates a substantial advantage of plugging an unknown parameter into the equations governing the system.

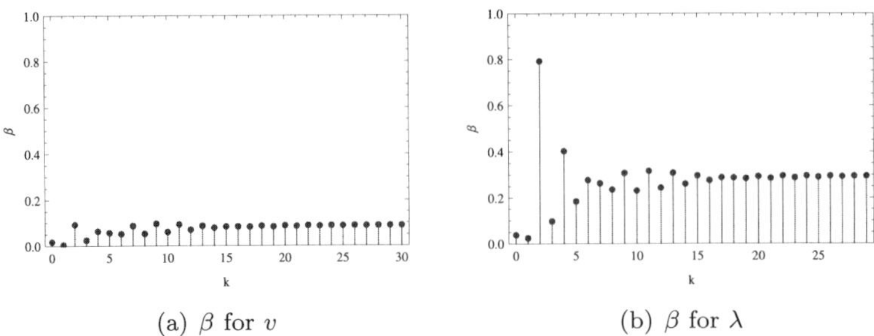

(a) β for v (b) β for λ

Fig. 5.24. List plots of the ratio β to reveal the convergence of the HAM solutions for equation (5.38).

5.5.4.4. *Undamped Duffing equation*

The second-order nonlinear undamped Duffing equation

$$u'' + u + \epsilon u^3 = 0, \quad u(t=0) = A, \quad u'(t=0) = 0, \tag{5.39}$$

governs the oscillator systems [23], where A represents the amplitude of the oscillations and ϵ denotes the strength of the nonlinearity. Using a suitable transformation $\tau = \omega\, t$, where ω is the frequency of the oscillations, equation (5.39) becomes

$$\omega^2 u'' + u + \epsilon u^3 = 0, \quad u(0) = A, \quad u'(0) = 0. \tag{5.40}$$

Note that ω is treated as unknown here, which depends on the embedding parameter p, as mentioned by Liao [2]. To gain the approximations of the solution of (5.40) in the frame of the HAM, we choose the initial guess, the auxiliary linear operator and the auxiliary function in the following manner

$$u_0(\tau) = A\cos\tau, \quad \mathcal{L} = \omega_0^2\Big(\frac{d^2}{d\tau^2} + 1\Big), \quad H(\tau) = 1,$$

where ω_0 and other approximations ω_i regarding

$$\omega = \sum_{i=0}^{\infty} \omega_i$$

are determined by avoiding the appearance of secular terms during the solution process of homotopy terms in (8.27). For more details, please refer to [24].

For the particular choices of the parameters $A = \epsilon = 1$, the interval of convergence at the 10th-order, 20th-order and 30th-order of approximation is computed using $u''(0)$ and ω for the h-curve, as shown in Figures 5.25

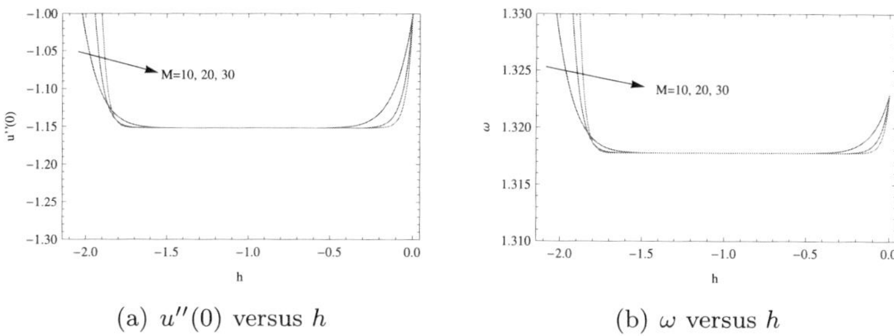

(a) $u''(0)$ versus h

(b) ω versus h

Fig. 5.25. Constant h-curves for equation (5.40).

(a–b), respectively. It seems that the interval of convergence is $[-1.8,0]$, which appears to narrow down from the left bound as the order of approximation increases. In fact, a justification is made using (5.17) for $u''(0)$ that results in an exact interval $[-1.8041,0]$ and also using (5.18) for ω that gives the interval $[-1.7903,0]$, where both are evaluated at the 31st-order of approximation. It is found that the optimal value of h calculated from direct numerical treatment of equations (5.20–5.22) (taking the integration interval as $[0,500]$) at the 31st-order of approximation is about $h = -1$, whose effects on the frequency and the absolute residual error at different orders of approximations are summarized in Table 5.14. It is clear that the higher the order of approximation, the better the approximation of frequency, and the faster the absolute residual error decays.

Table 5.14. The HAM approximations of ω and the corresponding Res of equation (5.40) in case of $A = \epsilon = 1$ when $h = -1$.

M	2	10	20	30
ω	1.317803946	1.317776065	1.317776065	1.317776065
Res[a]	4.7575×10^{-1}	2.0460×10^{-8}	3.9128×10^{-17}	1.0825×10^{-26}

[a]Equation (5.21)

A further assessment on the convergence of the homotopy series can be implemented by checking the ratios (5.15) for u and (5.18) for ω when $h = -0.98$, as shown in Figures 5.26 (a–b). Both ratios of β obviously tend to a finite value of approximately 0.15, which indicates that the homotopy approximations converge to the solution of the Duffing oscillator problem. In fact, Figures 5.26 (a–b) suggest that keeping track of the ratio of ω is sufficient. Moreover, it is unnecessary to evaluate the integrals for the squared residuals. This example hence approves the advantage of the present approach of using ratio, instead of the squared residual error.

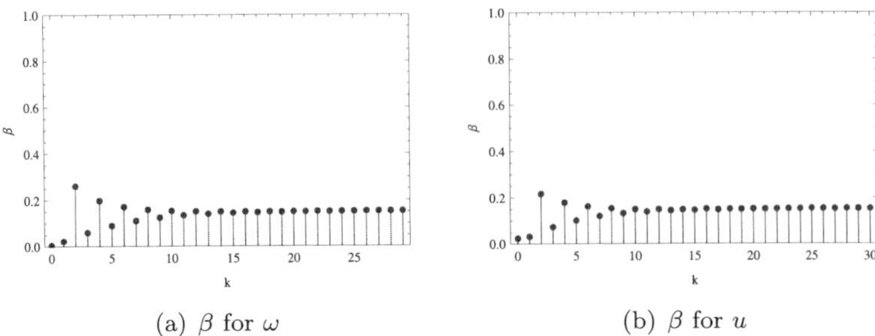

(a) β for ω (b) β for u

Fig. 5.26. List plots of the ratio β to reveal the convergence of the HAM solutions for equation (5.40) by means of $h = -0.98$.

In fact, making use of ω at different orders of approximation, optimum values of the convergence control parameters h can be gained easily by minimizing the ratio β in the homotopy expansion of ω in (8.27). This is implemented in Table 5.15. This strongly indicates that the ratio approach points to an optimal value close to $h = -1$.

Table 5.15. The optimum values of h evaluated by minimizing ratio (5.18) for ω.

M	6	12	16	20	24	30
h	-1.00096	-1.00056	-1.00044	-1.00036	-1.00031	-1.00025

A proper value of h can further be determined by restricting the domain of integration in (5.40) only to $[0,2\pi]$ (which might be sufficient for the desired physical solution), as supposedly taken in Liao's residual expressions (2.70) and (2.72) on page 41 in [2]. To illustrate the similar occurrence of optimums from both the residual and the ratio approach for the variable u, Figures 5.27 (a–b) are drawn only with the 20th-order of HAM approximation, with 100 equidistant points. The optimums are well predicted again by means of both methods. Making use of the discrete versions for both the residual and the ratio approach involving u, the optimum values of convergence control parameter h, the minimum ratio β and the corresponding values of CPU time are eventually sketched on Table 5.16. Consistent with the above findings, the optimal values limits to -1 for large M, the order of approximation. Furthermore, it is found once again that the ratio approach evaluates the optimums faster.

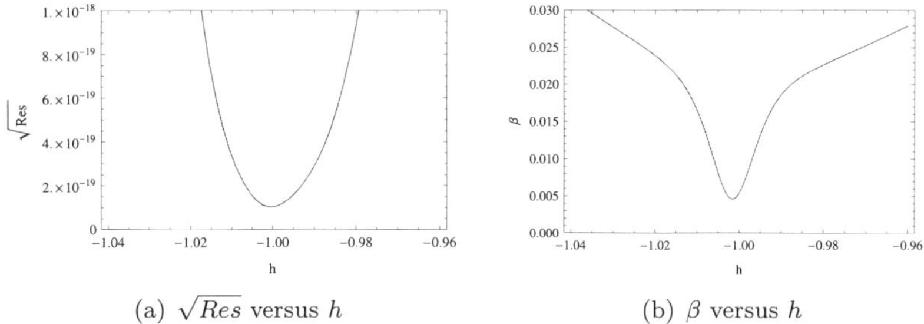

(a) $\sqrt{Res}$ versus h (b) β versus h

Fig. 5.27. Residual error and ratio for equation (5.40) at the 20th-order of HAM approaximation.

Table 5.16. The optimum values of h, the ratios β and the CPU times (seconds) from the residual and ratio approach for equation (5.40) at different orders of approximation M.

M	h^{a}	CPU times	h^{b}	β	CPU times
2	−1.0021	4.025	−1.0087	0.0002	1.357
5	−1.0010	15.12	−1.0041	0.0018	7.04
10	−1.0008	45.04	−1.0028	0.0029	23.34
20	−1.0004	78.20	−1.0015	0.0046	54.43
25	−1.0003	79.33	−1.0012	0.0053	61.85
30	−1.0003	82.90	−1.0011	0.0057	74.76

[a]Equation (5.23)
[b]Equation (5.26)

Finally, Figures 5.28 (a–d) show the uniform convergence of the homotopy approximations employed here, with the convergence control parameter $h = -1$ even for very moderate numbers of homotopy series truncation, compared to the numerical solution. It should be noticed that although the above analysis was implemented by restricting the domain to smaller intervals, the HAM is still able to extract uniformly valid solutions over reasonably larger intervals demonstrated in Figures 5.28 (a–d), such as $t \in [0, 30000]$.

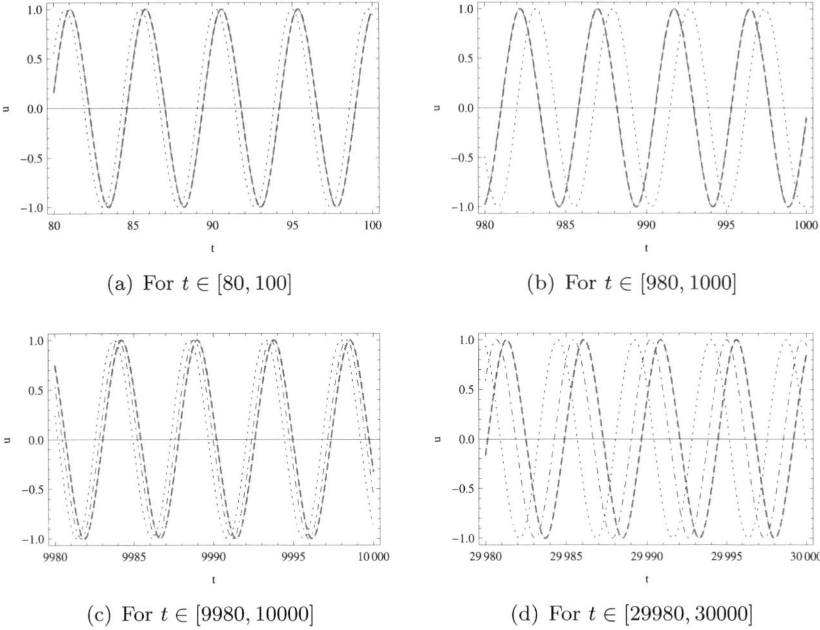

(a) For $t \in [80, 100]$

(b) For $t \in [980, 1000]$

(c) For $t \in [9980, 10000]$

(d) For $t \in [29980, 30000]$

Fig. 5.28. Snapshots from the solution of undamped Duffing equation (5.39). Solid line: the exact solution; Dashed-line: 5th-order homotopy approximation; Dashed-dotted line: 2nd-order homotopy approximation; Dotted line: the initial approximation.

5.5.4.5. *Damped Duffing equation*

Let us consider the damped Duffing equation

$$u'' + \alpha u' + \beta u + \epsilon u^3 = 0, \quad u(0) = A, \quad u'(0) = 0, \tag{5.41}$$

where α is the damping coefficient. Without loss of generality, let us consider the case $\alpha = 2$, $\beta = 5$, $A = \epsilon = 1$ in the following analysis. Different from the undamped case mentioned above, there is no need to transform the original variables. The auxiliary initial guess, the auxiliary linear operator and the auxiliary function corresponding to the homotopy (8.2) are as follows:

$$u_0(t) = \frac{1}{2}e^{-t}[2\cos(2t) + \sin(2t)], \quad \mathcal{L} = \frac{d^2}{dt^2} + 2\frac{d}{dt} + 5, \quad H(t) = 1.$$

Evaluating the ratio β in (5.17) for $u''(0)$ results in the exact formula $\beta = |1 + h|$, which means that the convergence control parameter h can

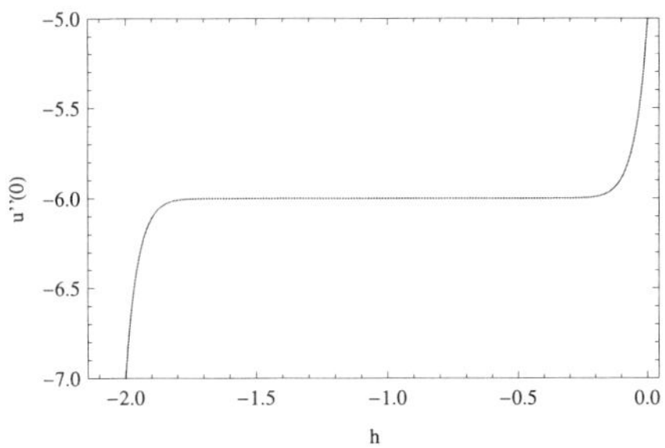

Fig. 5.29. Constant h-curve for equation (5.41).

be taken within the interval $[-2{,}0]$. Figure 5.29 also approves this reality, which is plotted at the 21st-order of HAM approximation.

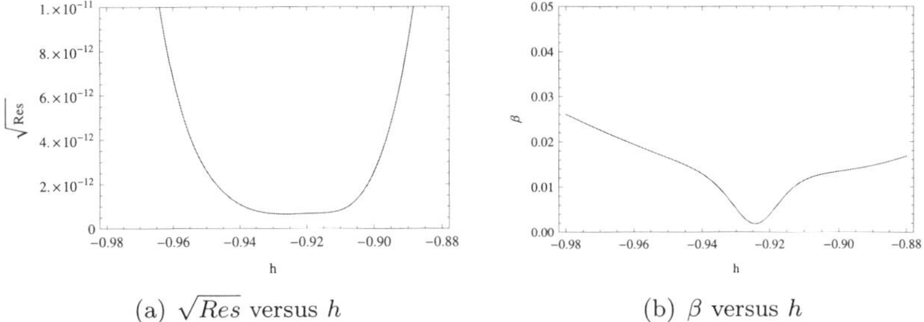

(a) $\sqrt{Res}$ versus h (b) β versus h

Fig. 5.30. Residual error and ratio for equation (5.41) at the 11st-order of the HAM approximation.

As shown in Figures 5.30 (a–b), the discrete residual computed from (5.23) and the ratio computed from (5.26) (with $p = 2$) at the 11th-order of approximation clearly indicate that the optimal convergence control parameter lies in the interval $h \in [-0.95, -0.90]$. Actually, the optimal values at each order of homotopy truncation are tabulated in Table 5.17. Optimums from both approaches are observed to asymptote to the same limiting value $h = -0.93$ as the approximation order increases. Moreover, the ratio β is so small that the convergence of the homotopy approximations given by choosing the optimum values of h from Table 5.17 is doubtless.

The convergence of the HAM approximations to the exact solution of the damped Duffing equation can be further visualized from Figure 5.31, which demonstrates the ratios for $h = -0.95$. In agreement with Table 5.17,

Table 5.17. The optimum values of h and the ratio β evaluated from the minimums of residual and ratio for equation (5.41).

M	2	6	10	14	20
h^{a}	-0.9472	-0.9315	-0.9251	-0.9296	-0.9282
h^{b}	-0.8771	-0.9056	-0.9243	-0.9161	-0.9253
β	0.01425	0.01446	0.00188	0.01056	0.00590

[a] From equation (5.23)
[b] From equation (5.26)

a very small limiting value of the ratio β shown in Figure 5.31 reveals the convergence of the HAM.

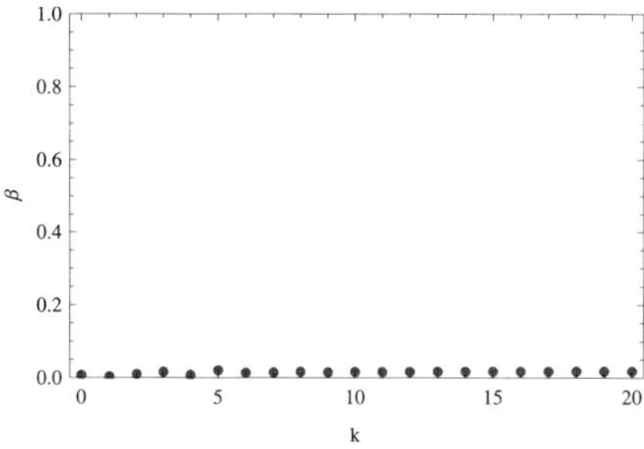

Fig. 5.31. A List plot of the ratio β to reveal the convergence of the HAM solutions for equation (5.41).

Figures 5.32 (a–b) exhibit the exact solution and the 21st-order HAM approximations given by $h = -0.92$, together with their absolute error. The convergence take places at a considerable rate even for the 5th-order HAM approximation, as revealed in Figure 5.32 (a). The uniform validity of the HAM for the present problem over the entire semi-infinite domain is also confirmed by the absolute error displayed in Figure 5.32 (b).

5.5.4.6. *Thomas–Fermi equation*

Let us consider now one of the most important nonlinear ordinary differential equation that occurs in mathematical physics, namely the Thomas–Fermi equation [27, 37]:

$$u'' = \sqrt{\frac{u^3}{x}}, \quad u(0) = 1, \quad u(\infty) = 0, \tag{5.42}$$

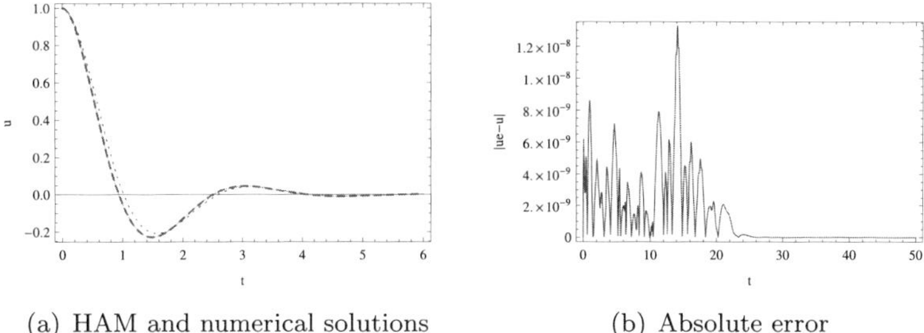

(a) HAM and numerical solutions (b) Absolute error

Fig. 5.32. (a) Solutions of damped Duffing equation (5.41). Solid line: the numerical solution; Dashed line: the 5th-order HAM approximation; Dotted line: the initial approximation. (b) Absolute error of the 21st-order HAM approximation.

which has a singularity at $x = 0$ since $u''(0) \to \infty$. So far, the accurate value of $u'(0)$ is -1.588071022611375313, given by [38] using Hankel–Padé method.

Although the HAM possesses several freedoms in it, choosing less useful auxiliary parameters may not always yield satisfactory series approximations to the exact solution. Thomas–Fermi equation is one such strong nonlinear differential equation that requires much care while benefiting from the advantages of the HAM. Otherwise, the accuracy of the approximations might be far from the desired, as mentioned by Liao [3]. Being alerted about this fact, in order to solve the Thomas–Fermi equation (5.42), we first convert it into more conventional form following Zhao *et al.* [37]. Employing the transformations

$$u(x) = g^2(t), \quad t = 1 + \lambda\sqrt{x}, \tag{5.43}$$

equation (5.42) turns out to be

$$\lambda^3 g g''' + 3\lambda^3 g' g'' - 6(t-1)g^2 g' - 4g^3 = 0, \tag{5.44}$$

subject to the boundary condition

$$g(1) = 1, \; g'(1) = g(\infty) = 0. \tag{5.45}$$

To approximate the solution of (5.42) and hence (5.44), the optimal initial guess, the auxiliary linear operator and the auxiliary function were supplied in [37], which we make use of them here as

$$g_0(t) = \frac{4+\epsilon}{t^3} - \frac{3+2\epsilon}{t^4} + \frac{\epsilon}{t^5},$$

$$\mathcal{L} = t^3 \frac{\partial^3}{\partial t^3} - 5t^2 \frac{\partial^2}{\partial t^2} - 100t \frac{\partial}{\partial t} - 180,$$

$$H(t) = \frac{t^4}{\lambda^3},$$

where $\lambda = 5/16$ and $\epsilon = -86/17$. These constants and the auxiliary linear operator were extracted by minimizing the squared residual of the governing equation at some low order of homotopy approximations, as mentioned by Zhao *et al.* [37] in details.

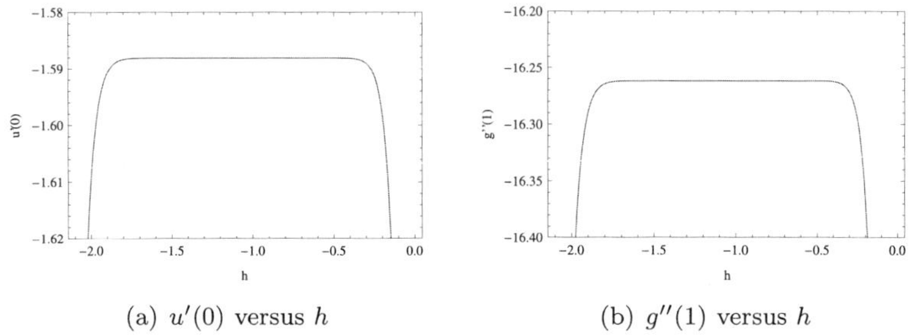

(a) $u'(0)$ versus h (b) $g''(1)$ versus h

Fig. 5.33. Constant h-curves for $u'(0)$ in (5.42) and $g''(1)$ in (5.44).

The h-curves of $u'(0)$ and $g''(1)$ at the 21st-order of the HAM approximation are shown in Figures 5.33 (a–b). The interval of convergence is anticipated to be about $h \in (-2, 0)$ from the Figures 5.33 (a–b). A better estimation of the convergence interval is obtained by solving the equation $|g''_{21}(1)/g''_{20}(1)| < 1$, which gives $h \in [-1.8738, 0)$.

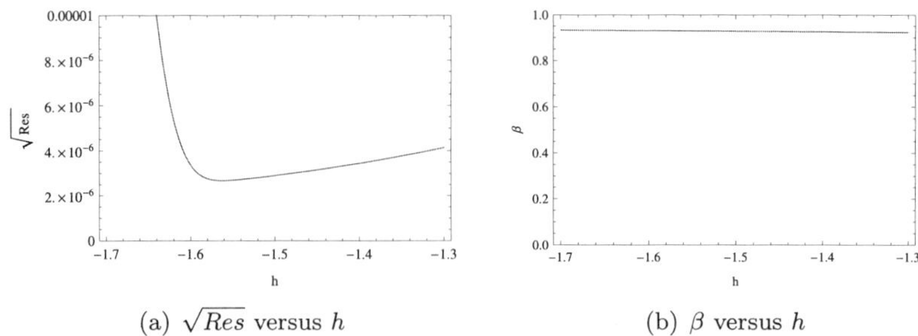

(a) $\sqrt{Res}$ versus h (b) β versus h

Fig. 5.34. Residual error and the ratio for equation (5.44) at the 21st-order of the HAM approximation.

In order to gain the optimum value of the convergence control parameter h, Figures 5.34 (a–b) are then sketched at the 21st-order of the HAM approximation. The residual error in Figure 5.34 (a) is calculated using the discrete formula (5.23) and the ratio in Figure 5.34(b) using (5.26) (with $p = 2$), respectively. Due to the strong nonlinearity of the Thomas–Fermi equation, more and more terms are needed to be evaluated. It is found from the discrete error that the optimum h is -1.5618, better than the value given in [37] with $h = -1.382$. This is indeed the case, since a relative error of 5.9×10^{-7} was obtained by $h = -1.382$, see Table 1 in [37], while our optimum value results in a better relative error 1.3×10^{-7}. It is most intriguing to anticipate from Figure 5.34 (b) that the ratio is almost constant for some values of h between -2 and 0, the constant being close to one but fortunately less, indicating that the homotopy series solution (8.27) converges within a sufficiently broad range of h. Furthermore, the optimum values of h and the values $u'(0)$ are summarized in Table 5.18 at different orders of homotopy approximations. From Table 5.18, the good performance of the HAM approach is apparent and the accuracy is greatly increased as the order of homotopy approximation is getting high. The Padé–approximant can increase the accuracy further, as mentioned by Zhao *et al.* [37].

Table 5.18. Values of $u'(0)$ and optimum h for equation (5.44).

M	2	5	10	15	21
$u'(0)$	-1.61316215	-1.58806931	-1.58807014	-1.58807132	-1.58807123
h^{a}	-1.38432974	-1.17961548	-1.35694215	-1.46696045	-1.56175956

[a]Equation (5.23)

The convergence of the HAM approximations when $h = -1.55$ for the current physical problem is also verified on Table 5.19 and in Figure 5.35. Despite the fact that the ratio approaches unity, it still remains less than 1, as shown in Figure 5.34 (b). Figure 5.36 well demonstrates that the HAM approximations of the Thomas–Fermi equation (5.42) converges uniformly to the exact solution over a large enough domain.

5.5.4.7. *Gelfand equation*

The Gelfand equation

$$u'' + \lambda e^{u} = 0, \quad u(0) = u(1) = 0, \tag{5.46}$$

Table 5.19. The ratio β evaluated with $h = -1.55$ at several order (M) of the HAM approximations.

M	5	10	20	30	35	40
β	0.69742	0.83023	0.92873	0.96356	0.97352	0.98099

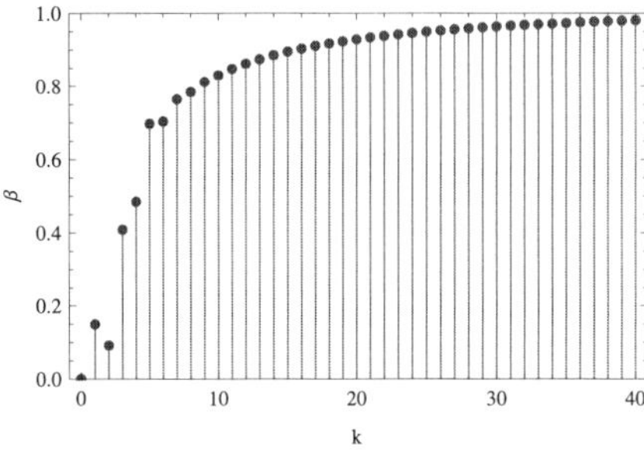

Fig. 5.35. A list plot of the ratio β to reveal the convergence of the HAM solutions for equation (5.44).

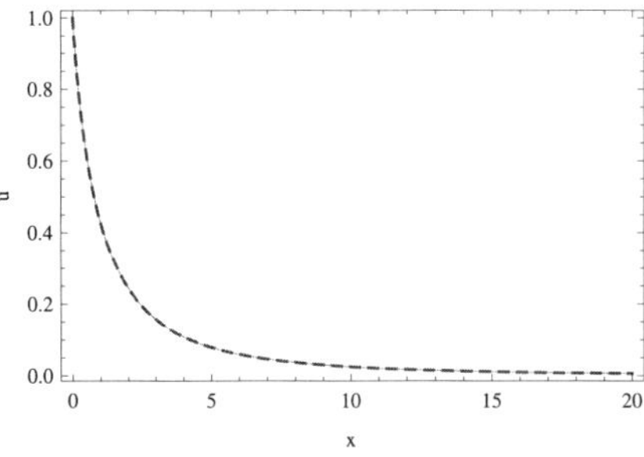

Fig. 5.36. Solutions of Thomas–Fermi equation (5.42): straight curve from the exact solution and dashed curve from the 21st-order ($M = 21$) homotopy solution.

presents an exponential type nonlinearity, where λ is a constant. The Gelfand equation (5.46) represents the steady state of diffusion and transfer of heat conduction of a thermal reaction process in combustion [39]. Li and Liao [40] solved this equation by first converting it into another kind of differential equation in which the exponential nonlinearity is no longer present. However, thanks to the theorem provided in [41], there is no need to transform the Gelfand equation (5.46), which can be treated as it is by the HAM. Having this in mind, the subsequent auxiliary variables

$$u_0(t) = 0, \quad \mathcal{L} = \frac{d^2}{dt^2}, \quad H(t) = 1$$

are used to approximate the exact solution of (5.46).

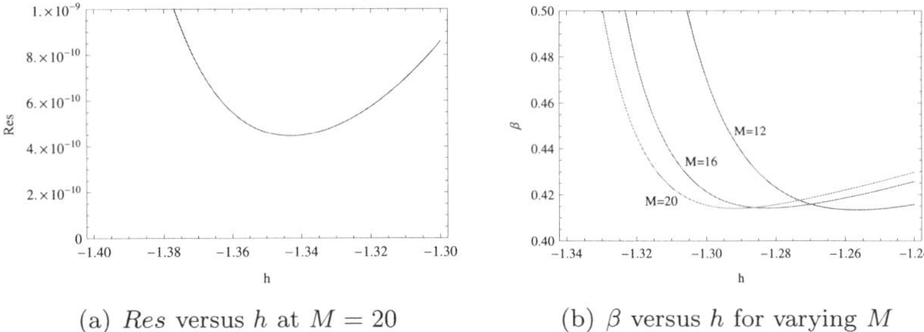

(a) *Res* versus h at $M = 20$

(b) β versus h for varying M

Fig. 5.37. Residual error and the ratio at Mth-order of HAM approximations for equation (5.46).

Taking $\lambda = 2$ as an example, at the 20th-order of approximation the homotopy (8.2) generates an optimum convergence control parameter $h = -1.3434$ from the residual (5.24) (see Figure 5.37 (a)) and $h = -1.2916$ from the ratio (5.25) with $p = 1$ (see Figure 5.37 (b)). Note that as M increases both approaches attain the nearly same optimum values of the convergence control parameter h. It should be reminded that equation (5.21) is not analytically integrable at all, owing to the strong nonlinearity of (5.46), hence a numerical integration is necessarily carried out with 500 integration points for the computation of the residual. On the other hand, equation (5.25) is analytically integrable to yield the exact values of the ratio β. Therefore, the advantage of the present approach deserves a special mention again here. Table 5.20 displays the values of $u'(0)$, the absolute residual errors *Res* and the ratio β at several order of approximations given by $h = -1.3$. It is found that the HAM approximations converge to its numerical value $u'(0) = 1.248217518$ fairly rapidly.

The interval of convergence computed from the constant h-curves using both $u'(0)$ and $u'(1)$ is revealed in Figures 5.38 (a–b). They clearly indicate that within the interval $[-2, 0)$ the convergence of the homotopy approximation is possible. Making use of the ratios $|u'_{20}(0)/u'_{19}(0)| < 1$ and $|u'_{20}(1)/u'_{19}(1)| < 1$, we find the interval of convergence as $[-2.0279, 0)$, that is consistent with Figures 5.38 (a–b). The limiting behavior of the ratio (together with $h = -1.2$) approximately as 0.45 revealed in Figure 5.39

Table 5.20. Values of $u'(0)$, Res and β for equation (5.46).

M	2	5	10	15	20
$u'(0)$	1.191666667	1.248024965	1.248210956	1.248217466	1.248217517
Res^{a}	8.997×10^{-2}	9.876×10^{-5}	9.807×10^{-6}	7.324×10^{-8}	8.634×10^{-10}
β	0.965476187	0.055553396	0.550683772	0.374199220	0.415464040

[a]Equation (5.21)

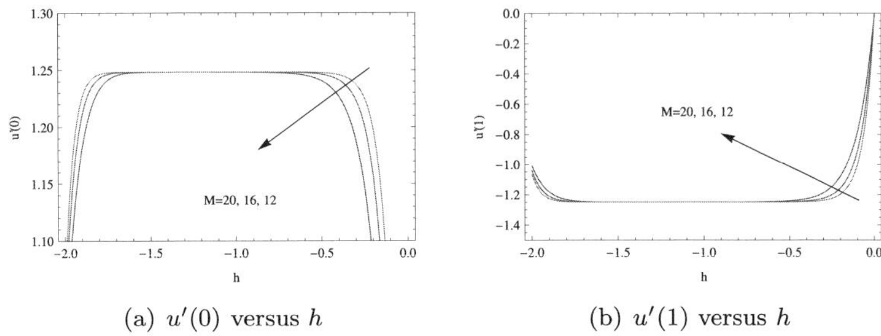

(a) $u'(0)$ versus h (b) $u'(1)$ versus h

Fig. 5.38. Constant h-curves for equation (5.46).

is sufficient in accordance with the Theorem 5.1 in §5.3 for the convergence of the HAM approximations of the Gelfand equation (5.46). A further support to the uniform convergence of the HAM approximations to the exact solution of (5.46) comes from Figures 5.40 (a–b).

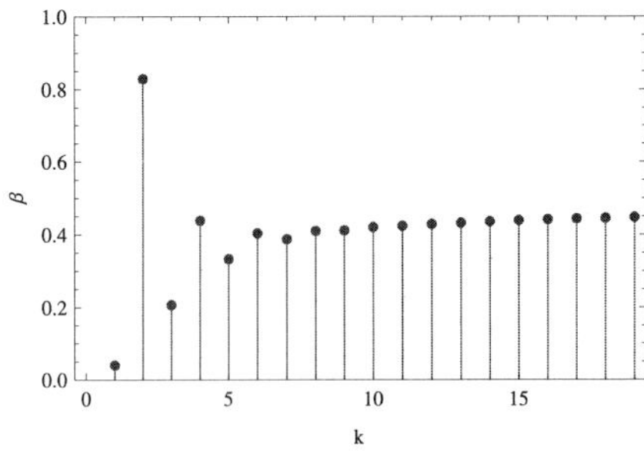

Fig. 5.39. A list plot of the ratio β to reveal the convergence of the HAM solutions for equation (5.46).

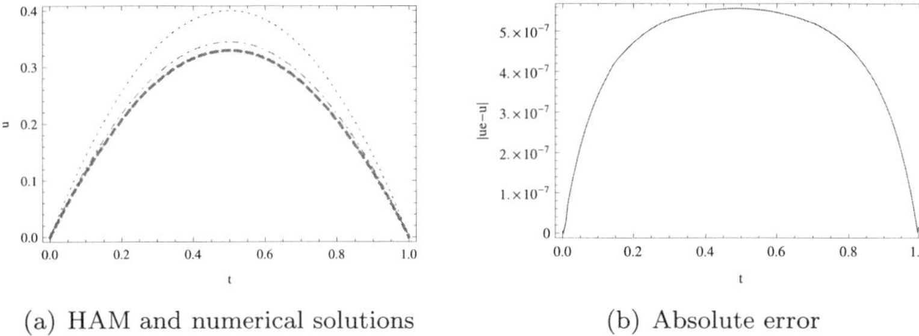

(a) HAM and numerical solutions (b) Absolute error

Fig. 5.40. (a) Solutions of Gelfand equation (5.46). Solid line: the exact solution; Dashed line: the 20th-order HAM approximation; Dot-dashed line: the 3rd-order HAM approximation; Dotted line: the 1st-order HAM approximation. (b) Absolute error.

5.5.4.8. *Uniform beam acted by axial load*

A strongly nonlinear eigenvalue problem arising from the uniform beam acted by axial load is expressed by the system (see, page 324 in [2])

$$u'' + \lambda \sin u = 0, \quad u'(0) = u'(\pi) = 0, \tag{5.47}$$

where λ is an eigenvalue corresponding to a compressive force. The system (5.47) is supplied with an extra normalizing condition $u(0) = \gamma$. Here, we consider the case of $\gamma = 1$.

The nonlinear eigenvalue problem (5.47) was treated by Liao [2] and so we closely follow that source in our HAM approach. Like Liao [2], we choose the following initial guess, the auxiliary linear operator and the auxiliary function

$$u_0(t) = \cos t, \quad \mathcal{L} = \frac{d^2}{dt^2} + 1, \quad H(t) = 1,$$

and the nonlinear operator is defined by

$$\mathcal{N}[u(t;p), \lambda(p)] = u''(t;p) + \lambda(p) \left\{ \frac{\sin[u(t;p)]}{p} \right\}.$$

It is known from the work of Liao [2] that the corresponding positive eigenvalue is $\lambda = 1.137069$. Making use of the above homotopy approach, the curves of $u''(0)$ and λ with respect to the convergence control parameter h are depicted in Figure 5.41 (a–b) at different orders of approximation M. The convergence regions shown in Figure 5.41 (a–b) are resolved exactly by means of the ratios (5.17) and (5.18) that result in

$$[-1.9704, 0), \quad [-1.9787, 0), \quad [-1.9827, 0),$$

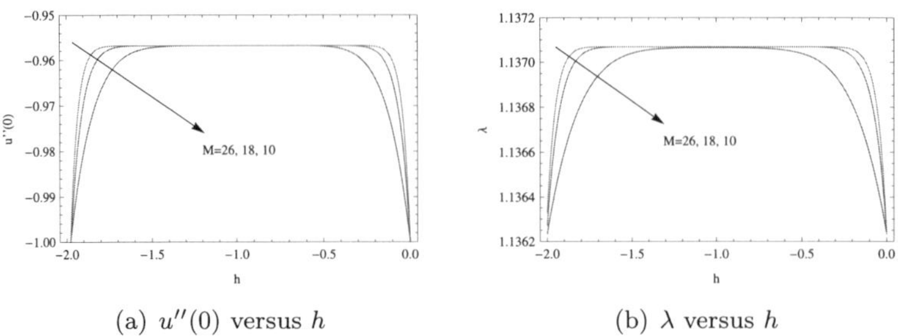

(a) $u''(0)$ versus h

(b) λ versus h

Fig. 5.41. Constant h-curves for equation (5.47).

from $u''(0)$ and

$$[-2.0035, 0), \quad [-2.0099, 0), \quad [-2.0145, 0),$$

from λ, at the 10th-order, 18th-order and 26th-order of approximations, respectively.

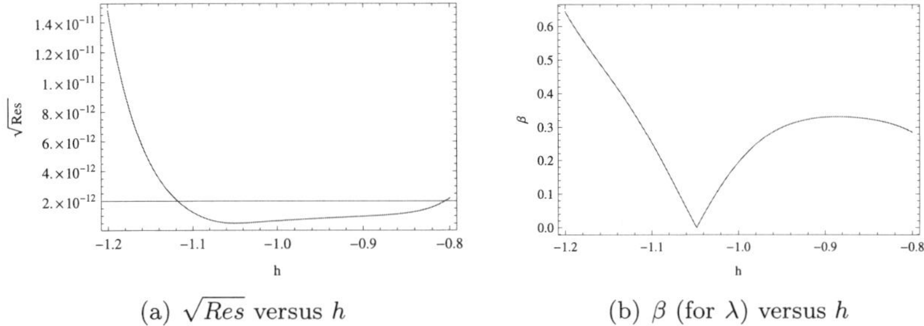

(a) $\sqrt{Res}$ versus h

(b) β (for λ) versus h

Fig. 5.42. Residual error and ratio for equation (5.47) at the 24th-order of the HAM approximation.

Next, the curve of squared residual of the governing equation (5.47) and also the curve of ratio for λ versus the convergence control parameter h are shown in Figures 5.42 (a–b). Both of them clearly indicate that the optimal convergence control parameter h at the 24th-order of approximation is close to -1: similar to the work in [2], $h = -1.0496$ from the discrete squared residual (5.23) and $h = -1.0478$ from the the ratio β associated with λ, respectively. Note that analytic integration for the squared residual is impossible at any order since the homotopy series terms are imbedded into the trigonometric function of (5.47). On the other hand, the obvious advantage here is to search for the optimum from the consecutive homotopy terms of the eigenvalue λ, which does not require cumbersome integrations, but only a minimization process. To illustrate the further and most important feature of the ratio approach, at the 24th-order of the HAM approximation, only 0.187 seconds CPU time is needed to evaluate minimum β of λ, whereas 49.485 seconds is consumed for the calculation of minimum of the discrete residual Res.

Table 5.21. Values of λ and optimum convergence control parameter h at different orders of approximation for equation (5.47).

M	3	5	10	20	26
λ	1.1250000000	1.1354166667	1.1370644139	1.1370687232	1.13706872330
h[a]	−1.058125369	−1.078444218	−1.129091502	−0.998348379	−1.0353924539

[a]Equation (5.23)

Besides the accuracy of the HAM approximations as demonstrated in Table 5.21, the behavior of ratio β related to λ given by different convergence control parameters is revealed eventually in Fgures 5.43 (a–c). The ratio seems to be oscillatory in nature for the convergence control parameters in the vicinity of the optimal value, smooth for the smaller ones. Thus, more homotopy terms are needed to better visualize the ratio for the h values close to -1, but unnecessary for small enough h.

5.5.5. *A nonlinear fractional differential equation*

Fractional calculus is an area of mathematics that is growing out of the traditional definitions of integral and derivative operators, see for instance the books [42, 43]. It has been often used in studies of viscoelastic materials, as well as in many fields of science and engineering including fluid flow, food science, rheology, diffusive transport, electrical networks, electromagnetic theory, control theory and probability, as mentioned in [44–47]. Here, let us consider one such nonlinear fractional differential equation

$$D^{\alpha}u + u^2 = 1, \quad u(0) = 0, \quad t \in [0, 1], \tag{5.48}$$

where $0 < \alpha < 1$ means the order of fractional derivative D^{α}. Without loss of generality, we concentrate on $\alpha = 1/2$ here. Note that in the limit α tends to 1, system (5.48) evolves into the traditional physical problem presented in § 5.5.4.2.

To solve the fractional differential initial-value problem (5.48), we use the Riemann–Liouville fractional integral operator J^{α} of order α and the

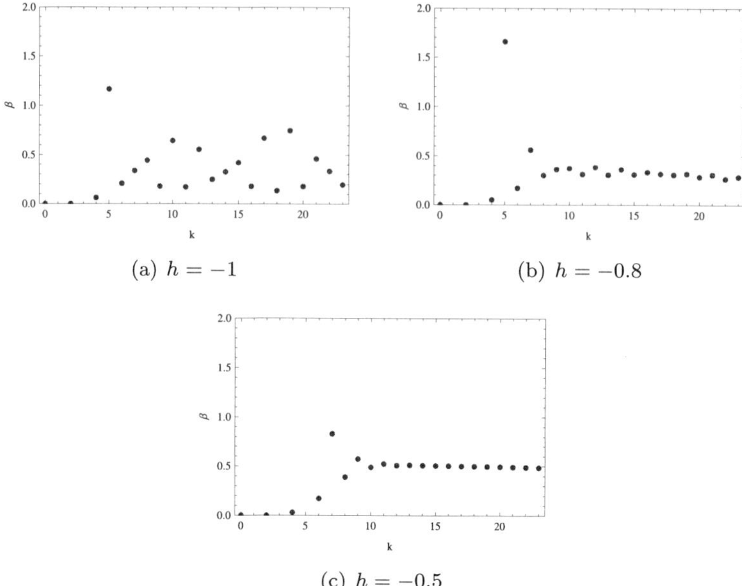

Fig. 5.43. List plots of the ratio β for λ to reveal the convergence of the HAM solutions for equation (5.47).

fractional derivative D^α in the Caputos sense, respectively, defined by

$$J^\alpha f(t) = \frac{1}{\Gamma(\alpha)} \int_0^t (t-\tau)^{\alpha-1} f(\tau) d\tau, \tag{5.49}$$

$$D^\alpha f(t) = \frac{1}{\Gamma(1-\alpha)} \int_0^t (t-\tau)^{-\alpha} f'(\tau) d\tau, \tag{5.50}$$

where Γ is the well-known Gamma function.

The HAM outlined in § 5.2 is not difficult to be adapted to gain approximate analytic solutions of (5.48) after taking into account (5.49) and (5.50) by means of the following auxiliary linear operator, initial guess and auxiliary function

$$\mathcal{L} = D^\alpha, \quad u_0(t) = c\, t^\alpha, \quad H(t) = 1,$$

where $c = 21/25$ is determined by optimizing the squared residual at the sixth-order of the HAM approximation.

The constant h-curves for the unknown $u(1)$ at the 8th-order, 14th-order and 20th-order of approximations are depicted in Figure 5.44. Note that the value of $u(1)$ converges to a constant close to 0.7 for the convergence

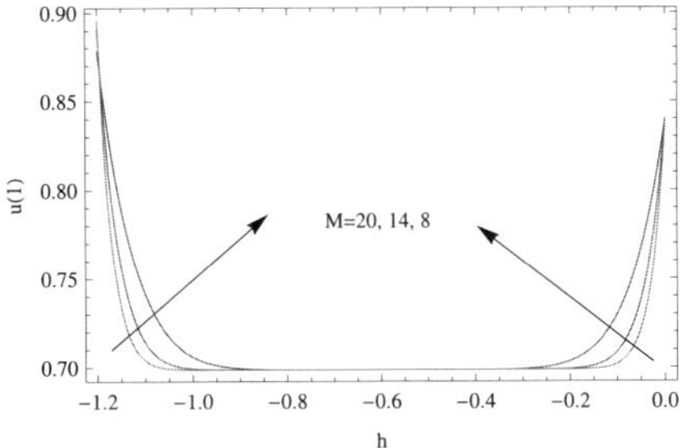

Fig. 5.44. Constant h-curves for equation (5.48).

control parameters $h \in [-1.1, 0.1]$. This convergence region of h indicated approximately by the graph can be approved by solving the inequality in (5.17) for $u(1)$, that gives exactly $h \in [-1.1946, 0)$ at the 14th-order of the HAM approximation.

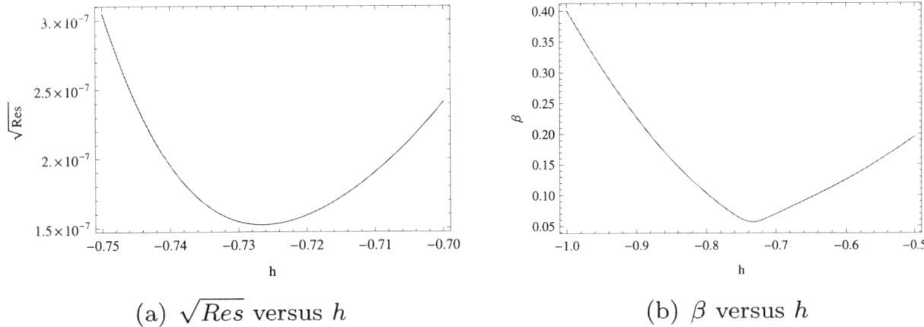

(a) $\sqrt{Res}$ versus h

(b) β versus h

Fig. 5.45. Residual error and ratio for equation (5.48) at the order of homotopy $M = 10$.

Making use of the exact residual (5.20) and the ratio (5.25) (with $p = 2$), the better interval of the convergence control parameter and further the optimal value of it are demonstrated in Figures 5.45 (a–b) at the 10th-order of approximation. Moreover, the optimum values of the convergence control parameter h given by the minimum of residual square and the ratio at different order of the HAM approximations are as shown in Table 5.22, together with the corresponding approximations of $u'(1)$, the ratio β and the CPU times in seconds. It is found that the HAM approximations converge quite rapidly by means of nearly the same optimum convergence control parameter h evaluated from both of the squared residual and the ratio approaches. Note that, the ratio approach is much more efficient in terms of the computational time, similar to the case for the previous examples. It can be conjectured here that for more complicated functions, the squared residual (5.20) might consume much more CPU time owing to (5.49), even it would be necessary to resort to the numerical integration, whereas the ratio β and its minimum can always be obtained from equation (5.25).

Table 5.22. The optimum values of h evaluated from the minimum of residual and the ratio for equation (5.48), together with the corresponding CPU times (in parenthesis), $u(1)$ and the ratio β.

M	4	6	8	10
h^a	−0.70132(31.77)	−0.71463(70.82)	−0.72196(130.56)	−0.72656(204.09)
$u(1)$	0.6988976632	0.6987489582	0.6987398224	0.6987392647
h^b	−0.71209(8.66)	−0.72522(15.95)	−0.73065(27.86)	−0.73360(48.76)
β	0.04445	0.05103	0.05442	0.05651

[a]Equation (5.20)
[b]Equation (5.25)

We eventually demonstrate the list plot of β in Figure 5.46 in case of the convergence control parameter $h = -0.73$. The convergence of the homotopy series is again confirmed for the nonlinear fractional differential equation (5.48), since β remains less than unity.

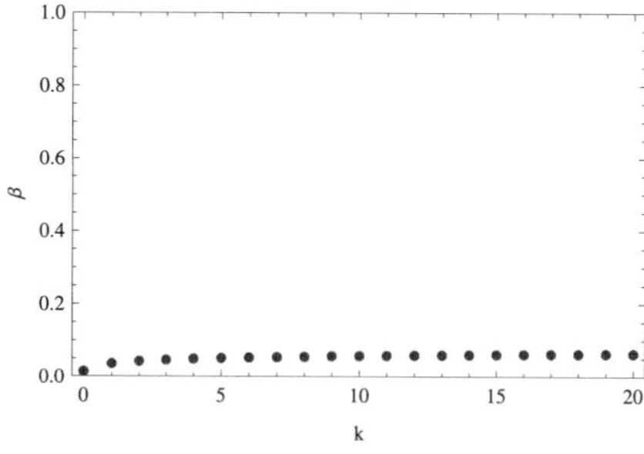

Fig. 5.46. A list plot of the ratio β to reveal the convergence of the HAM solutions for equation (5.48).

5.5.6. *Coupled nonlinear differential equations*

5.5.6.1. *Von Karman viscous flow*

We consider the steady laminar flow of a viscous, incompressible, electrically conducting and rotating unbounded fluid in the vicinity of the equator of a rotating sphere, governed by the set of highly nonlinear differential equations and boundary conditions [20]

$$\left\{\begin{aligned} F'' - HF' - F^2 - G^2 + 1 - mF &= 0, \\ G'' - HG' - m(G-1) &= 0, \\ F + H' &= 0, \\ F(0) = 0, \quad G(0) = \lambda, \quad H(0) &= s, \\ F(\infty) = 0, \quad G(\infty) &= 1, \end{aligned}\right. \tag{5.51}$$

where m, s and λ are physical parameters. For more information about the flow configuration, please refer to Turkyilmazoglu [20].

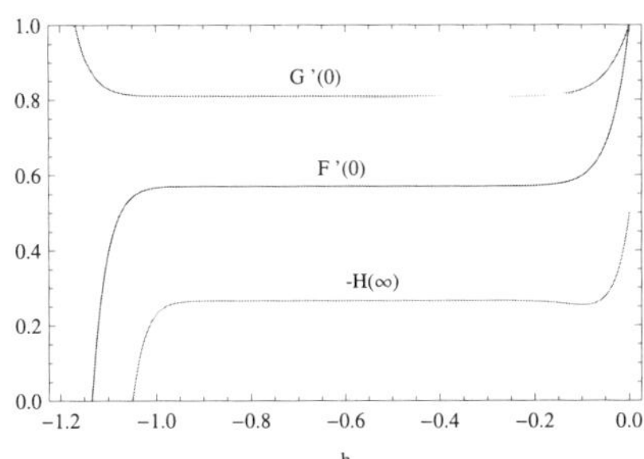

Fig. 5.47. Constant h-curves for the system (5.51).

For simplicity, we use the same auxiliary linear operator $\mathcal{L}$ and the same auxiliary convergence control parameter h to solve the coupled nonlinear ODEs. We choose the following initial guesses

$$F_0(\eta) = \eta e^{-\eta}, \quad G_0(\eta) = 1 + (\lambda - 1)e^{-\eta}, \quad H_0(\eta) = -1 + s + e^{-\eta}(1+\eta),$$

and the auxiliary linear operator

$$\mathcal{L} = \frac{d^2}{d\eta^2} - 1.$$

In the case of $s = 1/2$, $\lambda = 0$ and $m = 1$, at the 21st-order of the HAM approximation, the convergence interval for h is shown in Figure 5.47 by means of plotting the unknown values of $F'(0)$, $G'(0)$ and $-H(\infty)$ against

h. Dissimilar to the above examples, although the convergence domain varies in accordance with the considered quantity, the common interval of convergence is seen to be confined to approximately $[-1, -0.1]$. The intervals shown in Figure 5.47 are also verified via the ratios in (5.17) such that the ratios

$$\left|\frac{F'_{21}(0)}{F'_{20}(0)}\right|, \quad \left|\frac{G'_{21}(0)}{G'_{20}(0)}\right|, \quad \left|\frac{H_{21}(\infty)}{H_{20}(\infty)}\right|$$

give, respectively, the more accurate intervals

$$[-1.0061, 0), \quad [-1.0078, 0), \quad [-0.9968, 0).$$

This example illustrates that, for the system of equations like (5.51), better information can be accessed through the ratio β given in (5.17), rather than the classical constant h-curves in Figure 5.47. Even though, it may be discussed from Figures 5.47 that the interval of convergence seems to contain the value $h = -1$, but the ratio approach tells us that this is not true, as shown below.

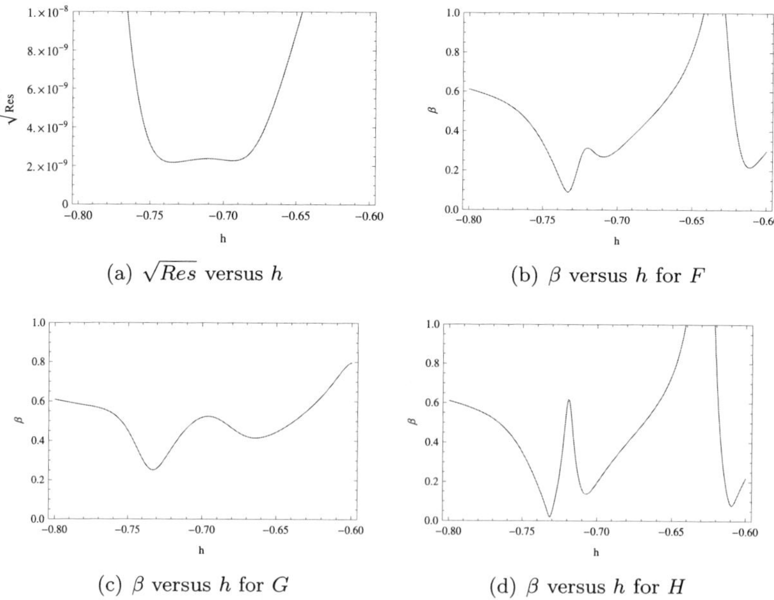

(a) $\sqrt{Res}$ versus h (b) β versus h for F

(c) β versus h for G (d) β versus h for H

Fig. 5.48. Residual error $\sqrt{Res}$ and ratios β for F, G and H at the 21st-order of the HAM approximation.

The optimal value of the convergence control parameter h is found to be about $h = -0.75$ at the 21st-order of approximation using the discrete squared residual

$$Res(h) = \int_0^{\infty} \left[g_1^2(\eta) + g_2^2(\eta) + g_3^2(\eta)\right] d\eta,$$

where

$$g_1 = F'' - HF' - F^2 - G^2 + 1 - mF, g_2 = G'' - HG' - m(G-1), g_3 = F + H'.$$

This optimum value is better visualized from the residual in Figure 5.48 (a) and further from the ratio (5.25) (with $p = 2$) in Figures 5.48 (b–d). Table 5.23 presents the values of $F'(0)$, $G'(0)$ and $H(\infty)$ using this optimum convergence control parameter at different order of the HAM approximations, M. The HAM approximations for $M \geq 20$ are the same, and thus, as compared to the previous examples, the convergence seems to take place at a considerably fast rate towards the exact solution for this highly nonlinear system.

Table 5.23. The Mth-order approximation of $F'(0)$, $G'(0)$ and $H(\infty)$ of (5.51) given by $h = -3/4$.

M	10	20	30	40	50
$F'(0)$	0.57044244	0.57043912	0.57043912	0.57043912	0.57043912
$G'(0)$	0.80969065	0.80969193	0.80969193	0.80969193	0.80969193
$-H(\infty)$	0.26539975	0.26540339	0.26540337	0.26540337	0.26540337

As also highlighted in [20], Figures 5.49 (a–c) demonstrate the ratios (5.15) in Theorem 5.1 for F, G and H, which are computed from the homotopy series (8.27) by means of $h = -0.75$. Even though sharp oscillations occur for small number of terms, as the number of terms increases, the oscillations settle down and the ratios tend to a limiting value of approximately 0.53. The convergence for the current physical system (5.51) is then confirmed, since the ratios remain less than unity.

Finally, we point to the fact that the homotopy perturbation method (HPM) proposed in 1998 is only a special case of the homotopy analysis method (HAM) introduced first by Liao in 1992 [1]. As pointed out by Liao [2], this is obvious by simply setting $h = -1$ in the homotopy equation (8.2). This means that the homotopy perturbation method (HPM) is only a special case of the HAM when $h = -1$, and not as flexible as the HAM so that divergence becomes inevitable for the HPM to most of strongly

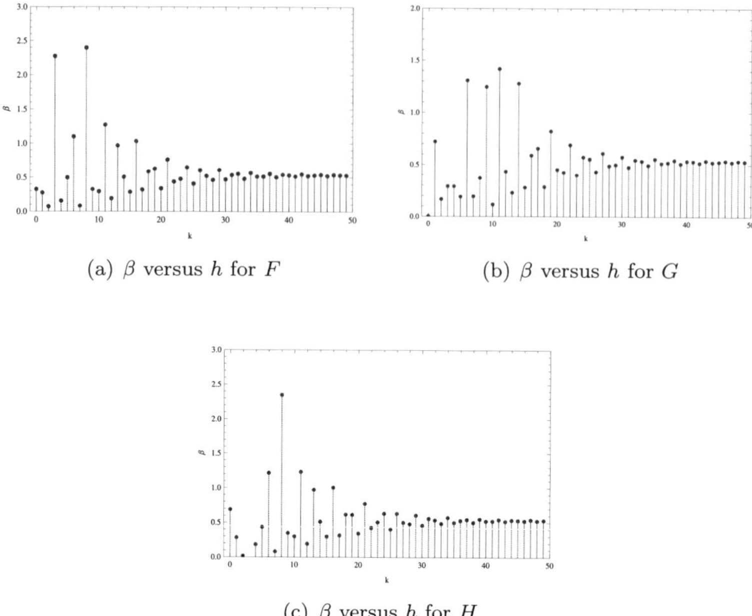

(a) β versus h for F

(b) β versus h for G

(c) β versus h for H

Fig. 5.49. List plots of the ratios to reveal the convergence of the HAM solutions for equation (5.51) when $h = -3/4$.

nonlinear problems. However, by means of choosing proper value of the convergence-control parameter h, the HAM provides us with a simple way to guarantee the convergence of approximation series. Thus, the convergence control parameter plays a very important role, which differs the HAM from all other analytic techniques. For example, as shown in Figures 5.50 (a–c), approximations given by the homotopy perturbation method (HPM), corresponding to the HAM in the special case of $h = -1$, do not converge to the exact solution, because the limit of the ratio obviously exceeds unity, reaching 1.04. Therefore, while making use of the so-called homotopy perturbation method (HPM), the convergence of the produced approximation series must be mathematically justified, which was unfortunately not adhered in most of the relevant studies in the literature.

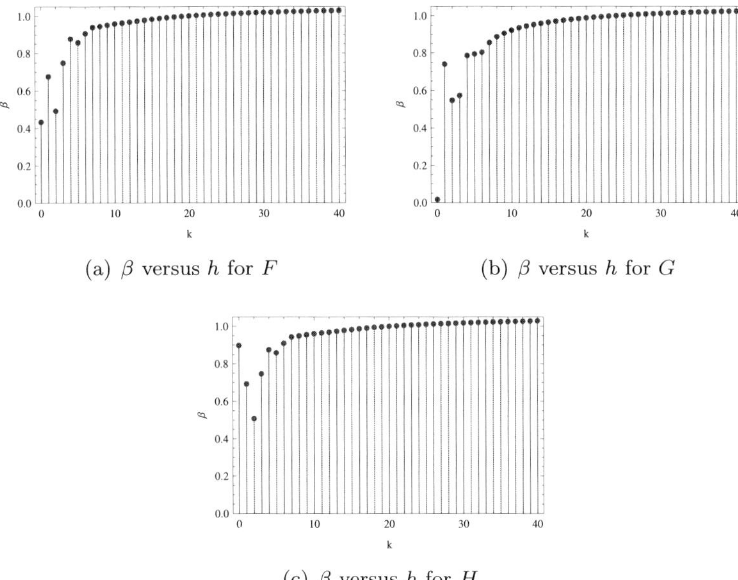

(a) β versus h for F

(b) β versus h for G

(c) β versus h for H

Fig. 5.50. List plots of the ratios to reveal the divergence of the HPM solutions for equation (5.51).

5.5.6.2. *A system of equations modeling the smokers habit in Spain*

Consider now the system of nonlinear differential equations [48]

$$\begin{cases} n'(t) = \mu(1-n) - \beta n(s+c), \\ s'(t) = \beta n(s+c) + \rho e + \alpha c - (\gamma + \lambda + \mu)s, \\ c'(t) = \gamma s - (\alpha + \delta + \mu)c, \\ e'(t) = \lambda s + \delta c - (\rho + \mu)e, \end{cases} \tag{5.52}$$

which govern the dynamics of the smokers in Spain. The necessary explanations on the variables and parameters are given in [48], together with the initial conditions. However, unlike [48], we use here

$$\mathcal{L} = \frac{d}{dt} + \frac{1}{4}$$

as the auxiliary linear operator. The so-called h-curves of $n'(0), s'(0), c'(0)$ and $e'(0)$ at the 25th-order of approximation are as shown in Figure 5.51. They suggest that the HAM approximations should be convergent when h is approximately in the interval $[-1.8, -0.2]$. Theoretical analysis of the ratios β for the four variables in equation (5.52) also results in the similar interval of the convergence-control parameter.

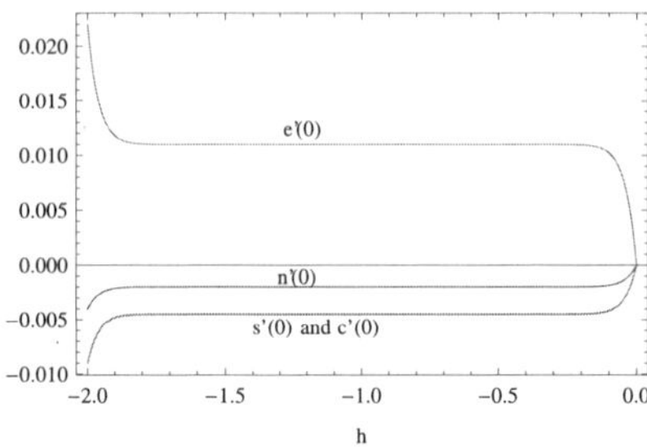

Fig. 5.51. Constant h-curves for (5.52) at the 25th-order of the HAM approximation.

The residual for the present system is defined by the integral

$$Res(h) = \int_0^{\infty} [g_1^2(t) + g_2^2(t) + g_3^2(t) + g_4^2(t)]dt,$$

where

$$\begin{aligned} g_1 &= n' - \mu(1-n) + \beta n(s+c), \\ g_2 &= s' - \beta n(s+c) - \rho e - \alpha c + (\gamma + \lambda + \mu)s, \\ g_3 &= c' - \gamma s + (\alpha + \delta + \mu)c, \quad g_4 = e' - \lambda s - \delta c + (\rho + \mu)e. \end{aligned}$$

However, unlike the previous example, the ratio β is defined here as follows

$$\beta = \frac{1}{4}\left[\frac{\|n_{k+1}(t)\|}{\|n_k(t)\|} + \frac{\|s_{k+1}(t)\|}{\|s_k(t)\|} + \frac{\|c_{k+1}(t)\|}{\|c_k(t)\|} + \frac{\|e_{k+1}(t)\|}{\|e_k(t)\|}\right],$$

where the norm is assumed as L^2. Their residual error Res and ratio β at the 25th-order of the HAM approximation are shown in Figures 5.52 (a–b). Both of them suggest the optimum value $h \approx -1.2$. Moreover, the corresponding optimum values of h and the CPU times of the two approaches at different orders of the HAM approximation are given in Table 5.24. Although the CPU times of the two approaches are close, we still observe the better performance of the ratio approach.

The convergence of the HAM approximation in case of $h = -1.18$ for the current smoking habit problem is also verified in Table 5.25 and Figure 5.53. Despite the fact that the ratio keeps oscillating, the limit is likely to be about 0.07. As shown in Figure 5.54, the 25th-order HAM approximation in case of $h = -1.18$ agrees well with the exact solution. The present

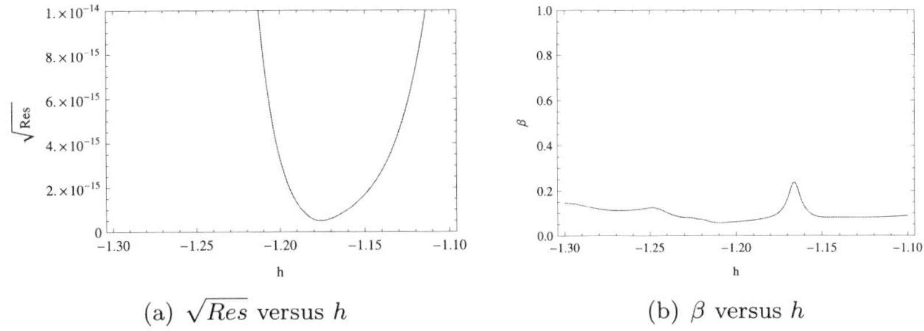

(a) $\sqrt{Res}$ versus h (b) β versus h

Fig. 5.52. Residual error $\sqrt{Res}$ and ratio for equation (5.52) at the 25th-order of approximation.

Table 5.24. The optimum values h, ratios β and CPU times for equation (5.52) at different orders of approximation M.

M	h^{a}	CPU time	h^{b}	β	CPU time
2	−1.5036	2.67	−1.4256	0.0833	1.90
5	−1.2474	8.36	−1.2308	0.0765	5.89
10	−1.1803	26.50	−1.0252	0.0670	17.91
20	−1.1761	72.66	−1.1511	0.0617	73.68
24	−1.1759	94.89	−1.2094	0.0570	88.89

[a]Equation (5.23)
[b]Equation (5.26)

HAM approximations are much better than those given in [48]. Note that Guerrero *et al.* [48] employed the auxiliary operator $\mathcal{L} = \frac{d}{dt}$. It should be emphasized that the homotopy Padé approximation had to be employed in [48] so as to get convergent results in a large time interval, which are already well predicted from the present HAM approximations. This illustrates the importance of choosing a better auxiliary linear operator in the frame of the HAM.

Table 5.25. The ratio β evaluated with the 50 homotopy terms in case of $h = -1.18$.

M	5	10	20	30	40	50
β	0.12458	0.57251	0.07176	0.07933	0.07528	0.07185

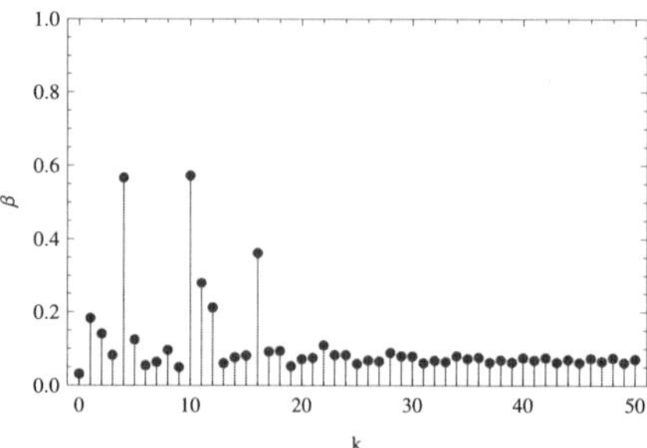

Fig. 5.53. A list plot of the ratio β to reveal the convergence of the HAM solutions for equation (5.52).

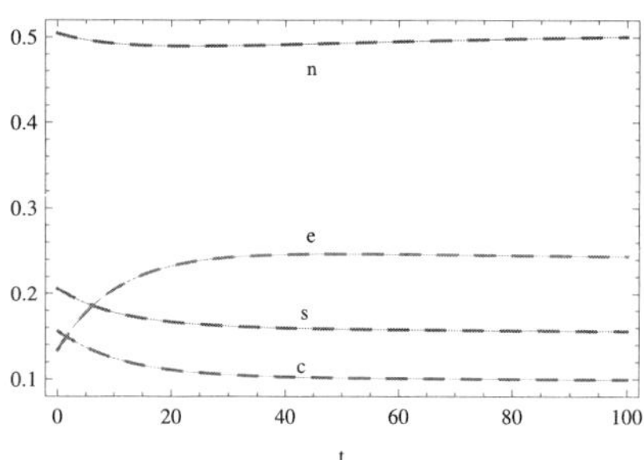

Fig. 5.54. The exact (unbroken) and the HAM (broken) solutions for equation (5.52).

5.5.7. *Partial differential equations*

5.5.7.1. *Burger's equation*

The famous Burger's equation

$$u_t + uu_x = u_{xx} \tag{5.53}$$

describes various kind of phenomena, such as the turbulence and the shock wave traveling in a viscous fluid [49]. Equation (5.53) under the initial condition

$$u(x,0) = 2x, \tag{5.54}$$

admits an exact solution

$$u(x,t) = \frac{2x}{1+2t}. \tag{5.55}$$

To gain the solution (5.55) by means of the HAM, we choose the following initial guess, auxiliary linear operator and auxiliary function

$$u_0(x,t) = 2x, \quad \mathcal{L} = \frac{\partial}{\partial t}, \quad H(x,t) = 1.$$

As a result, the homotopy (8.2) turns out to be

$$\begin{cases} \left(1 - \frac{1-p}{hp}\right) u_t(x,t,p) + u(x,t,p)u_x(x,t,p) - u_{xx}(x,t,p) = 0, \\ u(x,0,p) = 2x. \end{cases} \tag{5.56}$$

The corresponding homotopy series solution of (5.53) reads

$$\begin{aligned} u(x,t) = &\ 2x + 4htx + 4ht(1+h+2ht)x + 4ht(1+h+2ht)^2 x \\ &+ 4ht(1+h+2ht)^3 x + 4ht(1+h+2ht)^4 x + \cdots, \end{aligned} \tag{5.57}$$

which is convergent, if

$$\lim_{n\to\infty} \frac{|u_{n+1}|}{|u_n|} = |1 + h(1+2t)| < 1,$$

where the norm is in the sense of absolute value. Thus, according to Theorem 5.1 in § 5.3, this holds exactly when

$$-\frac{1}{2} < t < -\frac{2+h}{2h}, \quad -2 < h < 0.$$

Note that, in case of $h = -1$ (corresponding to the so-called homotopy perturbation method), the series is convergent only in the interval $-\frac{1}{2} < t < \frac{1}{2}$. However, as $h \to 0$, the HAM approximations converge to the exact solution in an infinite interval $0 \le t < +\infty$.

5.5.7.2. *Flexible beam equation*

Consider the following fourth-order parabolic partial differential equation arising in the study of the transverse vibrations of a uniform flexible beam [50]

$$\begin{cases} u_{tt} + \left(\frac{y+z}{2\cos x} - 1\right) u_{xxxx} + \left(\frac{z+x}{2\cos y} - 1\right) u_{yyyy} \\ \qquad + \left(\frac{x+y}{2\cos z} - 1\right) u_{zzzz} = 0, \\ u(x,y,z,0) = -u_t(x,y,z,0) = x + y + z - (\cos x + \cos y + \cos z), \end{cases} \tag{5.58}$$

whose exact solution reads

$$u(x,t) = (x + y + z - \cos x - \cos y - \cos z)e^{-t}. \tag{5.59}$$

The above nonlinear PDE can be solved in the frame of the HAM using the following initial guess, auxiliary linear operator and auxiliary function

$$u_0(x,t) = (x+y+z-\cos x-\cos y-\cos z)(1-t), \quad \mathcal{L} = \frac{\partial^2}{\partial t^2}, \quad H(x,t) = 1.$$

Then, the homotopy (8.2) for $h = -1$ generates the subsequent closed-form homotopy series

$$u(x,t) = \sum_{n=0}^{\infty}(x+y+z-\cos x-\cos y-\cos z)\left[\frac{t^{2n}}{(2n)!} - \frac{t^{2n+1}}{(2n+1)!}\right], \tag{5.60}$$

which leads, using the ratio in (5.15), to

$$\lim_{n\to\infty}\frac{|u_{n+1}|}{|u_n|} = 0.$$

Therefore, according to Theorem 5.1 in § 5.3, the homotopy series (8.2) converges to the exact solution (5.58) for *all* t.

5.5.7.3. *A linear partial differential equation of high order*

Let us consider the following initial value problem governed by the linear partial differential equation

$$u_t + u_x - 2u_{xxt} = 0, \quad u(x,0) = e^{-x}, \tag{5.61}$$

whose exact solution reads

$$u(x,t) = e^{-x-t}. \tag{5.62}$$

The zeroth-order deformation equation (8.2) then becomes

$$\begin{cases} (1-p)u_t(x,t,p) - ph(u_t(x,t,p) + u_x(x,t,p) - 2u_{xxt}(x,t,p)) = 0, \\ u(x,0,p) = e^{-x}, \end{cases} \tag{5.63}$$

where the following initial guess, auxiliary linear operator and auxiliary function

$$u_0(x,t) = e^{-x}, \quad L = \frac{\partial}{\partial t}, \quad H(x,t) = 1$$

are used.

When $h = -1$, it appears that the homotopy series solution is convergent only for $t \le 0$, that is out of physical interest, since t denotes time in equation (5.61). On the other hand, when $h = 1$, the successive ratio in (5.15) yields

$$\left|\frac{u_{n+1}(x,t)}{u_n(x,t)}\right| = \frac{|t|}{n+1},$$

whose limit as n tends to infinity is simply zero. As mentioned before, the homotopy perturbation method (HPM) is a special case of the homotopy analysis method (HAM) when $h = -1$. This well explains why the so-called HPM is unable to derive convergent homotopy series, and further explains why the homotopy series solution (8.27) with the convergence control parameter $h = 1$ represents the real physical solution over $t \geq 0$, as discussed by Liang *et al.* [10] in details.

5.5.7.4. *Korteweg-de Vries Burgers (KdVB) equations*

The KdVB equation has the form

$$u_t + \epsilon u u_x - \nu u_{xx} + \mu u_{xxx} = 0, \tag{5.64}$$

where ϵ, ν and μ are constants. This nonlinear partial differential equation, which was originally derived by [51], is a cornerstone for some significant physical problems in mathematical physics. Unlike the above examples, we are interested in the non-separable solutions of (5.64). Hence, we consider (5.64) in a special case of $\epsilon = -6$, $\nu = 0$ and $\mu = 1$, together with the initial condition $u(x,0) = -2\text{sech}^2 x$.

Substituting the following initial guess, auxiliary linear operator and auxiliary function

$$u_0(x,t) = -2\text{sech}^2 x, \quad \mathcal{L} = \frac{\partial}{\partial t}, \quad H(x,t) = 1$$

in the zeroth-order deformation equation (8.2), the homotopy approximations at any order can be easily succeeded. A few of them are given by

$$\begin{aligned}
u_1(x,t) &= 16ht\text{sech}^2 x \tanh x \\
u_2(x,t) &= -8ht\text{sech}^4 x(-8ht + 4ht\cosh(2x) - (1+h)\sinh(2x)) \\
&\vdots
\end{aligned}$$

Taking into account the norm defined by

$$||u(x,t)|| = \int_0^1 \int_0^\infty u^2(\zeta,\tau)d\zeta d\tau, \tag{5.65}$$

the ratio given in (5.15) guaranties the convergence of the homotopy series (8.27) within the time domain [0,1] and space domain $[0,\infty)$, provided that $\beta < 1$. For instance, the successive ratios

$$\left|\left|\frac{u_2}{u_1}\right|\right|, \quad \left|\left|\frac{u_4}{u_3}\right|\right|, \quad \left|\left|\frac{u_6}{u_5}\right|\right|$$

yield the convergence control parameters residing in the intervals, respectively

$$[-0.2545, 0], \quad [-0.0786, 0], \quad [-0.0472, 0].$$

This clearly shows that a sufficiently small convergence control parameter should be chosen to ensure the convergence within the mentioned domain of definition. In addition to this, the above ratios indicate that $h = -1$ is not inside the valid region of convergence control parameter, thus, the so-called homotopy perturbation method may not succeed for the KdVB problem (5.64). In fact, in order to understand what happens when h is set to -1, let us consider the norm

$$\| u(x,t) \| = \left| \int_0^\infty u^2(\zeta, t) d\zeta \right|, \tag{5.66}$$

to comply with the Corollary 2. Then, the successive ratios

$$\left\| \frac{u_1}{u_0} \right\|, \quad \left\| \frac{u_6}{u_5} \right\|, \quad \left\| \frac{u_{31}}{u_{30}} \right\|, \quad \left\| \frac{u_{51}}{u_{50}} \right\|, \quad \left\| \frac{u_{71}}{u_{70}} \right\|$$

result in the time intervals

$$[0, 0.2795], \quad [0, 0.3493], \quad [0, 0.3834], \quad [0, 0.3870], \quad [0, 0.3886],$$

pointing to the reality that the validity region of time domain of the homotopy series solution for $h = -1$ is close to [0,0.4), which explains the above failure of the homotopy perturbation method (HPM) over the interval $t \in [0, 1]$.

5.5.7.5. *Coupled nonlinear solitary wave equations*

Consider two coupled nonlinear equations

$$\begin{cases} u_t - uu_x - v_x + u_{xx}/2 = 0, \\ v_t - uv_x - vu_x - v_{xx}/2 = 0, \end{cases} \tag{5.67}$$

subject to the initial conditions

$$u(x, 0) = 1 + \tanh x, \quad v(x, 0) = \operatorname{sech}^2 x, \tag{5.68}$$

for the long solitary waves in deep water [52–54].

For the sake of simplicity, we use the same auxiliary linear operator $\mathcal{L} = \frac{\partial}{\partial t}$ and the same convergence control parameter h for the two partial differential equations in (5.67), and the initial guesses

$$u_0(x, t) = 1 + \tanh x, \quad v_0(x, t) = \operatorname{sech}^2 x.$$

Taking into consideration the norm from equation (5.66), the successive ratios concerning u

$$\left\|\frac{u_2}{u_1}\right\|, \quad \left\|\frac{u_6}{u_5}\right\|, \quad \left\|\frac{u_{10}}{u_9}\right\|, \quad \left\|\frac{u_{15}}{u_{14}}\right\|$$

produce the valid regions

$$[0, 2.2361], \quad [0, 1.6413], \quad [0, 1.6116], \quad [0, 1.5977],$$

and the successive ratios concerning v

$$\left\|\frac{v_2}{v_1}\right\|, \quad \left\|\frac{v_6}{v_5}\right\|, \quad \left\|\frac{v_{10}}{v_9}\right\|, \quad \left\|\frac{v_{15}}{v_{14}}\right\|$$

produce the valid regions

$$[0, 1.1832], \quad [0, 1.3974], \quad [0, 1.4616], \quad [0, 1.4962],$$

for time t, in case of the convergence control parameter $h = -1$. The corresponding 10th-order HAM approximation reads

$$\begin{aligned}
u(x,t) = {} & 1 + \tanh x + t^9 \operatorname{sech}^{10} x(1 - t\tanh x) \\
& + \frac{1}{3} t^7 \operatorname{sech}^8 x \left(-3 - 5t^2 + t\left(3 + 4t^2\right)\tanh x\right) \\
& + \frac{1}{45} t^5 \operatorname{sech}^6 x \left(5\left(9 + 12t^2 + 7t^4\right) - 3t\left(15 + 15t^2 + 7t^4\right)\tanh x\right) \\
& + \frac{1}{945} t^3 \operatorname{sech}^4 x(-945 - 945t^2 - 378t^4 - 85t^6 \\
& + t\left(945 + 630t^2 + 189t^4 + 34t^6\right)\tanh x) \\
& + \frac{1}{14175} t \operatorname{sech}^2 x(5\left(2835 + 2t^2\left(945 + 189t^2 + 18t^4 + t^6\right)\right) \\
& - t\left(14175 + 4725t^2 + 630t^4 + 45t^6 + 2t^8\right)\tanh x), \\
v(x,t) = {} & \frac{1}{14175} \operatorname{sech}^2 x(14175 + 28350t^2 + 9450t^4 + 1260t^6 + 90t^8 + 4t^{10} \\
& + t\operatorname{sech} x(-155925 t^9 \operatorname{sech}^9 x \\
& - 10(2835 + 2t^2\left(945 + 189t^2 + 18t^4 + t^6\right))\sinh x \\
& + 14175 t^7 \operatorname{sech}^7 x\left(9 + 22t^2 - 10t\tanh x\right) \\
& - 945 t^5 \operatorname{sech}^5 x(105 + 225t^2 + 209t^4 - 40t\left(3 + 5t^2\right)\tanh x) \\
& + 15 t^3 \operatorname{sech}^3 x(4725 + 8820t^2 + 6615t^4 + 2816t^6 \\
& - 630t\left(9 + 12t^2 + 7t^4\right)\tanh x) \\
& + 3t\operatorname{sech} x(-14175 - 23625t^2 - 13230t^4 - 3825t^6 - 682t^8 \\
& + 20t\left(945 + 945t^2 + 378t^4 + 85t^6\right)\tanh x))).
\end{aligned}$$

5.5.7.6. *A nonlinear age-structured population model*

The following nonlinear partial differential equation

$$u_t(x,t) + u_x(x,t) = -\left(1 + \int_0^\infty u(x,t)dx\right) u(x,t), u(x,0) = \frac{e^{-x}}{2} \quad (5.69)$$

models the age-structured populations [55, 56], where x and t denote age and time, and $u(x,t)$ is the age-specific density of individuals of age x at time t, respectively. The maximum age that an individual of the population may receive is assumed to grow indefinitely so that $0 \leq x < \infty$ and we confine the time to $t \in [0,1]$. In the frame of the HAM, we use the following auxiliary linear operator, initial guess and auxiliary linear operator

$$\mathcal{L} = \frac{\partial}{\partial t}, \quad u_0(x,t) = e^{-x}/2, \quad H(x,t) = 1.$$

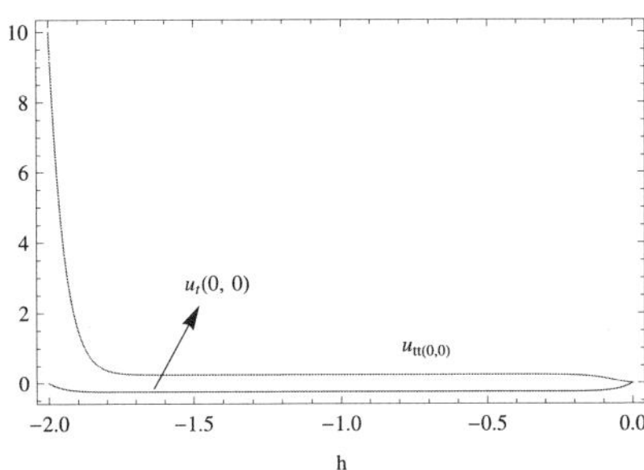

Fig. 5.55. Constant h-curves for equation (5.69).

Similarly, the constant h-curves of $u_t(0,0)$ and $u_{tt}(0,0)$ at the 20th-order of approximation are as shown in Figure 5.55, which suggests that the convergence control parameter h for convergent series should be within the interval [−2,0]. The better estimation of the interval is recovered from the ratio (5.17), which gives [−2,0] for $u_t(0,0)$ and [−1.9474,−0.0526] for $u_{tt}(0,0)$, respectively. However, as commented before, the obtained intervals from the constant h-curves are locally good approximates. Better convergence control parameter can be gained by means of the ratio (5.15) together with the norm given in (5.65). Hence, the ratios

$$\left\|\frac{u_2}{u_1}\right\|, \quad \left\|\frac{u_6}{u_5}\right\|, \quad \left\|\frac{u_{10}}{u_9}\right\|, \quad \left\|\frac{u_{20}}{u_{19}}\right\|, \quad \left\|\frac{u_{30}}{u_{29}}\right\|$$

result in the the subsequent intervals of the convergence control parameter h: $[-1.4474, 0]$, $[-1.3849, 0]$, $[-1.3655, 0]$, $[-1.3498, 0]$, $[-1.3443, 0]$. As compared to the intervals given by the constant h-curves, the above intervals are better since they are computed from norms over the entire regions, unlike the local ones obtained by means of the h-curves in Figure 5.55.

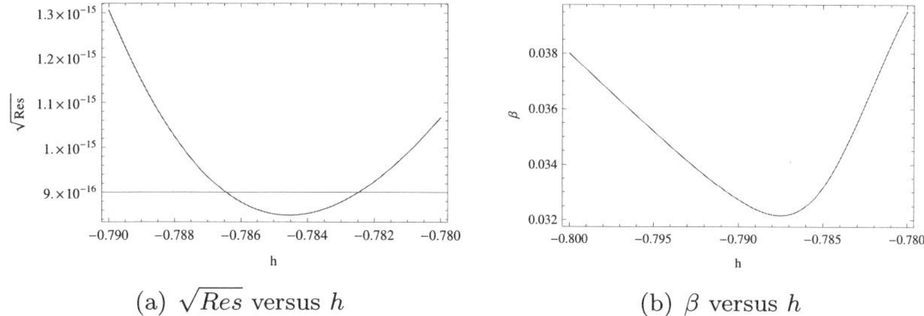

(a) $\sqrt{Res}$ versus h (b) β versus h

Fig. 5.56. Residual error and ratio for equation (5.69) at the order of homotopy $M = 20$.

Figures 5.56 (a–b) demonstrate the residual and the ratio at the 20th-order of the HAM approximation, using the exact integrations based on the norm (5.65). Moreover, Table 5.26 presents the optimal values of h, the ratios β and the used CPU times. Similar to the ordinary differential equations mentioned before, the ratio approach is highly successful for this nonlinear partial differential equation in obtaining the optimal values of the convergence control parameters h, which are very close to those obtained by means of the squared residual approach that however unfortunately needs more computational times.

Table 5.26. The optimum values h, ratios β and CPU times for equation (5.69) at different orders of approximation M.

M	h^{a}	$\sqrt{Res}$	CPU time	h^{b}	β	CPU time
2	−0.7259	4.6028×10^{-3}	0.98	−0.7237	0.0049	0.33
5	−0.7586	3.0671×10^{-5}	4.34	−0.7649	0.0198	0.98
10	−0.7743	8.9322×10^{-9}	8.64	−0.7791	0.0272	3.12
20	−0.7846	8.4875×10^{-16}	37.56	−0.7875	0.0322	17.36

[a]Equation (5.20)
[b]Equation (5.25)

The convergence of the HAM approximation in case of $h = -0.79$ to the exact solution of the nonlinear age-structured population model (5.69) is further confirmed by Figure 5.57, which clearly demonstrates the ratio less than unity. Therefore, this example once again validates the ratio approach even for strongly nonlinear partial differential equations.

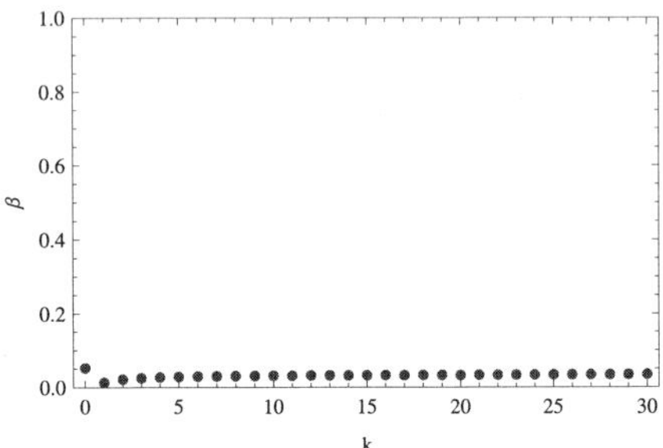

Fig. 5.57. A list plot of the ratio β to reveal the convergence of the HAM solutions for equation (5.69).

5.6. Concluding remarks

In this chapter, the homotopy analysis method (HAM) has been investigated with an aim to study the convergence conditions of the homotopy series solutions of the nonlinear algebraic equations as well as the nonlinear differential-difference, integro-differential, fractional differential and ordinary/partial differential equations, and also system of differential equations. In the context of the theorems and their corollaries, we illustrate that if the convergence control parameter, the auxiliary linear operator, the initial guess and so on are properly chosen in the frame of the HAM, the homotopy analysis results may indeed converge to the exact solution of considered nonlinear problems.

The concept of the convergence control parameter employed in the HAM technique has been thoroughly reviewed. A novel definition for the optimal convergence control parameter is given in this chapter, which is distinct from the well-known optimum approach based on minimizing the squared residual. The new optimum approach for the convergence control parameter is based on the minimization of the ratio of the homotopy series terms. It often gives optimum convergence control parameter close to those gained by the squared residual approach at a finite order of approximations. When the order of approximation tends to infinity, the majority of the considered examples clearly imply the coincidence of the optimal convergence control parameters given by the two approaches. Different types of nonlinear equations have been used in this chapter to verify the theorems and the corollaries proved in this chapter.

By means of the ratio approach, not only the convergence of the homotopy series is guaranteed, but also the interval of the convergence control parameter ensuring the convergence of the HAM results is analytically determined. In particular, the interval of h for convergence estimated by the traditional constant h-curves is often worked out by means of the ratio method. Being able to better determine the bounds of the interval of the convergence control parameter h, the ratio approach well explains why the so-called homotopy perturbation method (HPM), which is actually a special case of the HAM, might fail or only correspond to some restricted intervals of the space and time variables of considered physical problems.

The ratio approach seems promising, since it takes considerably less computational time than the traditional ways. Our examples illustrate that this is indeed the case, if both of the traditional squared residual and the ratio approaches are used to estimate the interval of the convergence control parameters. For some strongly nonlinear problems, like the Gelfand problem, the eigenvalue problems involving trigonometric functions and so on, exact evaluation of the squared residual is not possible in most cases, whereas the ratio concerning the homotopy terms can be computed up to any order of approximation. Particularly, for fractional differential equations, the advantage of the ratio approach can not be disregarded.

The minimization of the ratio approach to gain optimal h still requires less CPU times than the minimization of the squared residual approach. This is also generally true if discrete squared residual versus the exact ratio is concerned. Furthermore, for problems involving an unknown parameter, such as the undamped Duffing oscillator, the eigenvalue problems, the parameterized differential equations and so on, the ratio approach has the advantage that both the optimal value of convergence control parameter and the convergence analysis can be pursued only from the ratios concerning the parameter, leaving aside the need of using residuals for other variables, and thus avoiding heavy integrations. Even if there is no such kind of unknown parameter in the differential equation, a pseudo-like parameter can always be introduced using an unknown boundary condition together with a suitable scaling, so that the advantage of new method can always be benefited.

Our examples clearly show that the proposed ratio approach generates almost the same optimal convergence control parameters as those given by the squared residual approach, as long as the order of HAM approximation is high enough. A rigorous mathematical analysis explaining why this is so warrants a further work. Also, in spite of the fact that the HAM in general has to be considered (due to computational restrictions) by truncating the homotopy series at a finite value, the behavior in the limit of larger order approximations should be examined which will also enable the use of ratio approach to better understand the uniform convergence of the HAM. Moreover, the present ratio approach can be employed in combination with the iterative HAM so that the fast convergence feature of such an iteration technique is enlightened. Finally, the extension of the presented ratio approach to higher order standard or fractional partial differential equations and how to evaluate the ratio for these equations in practice deserve further investigation.

References

[1] S. J. Liao. *The proposed homotopy analysis technique for the solution of nonlinear problems.* PhD thesis, Shanghai Jiao Tong University (1992).

[2] S. J. Liao, *Homotopy Analysis Method in Nonlinear Differential Equations.* Springer-Verlag (2012).

[3] S. J. Liao, *Beyond perturbation: introduction to homotopy analysis method.* Chapman & Hall/CRC (2003).

[4] S. Abbasbandy, The application of the homotopy analysis method to nonlinear equations arising in heat transfer, *Physics Letters A.* **360**, 109–113 (2006).

[5] S. P. Zhu, An exact and explicit solution for the valuation of American put, *Quant. Finan.* **6**, 229–242 (2006).

[6] M. Sajid and T. Hayat, Comparison of ham and hpm methods in nonlinear heat conduction and convection equations, *Nonlinear Analysis: Real World Applications.* **9**, 2296–2301 (2008).

[7] R. A. VanGorder and K. Vajravelu, Analytic and numerical solutions to the lane-emden equation, *Physics Letters A.* **372**, 6060–6065 (2008).

[8] F. T. Akyildiz and K. Vajravelu, Magnetohydrodynamic flow of a viscoelastic fluid, *Physics Letters A.* **372**, 3380–3384 (2008).

[9] Y. Y. Wu and K. F. Cheung, Homotopy solution for nonlinear differential equations in wave propagation problems, *Wave Motion.* **46**, 1–14 (2009).

[10] S. Liang and D. J. Jeffrey, Comparison of homotopy analysis method and homotopy perturbation method through an evolution equation, *Commun. Nonlinear Sci. Numer. Simulat.* **14**, 4057–4064 (2009).

[11] A. Molabahrami and F. Khani, The homotopy analysis method to solve the burgershuxley equation, *Nonlinear Anal. Real World Appl.* **10**, 589–600 (2009).

[12] V. I. Zhuk and O. S. Ryzhov, A closed-form solution to american options under general diffusion, *Quant. Finan. online* (2009).

[13] S. J. Liao, A new branch of solutions of boundary-layer flows over a permeable stretching plate, *Int. J. Non-Linear Mech.* **42**, 819–930 (2007).
[14] M. Turkyilmazoglu, A homotopy treatment of analytic solution for some boundary layer flows, *International Journal of Nonlinear Sciences and Numerical Simulation.* **10**(7), 885–889 (2009).
[15] M. Turkyilmazoglu, On the purely analytic computation of laminar boundary layer flow over a rotating cone, *International Journal of Engineering Science.* **47**(9), 875–882 (2009).
[16] M. Turkyilmazoglu, Purely analytic solutions of the compressible boundary layer flow due to a porous rotating disk with heat transfer, *Physics of Fluids.* **21**(10) (2009).
[17] M. Turkyilmazoglu, Purely analytic solutions of magnetohydrodynamic swirling boundary layer flow over a porous rotating disk, *Computers and Fluids.* **39**(5), 793–799 (2010).
[18] M. Turkyilmazoglu, Analytic approximate solutions of rotating disk boundary layer flow subject to a uniform suction or injection, *International Journal of Mechanical Sciences.* **52**(12), 1735–1744 (2010).
[19] M. Turkyilmazoglu, Analytic approximate solutions of rotating disk boundary layer flow subject to a uniform vertical magnetic field, *Acta Mechanica.* **218**(3-4), 237–245 (2011).
[20] M. Turkyilmazoglu, Numerical and analytical solutions for the flow and heat transfer near the equator of an mhd boundary layer over a porous rotating sphere, *International Journal of Thermal Sciences.* **50**, 831–842 (2011).
[21] M. Turkyilmazoglu, Series solution of nonlinear two-point singularly perturbed boundary layer problems, *Computers and Mathematics with Applications.* **60**(7), 2109–2114 (2010).
[22] M. Turkyilmazoglu, Analytic approximate solutions of parameterized unperturbed and singularly perturbed boundary value problems, *Applied Mathematical Modelling.* **35**(8), 3879–3886 (2011).
[23] M. Turkyilmazoglu, The selection of proper auxiliary parameters in the homotopy analysis method. a case study: Uniform solutions of undamped duffing equation, *Applied Mathematics.* **2**, 783–790 (2011).
[24] M. Turkyilmazoglu, An effective approach for approximate analytical solutions of the damped duffing equation, *Physica Scripta.* **86**, 015301 (2012).
[25] M. Turkyilmazoglu, An optimal analytic approximate solution for the limit cycle of duffing-van der pol equation, *Journal of Applied Mechanics, Transactions ASME.* **78**(2) (2011).
[26] M. Turkyilmazoglu, Accurate analytic approximation to the nonlinear pendulum problem, *Physica Scripta.* **84**(1) (2011).
[27] M. Turkyilmazoglu, Solution of the thomas-fermi equation with a convergent approach, *Commun. Nonlinear Sci. Numer. Simulat.* **11**, 4097–4103 (2012).
[28] M. Turkyilmazoglu, The airy equation and its alternative analytic solution, *Physica Scripta.* **86**, 055004 (2012).
[29] M. Turkyilmazoglu, An analytic shooting-like approach for the solution of nonlinear boundary value problems, *Mathematical and Computer Modelling.* **53**(9-10), 1748–1755 (2011).

[30] S. J. Liao, An optimal homotopy-analysis approach for strongly nonlinear differential equations, *Commun. Nonlinear Sci. Numer. Simulat.* **15**, 2003–2016 (2010).
[31] S. J. Liao, Notes on the homotopy analysis method: Some definitions and theorems, *Commun. Nonlinear Sci. Numer. Simulat.* **14**, 983–997 (2009).
[32] M. Turkyilmazoglu, Some issues on hpm and ham methods: A convergence scheme, *Mathematical and Computer Modelling.* **53**, 1929–1936 (2011).
[33] V. Marinca and N. Herisanu, Application of optimal homotopy asymptotic method for solving nonlinear equations arising in heat transfer, *Int. Commun. Heat Mass. Trans.* **35**, 710–715 (2008).
[34] K. Yabushita, M. Yamashita, and K. Tsuboi, An analytic solution of projectile motion with the quadratic resistance law using homotopy analysis method, *J. Phys. A-Math. Theor.* **40**, 8403–8416 (2007).
[35] Z. Wang, L. Zou, and H. Zhang, Applying homotopy analysis method for solving differential-difference equation, *Physics Letters A.* **369**, 77–84 (2007).
[36] H. Temimia, A. R. Ansari, and A. M. Siddiqui, An approximate solution for the static beam problem and nonlinear integro-differential equations, *Computers and Mathematics with Applications.* **62**, 3132–3139 (2011).
[37] Y. Zhao, Z. Lin, Z. Liu, and S. J. Liao, The improved homotopy analysis method for the Thomas Fermi equation, *Appl. Math. Comput.* **218**, 8363–8369 (2012).
[38] F. Fernandez, Rational approximation to the thomasfermi equations, *Appl. Math. Comput.* **217**, 6433–6436 (2011).
[39] I. M. Gelfand, Some problems in the theory of quasi-linear equations, *Am. Math. Soc. Transl. Ser.* **29**, 295381 (1963).
[40] S. Li and S. J. Liao, An analytic approach to solve multiple solutions of a strongly nonlinear problem, *Appl. Math. Comput.* **169**, 854–865 (2005).
[41] M. Turkyilmazoglu, A note on the homotopy analysis method, *Applied Mathematics Letters.* **23**, 1226–1230 (2010).
[42] R. Gorenflo and F. Mainardi, *Fractional calculus: integral and differential equations of fractional order. In: Fractals and fractional calculus in continuum mechanics.* Springer-Verlag (1997).
[43] I. Podlubny, *Fractional differential equations.* Academic Press (1999).
[44] Z. E. A. Fellah and C. Depollier, Application of fractional calculus to the sound waves propagation in rigid porous materials: Validation via ultrasonic measurement, *Acta Acustica.* **88**, 34–39 (2002).
[45] V. V. Anh and R. Mcvinish, Fractional differential equations diven by levy noise, *J. of Appl. Math. and Stoch. Anal.* **16**, 97–119 (2003).
[46] J. F. Douglas, Some applications of fractional calculus to polymer science, *Advances in chemical physics.* **102** (2003).
[47] A. Kaur, P. S. Takhar, D. M. Smith, J. E. Mann, and M. M. Brashears, Fractional differential equations based modeling of microbial survival and growth curves: model development and experimental validation, *J. Food Sci.* **8**, 403–414 (2008).
[48] F. Guerrero, F. J. Santonja, and R. J. Villanueva, Solving a model for the evolution of smoking habit in spain with homotopy analysis method, *Non-*

linear Analysis: Real World Applications. **14**, 549–558 (2012).

[49] B. E. R., Some new exact, viscous, nonsteady solutions of burgers' equation, *Phys. Fluids.* **9**, 1247–1248 (1966).

[50] D. J. Gorman, *Free vibrations analysis of beams and shafts.* Wiley (1975).

[51] C. H. Su and C. S. Gardner, Derivation of the korteweg de-vries and burgers equation, *J. Math. Phys.* **10**, 536–539 (1967).

[52] G. B. Whitham, Variational methods and applications to water wave, *Proceedings of the Royal Society A.* **299**, 6–25 (1967).

[53] L. T. F. Broer, Approximate equations for long water waves, *Applied Scientific Research.* **31**, 377–395 (1975).

[54] M. L. Wang, Application of a homogeneous balance method to exact solutions of nonlinear equations in mathematical physics, *Physics Letter A.* **6**, 67–75 (1996).

[55] M. G. Cui and C. Chen, The exact solution of nonlinear age-structured population model, *Nonlinear Anal. Real World Appl.* **8**, 1096–1112 (2007).

[56] O. Angulo, J. C. Lopez-Marcos, M. A. Lopez-Marcos, and F. A. Milner, A numerical method for nonlinear age-structured population models with finite maximum age, *J. Math. Anal. Appl.* **361**, 150160 (2010).

Chapter 6

Homotopy Analysis Method for Some Boundary Layer Flows of Nanofluids

Tasawar Hayat

Department of Mathematics, Quaid-I-Azam University 45320, Islamabad 44000, Pakistan
pensy_t@yahoo.com

Meraj Mustafa

Research Centre for Modeling and Simulation (RCMS), National University of Sciences and Technology (NUST), Islamabad 44000, Pakistan

This chapter describes the application of homotopy analysis method (HAM) to the nonlinear boundary value problems arising in nanofluid dynamics. The series solutions for some fundamental problems in fluid mechanics (including axisymmetric flow past a stretching sheet, unsteady boundary layer flow developed due to the impulsive motion of an extensible surface, squeezing flow between parallel disks and boundary layer flow of second grade fluid) involving nanofluids are computed and discussed.

Contents

6.1. Background 260
6.2. Unsteady boundary-layer flow of nanofluid past an impulsively stretching sheet 263
6.3. Axisymmetric flow of nanofluid over a radially stretching sheet with convective boundary conditions 271
6.4. Squeezing flow of nanofluid between parallel disks 275
6.5. Boundary layer flow of non-Newtonian nanofluid over a stretching sheet . . . 280
References 288

6.1. Background

Fluid flow problems characterizing either viscous or non-Newtonian fluids are usually modeled in terms of highly nonlinear boundary value problems. Obviously one requires an efficient and easy to use analytical technique like HAM to solve such problems and deepen our understanding to these complicated nonlinear phenomena. The researchers have therefore widely applied HAM to solve various nonlinear problems arising in fluid mechanics. For instance, an approximate analytic solution of the two-dimensional laminar viscous flow over a semi-infinite plate has been obtained by Liao [1]. The convergence of the solutions was ensured through the proper selection of the auxiliary parameter. The obtained solution was in an excellent agreement with the Howarth's numerical solution. Liao and Campo [2] used HAM to find the series solution of temperature distribution in a Blasius flow. The dimensionless heat transfer rate at the plate was also calculated. The analytic results were found in complete agreement with the obtained numerical solution for reasonably large values of the Prandtl number. Liao [3] computed series solutions for magnetohydrodynamic (MHD) flow of power-law fluid past a stretching sheet. The analytic solutions were obtained for real power law index in terms of recursive formulas of constant coefficients and these were in decent agreement with the numerical solution for all the values of magnetic parameter. He also obtained a simple analytic formula for calculating skin friction coefficient on a moving sheet to find wide applications in metallurgical processes. Homotopy solutions for unsteady boundary layer flow due to impulsively stretching plate have been provided by Liao [4]. Different from the previous perturbation solutions, the HAM solutions were convergent for all values of the dimensionless time. The validation to the findings was given through the obtained numerical solution in several different cases. Sajid and Hayat [5] obtained the local similarity solutions for the two-dimensional flow of Maxwell fluid above a continuously moving flat plate in a quiescent ambient fluid (the so-called Sakiadis flow) by HAM. The convergence was guaranteed for a fixed range of the auxiliary parameter by displaying the so called $\hbar$-curves. The convergent numerical results for various parametric values were also given. In another paper Sajid and Hayat [6] further interpreted that second order or second grade fluids are inadequate in describing the shear thinning or thickening behavior of various polymeric liquids. For this purpose they considered a flow of fourth grade fluid down a vertical cylinder. The modeled nonlinear problem was solved for the homotopy analytic solution. They also proved that obtained solution in case of homotopy perturbation method (HPM) is a special case of HAM and it offers divergent solutions for strong nonlinearities. Series solution for unsteady axisymmetric flow over a stretching sheet was provided by Sajid *et al.* [7]. A valid range of the auxiliary parameter was identified through $\hbar$-curves. Hayat *et al.* [8] investigated the viscoelastic effects in the Blasius flow by HAM. Different from Liao and Campo [2] the viscous dissipation effects were also considered. Similar initial guesses and the linear operators for the velocity and temperature functions were selected as in Liao [1], and Liao and Campo [2]. Local similarity solutions were obtained and convergence was guaranteed for various values of viscoelastic parameter. Homotopy solutions for Falkner–Skan flow have been reported by Abbasbandy and Hayat [9]. They have obtained the numerical values of dimensionless velocity gradient on the plate by Crocco's transformation. Based on the idea given in Liao [1], in addition to the auxiliary parameter $\hbar$ another parameter γ was introduced in the initial guess. Proper values of $\hbar$ and γ were identified so that the series solution would converge for all the values of embedded parameters. Homotopy-Pade method was used to compute the numerical values of skin friction coefficient and these values were shown to be in a very good agreement with those obtained numerically and by Crocco's transformation. In another paper, Abbasbandy and Hayat [10] considered the unsteady flow of special third grade fluid over a porous unsteady stretching sheet. The velocity distribution outside the boundary layer was assumed to be time dependent. The developed differential system was solved for the homotopy solutions. For accelerating the convergence, homotopy-Pade method was adopted and numerical values of velocity gradient on the sheet were computed for various parametric values. These values were compared with the numerical solution obtained through the computational software Mathematica and were shown to be in decent agreement. Xu *et al.* [11] explored the multiple solutions of the Navier–Stokes equations representing the viscous flow in a porous channel with moving walls by optimal homotopy analysis approach. They predicted the multiple solutions of the boundary layer equations by adjusting the auxiliary parameter. Rashidi *et al.* [12] provided the homotopy analytic solutions for heat transfer in the flow of micropolar fluid embedded in a porous medium. The thermal radiation effect was also considered. They defined the averaged residual errors of the corresponding differential equations and using the idea given by Liao [13] the optimal values of the auxiliary parameter were determined. They concluded that, by means of the convergence control (auxiliary) parameter, HAM can be applied to both strongly and weakly nonlinear problems which is a fundamental qualitative difference between HAM and other analytical methods.

Nanofluid is a liquid in which nanometer-sized particles (called nanoparticles) made up of metals, oxides, carbides or carbon nanotubes are suspended in the base fluid such as oils, water, ethylene glycol etc. Choi [14] experimentally found that addition of these nanoparticles in the base fluid appreciably enhances the effective thermal conductivity of the fluid. Non-homogeneous equilibrium model proposed by Buongiorno [15] reveals that such massive increase in the thermal conductivity occurs due to the presence of two main effects namely the Brownian diffusion and the thermophoretic diffusion of nanoparticles. Review studies on nanofluids have been made by Daungthongsuk and Wongwises [16], Wang and Mujumdar [17, 18], Eastman *et al.* [19] and Kaka and Pramuanjaroenkij [20]. Kuznetsov and Nield [21] have reported the pioneering work on the boundary layer flow behavior of nanofluids utilizing this Buongiorno's model. They formulated the momentum, transport and energy equations valid for any nanofluid. The equations were firstly simplified under usual boundary layer assumptions and then solved numerically for boundary layer flow past a vertical plate. Based on the idea developed by Kuznetsov and Nield [21] various nanofluid boundary layer flow problems were considered by the different researchers and their solutions were obtained by traditional numerical methods including explicit/implicit finite difference schemes, shooting method with Runge-Kutta integration techniques, finite element method etc. Despite of the complex nature of these equations, the analytic solution became an excessive challenge to the researchers from different quarters. Mustafa *et al.* [22] were probably the first to compute homotopy analytic solutions for stagnation-point flow of an incompressible nanofluid towards a stretching sheet. The results were compared with the previous numerical studies and were shown to be in complete agreement. Homotopy based analytic solution for mixed convection flow of nanofluid past a vertical cylinder has been provided by Dinarvand *et al.* [23] . Series solutions for flow of nanofluid past an exponentially stretching sheet were computed by Nadeem *et al.* [24]. Mustafa *et al.* [25] provided analytic solutions for squeezing flow of nanofluid between parallel disks. The resulting coupled nonlinear boundary values problem was formulated and solutions corresponding to the optimal values of the auxiliary parameter were presented. Ellahi *et al.* [26] developed the homotopy analytic solutions for flow of third grade nanofluid with temperature dependent viscosity. Non-orthogonal stagnation-point flow of second grade nanofluid towards a stretching surface has been examined by Nadeem *et al.* [27]. The HAM solution for flow of viscous nanofluid with porous medium and slip condition has been constructed by Zheng *et al.* [28]. The unsteady film flow of nanofluid induced by a stretching surface is analyzed by Xu *et al.* [29]. HAM is implemented for the development and analysis of series solution. In the subsequent sections we will discuss some interesting boundary layer flows of nanofluids in which HAM is applied to yield explicit series solutions of a variety of nonlinear problems.

6.2. Unsteady boundary-layer flow of nanofluid past an impulsively stretching sheet

Mustafa *et al.* [30] considered the unsteady incompressible flow of nanofluid over a stretching sheet (situated at $y = 0$ with a velocity $u_w(x) = ax$, temperature T_w and nanoparticles concentration C_w . The x- and y- axes are taken along and perpendicular to the sheet respectively and the flow is confined to $y \geq 0$. We denote T_∞ and C_∞ the ambient temperature and nanoparticles concentration respectively. The equations governing the unsteady flow of an incompressible nanofluid can be simplified under usual boundary layer assumptions as (see Kuznetsov and Nield [21])

$$\frac{\partial u}{\partial x} + \frac{\partial v}{\partial y} = 0, \tag{6.1}$$

$$\frac{\partial u}{\partial t} + u\frac{\partial u}{\partial x} + v\frac{\partial u}{\partial y} = \nu\frac{\partial^2 u}{\partial y^2}, \tag{6.2}$$

$$\frac{\partial T}{\partial t} + u\frac{\partial T}{\partial x} + v\frac{\partial T}{\partial y} = \alpha_m\frac{\partial^2 T}{\partial y^2} + \tau^*\left[D_B\frac{\partial C}{\partial y}\frac{\partial T}{\partial y} + \frac{D_T}{T_\infty}\left(\frac{\partial T}{\partial y}\right)^2\right], \tag{6.3}$$

$$\frac{\partial C}{\partial t} + u\frac{\partial C}{\partial x} + v\frac{\partial C}{\partial y} = D_B\frac{\partial^2 C}{\partial y^2} + \frac{D_T}{T_\infty}\frac{\partial^2 T}{\partial y^2}. \tag{6.4}$$

The considered boundary conditions are

$$\begin{cases} t < 0 : v = 0,\ u = 0,\ T = T_\infty,\ C = C_\infty & \text{for any} \quad x, y \\ \qquad u = ax,\ v = 0,\ T = T_w,\ C = C_w & \text{at} \quad y = 0, \\ t \geq 0 : u, v \to 0,\ T \to T_\infty,\ C \to C_\infty & \text{as} \quad y \to \infty. \end{cases} \tag{6.5}$$

Here u and v are the velocity components along x- and y-directions respectively, ν the kinematic viscosity, α_m the thermal diffusivity, D_B the Brownian motion coefficient, D_T the thermophoretic diffusion coefficient and $\tau^* = (\rho c)_p/(\rho c)_f$ the ratio of effective heat capacity of the nanoparticle material to heat capacity of the fluid. Following Liao [4] we seek the similarity solution of Eqs. (7.1)–(7.5) of the following form

$$\begin{cases} \psi = x\sqrt{a\nu\xi}f(\xi,\eta), \eta = \sqrt{\frac{a}{\nu\xi}}y, \xi = 1-\exp(-\tau), \tau = at, \\ \theta(\xi,\eta) = \frac{T-T_\infty}{T_w-T_\infty}, \phi(\xi,\eta) = \frac{C-C_\infty}{C_w-C_\infty}, \end{cases} \tag{6.6}$$

where ψ is the stream function, τ the dimensionless time and ξ and η the similarity variables. Substitution of Eq. (7.6) into Eqs. (7.2)–(7.5) yields the following differential equations

$$\frac{\partial^3 f}{\partial\eta^3} + \xi\left(f\frac{\partial^2 f}{\partial\eta^2} - \left(\frac{\partial f}{\partial\eta}\right)^2\right) + (1-\xi)\left(\frac{\eta}{2}\frac{\partial^2 f}{\partial\eta^2} - \xi\frac{\partial^2 f}{\partial\eta\partial\xi}\right) = 0, \tag{6.7}$$

$$\frac{1}{Pr}\frac{\partial^2\theta}{\partial\eta^2} + \xi f\frac{\partial\theta}{\partial\eta} + (1-\xi)\left(\frac{\eta}{2}\frac{\partial\theta}{\partial\eta} - \xi\frac{\partial\theta}{\partial\xi}\right) + Nb\theta'\phi' + Nt\theta'^2 = 0, \tag{6.8}$$

$$\frac{\partial^2\phi}{\partial\eta^2} + Sc\xi f\frac{\partial\phi}{\partial\eta} + Sc(1-\xi)\left(\frac{\eta}{2}\frac{\partial\phi}{\partial\eta} - \xi\frac{\partial\phi}{\partial\xi}\right) + \frac{Nt}{Nb}\frac{\partial^2\theta}{\partial\eta^2} = 0, \tag{6.9}$$

subject to the boundary conditions

$$f(\xi,0) = 0, \quad \left.\frac{\partial f}{\partial\eta}\right|_{\eta=0} = 1, \quad \theta(\xi,0) = 1, \quad \phi(\xi,0) = 1,$$

$$\left.\frac{\partial f}{\partial\eta}\right|_{\eta=+\infty} \to 0, \quad \theta(\xi,+\infty) \to 0, \quad \phi(\xi,+\infty) \to 0, \tag{6.10}$$

where $Pr = \nu/\alpha$ is the Prandtl number, $Sc = \nu/D_B$ the Schmidt number,

$$Nb = \frac{(\rho c)_p D_B(C_w - C_\infty)}{(\rho c)_f\nu}, \quad Nt = \frac{(\rho c)_p D_T(T_w - T_\infty)}{(\rho c)_f T_\infty\nu}$$

are the Brownian motion parameter and the thermophoresis parameter, respectively.

The skin friction coefficient C_f^x, the local Nusselt number Nu_x and the local Sherwood number Sh are given by

$$C_f^x = \frac{\tau_w}{\rho u_w^2}, Nu_x = \frac{xq_w}{k(T_w - T_\infty)}, Sh = \frac{xj_w}{D_T(T_w - T_\infty)}, \tag{6.11}$$

where the wall shear stress (τ_w) , wall heat flux (q_w) and the mass flux from the sheet (j_w) are defined by

$$\tau_w = \mu\left(\frac{\partial u}{\partial y}\right)_{y=0}, \; q_w = -k\left(\frac{\partial T}{\partial y}\right)_{y=0}, \; j_w = -D_B\left(\frac{\partial C}{\partial y}\right)_{y=0}. \tag{6.12}$$

In terms of dimensionless quantities (7.6) one obtains

$$\xi^{1/2} Re_x^{1/2} C_f^x = \left.\frac{\partial^2 f(\xi,\eta)}{\partial \eta^2}\right|_{\eta=0},$$

$$Nur = Nu_x/\xi^{1/2} Re_x^{1/2} = -\left.\frac{\partial \theta(\xi,\eta)}{\partial \eta}\right|_{\eta=0},$$

$$Shr = Sh/\xi^{1/2} Re_x^{1/2} = -\left.\frac{\partial \phi(\xi,\eta)}{\partial \eta}\right|_{\eta=0},$$

where Nur and Shr denote the reduced Nusselt and Sherwood numbers characterizing the rate of heat and nanoparticles transfer at the sheet. We now consider the following special cases:

(i) *Initial unsteady flow.* This solution corresponds to $\xi = 0$ ($\tau = 0$), where $f(0,\eta) = f_u(\eta)$, $\theta(0,\eta) = \theta_u(\eta)$ and $\phi(0,\eta) = \phi_u(\eta)$.Thus Eqs.(7.7)–(7.10) reduce to

$$f_u''' + \frac{1}{2}\eta f_u'' = 0, \tag{6.13}$$

$$\frac{1}{Pr}\theta_u'' + \frac{1}{2}\eta\theta_u' + Nb\theta_u'\phi_u' + Nt\theta_u'^2 = 0, \tag{6.14}$$

$$\phi_u'' + \frac{1}{2}Sc\eta\phi_u' + \frac{Nt}{Nb}\theta_u'' = 0, \tag{6.15}$$

subject to the boundary conditions

$$\begin{aligned} &f_u(0) = 0, \quad f_u'(0) = 1, \quad f_u'(\infty) = 1, \\ &\theta_u(0) = \phi_u(0) = 1, \quad \theta_u(\infty) = \phi_u(\infty) = 0, \end{aligned} \tag{6.16}$$

where prime denotes differentiation with respect to η. Eq. (7.15) subject to the boundary conditions (7.18) admits the exact solution (see Liao [4])

$$f_u(\eta) = \eta\left(1 - \operatorname{erf}\left(\frac{\eta}{2}\right)\right) + \frac{2}{\sqrt{\Pi}}\left(1 - \exp\left(\frac{-\eta^2}{4}\right)\right), \tag{6.17}$$

where erf is the error function defined as

$$\operatorname{erf(s)} = \frac{2}{\sqrt{\Pi}}\int_0^s e^{-z^2}dz. \tag{6.18}$$

Thus skin friction coefficient in the initial unsteady flow becomes

$$\xi^{1/2} Re_x^{1/2} C_f^x = -\frac{1}{\sqrt{\Pi}}. \tag{6.19}$$

(ii) *Final steady-state flow.* For this solution $\xi = 1$ as $\tau \to \infty$, where $f(1,\eta) = f_s(\eta)$, $\theta(1,\eta) = \theta_s(\eta)$ and $\phi(1,\eta) = \phi_s(\eta)$. Eqs. (7.7)–(7.9) now become

$$f_s^{'''} + f_s f_s^{''} - f_s^{'2} = 0, \tag{6.20}$$

$$\frac{1}{\Pr}\theta_s^{''} + f_s\theta_s^{'} + Nb\theta_s^{'}\phi_s^{'} + Nt\theta_s^{'2} = 0, \tag{6.21}$$

$$\phi_s^{''} + \mathrm{Sc} f_s\phi_s^{'} + \frac{Nt}{Nb}\theta_s^{''} = 0, \tag{6.22}$$

subject to the same boundary conditions (7.18) with index s. Here Eq. (7.22) with the relevant boundary conditions has the exact solution $f_s = 1 - \exp(-\eta)$. Based on the rule of solution expression and the involved boundary conditions, the initial guesses f_0, θ_0 and ϕ_0 of $f(\eta)$, $\theta(\eta)$ and $\phi(\eta)$ are chosen as

$$f_0(\xi,\eta) = 1 - \exp(-\eta), \quad \theta_0(\xi,\eta) = \phi_0(\xi,\eta) = \exp(-\eta), \tag{6.23}$$

and the auxiliary linear operators are selected as

$$\mathcal{L}_f(f) = \frac{\partial^3 f}{\partial\eta^3} - \frac{\partial f}{\partial\eta}, \quad \mathcal{L}_\theta(\theta) = \frac{\partial^2\theta}{\partial\eta^2} - \theta, \quad \mathcal{L}_\phi(\phi) = \frac{\partial^2\phi}{\partial\eta^2} - \phi. \tag{6.24}$$

If $q \in [0,1]$ is an embedding parameter and $\hbar$ denotes the non-zero auxiliary parameters then the generalized homotopic equations are constructed as follows:

$$(1-q)\mathcal{L}_f[\hat{f}(\xi,\eta;q) - f_0(\eta)] \\ = q\,\hbar\,\mathcal{N}_f\left[\hat{f}(\xi,\eta;q)\right], \tag{6.25}$$

$$(1-q)\mathcal{L}_\theta[\Theta(\xi,\eta;q) - \theta_0(\eta)] \\ = q\,\hbar\,\mathcal{N}_\theta\left[\hat{f}(\xi,\eta;q), \Theta(\xi,\eta;q), \Phi(\xi,\eta;q)\right], \tag{6.26}$$

$$(1-q)\mathcal{L}_\phi[\Phi(\xi,\eta;q) - \phi_0(\eta)] \\ = q\,\hbar\,\mathcal{N}_\phi\left[\hat{f}(\xi,\eta;q), \Theta(\xi,\eta;q), \Phi(\xi,\eta;q)\right], \tag{6.27}$$

subject to the boundary conditions

$$\hat{f}(\xi,\eta;q)\Big|_{\eta=0} = 0, \ \frac{\partial\hat{f}}{\partial\eta}(\xi,\eta;q)\Big|_{\eta=0} = 1, \ \frac{\partial\hat{f}}{\partial\eta}(\xi,\eta;q)\Big|_{\eta\to\infty} = 0, \tag{6.28}$$

$$\Theta(\xi,\eta;q)|_{\eta=0} = 1, \quad \Theta(\xi,\eta;q)|_{\eta\to\infty} = 0, \tag{6.29}$$

$$\Phi(\xi,\eta;q)|_{\eta=0} = 1, \quad \Phi(\xi,\eta;q)|_{\eta\to\infty} = 0, \tag{6.30}$$

in which the non-linear operators $\mathcal{N}_f, \mathcal{N}_\theta$ and $\mathcal{N}_\phi$ are

$$\begin{aligned}
&\mathcal{N}_f\left[\hat{f}(\xi,\eta;q)\right]\\
&= \frac{\partial^3 \hat{f}(\xi,\eta;q)}{\partial\eta^3}\\
&\quad + \xi\left\{\hat{f}(\xi,\eta;q)\frac{\partial^2 \hat{f}(\xi,\eta;q)}{\partial\eta^2} - \left(\frac{\partial \hat{f}(\xi,\eta;q)}{\partial\eta}\right)^2\right\}\\
&\quad + (1-\xi)\left(\frac{\eta}{2}\frac{\partial^2 \hat{f}(\xi,\eta;q)}{\partial\eta^2} - \xi\frac{\partial^2 \hat{f}(\xi,\eta;q)}{\partial\eta\partial\xi}\right),
\end{aligned} \tag{6.31}$$

$$\begin{aligned}
&\mathcal{N}_\theta\left[\hat{f}(\xi,\eta;q),\Theta(\xi,\eta;q),\Phi(\xi,\eta;q)\right]\\
&= \frac{1}{Pr}\frac{\partial^2\Theta(\xi,\eta;q)}{\partial\eta^2}\\
&\quad + \xi\hat{f}(\eta,p)\frac{\partial\Theta(\xi,\eta;q)}{\partial\eta} + Nb\frac{\partial\Theta(\xi,\eta;q)}{\partial\eta}\frac{\partial\Phi(\xi,\eta;q)}{\partial\eta}\\
&\quad + Nt\left(\frac{\partial\Theta(\xi,\eta;q)}{\partial\eta}\right)^2 + (1-\xi)\left(\frac{\eta}{2}\frac{\partial\Theta(\xi,\eta;q)}{\partial\eta} - \xi\Theta\right),
\end{aligned} \tag{6.32}$$

$$\begin{aligned}
&\mathcal{N}_\phi\left[\hat{f}(\xi,\eta;q),\Theta(\xi,\eta;q),\Phi(\xi,\eta;q)\right]\\
&= \frac{\partial^2\Phi(\xi,\eta;q)}{\partial\eta^2}\\
&\quad + Sc\hat{f}(\xi,\eta;q)\frac{\partial\Phi(\xi,\eta;q)}{\partial\eta} + \frac{Nt}{Nb}\frac{\partial^2\Theta(\xi,\eta;q)}{\partial\eta^2}\\
&\quad + Sc(1-\xi)\left(\frac{\eta}{2}\frac{\partial\Phi(\xi,\eta;q)}{\partial\eta} - \xi\Phi\right).
\end{aligned} \tag{6.33}$$

By Taylor's series

$$\hat{f}(\xi,\eta;q) = f_0(\xi,\eta) + \sum_{m=1}^{\infty} f_m(\xi,\eta)q^m, \tag{6.34}$$

$$\Theta(\xi,\eta;q) = \theta_0(\xi,\eta) + \sum_{m=1}^{\infty} \theta_m(\xi,\eta)q^m, \tag{6.35}$$

$$\Phi(\xi,\eta;q) = \phi_0(\xi,\eta) + \sum_{m=1}^{\infty} \phi_m(\xi,\eta)q^m, \tag{6.36}$$

where

$$f_m(\xi,\eta) = \frac{1}{m!}\left.\frac{\partial^m \hat{f}(\xi,\eta;q)}{\partial \eta^m}\right|_{p=0},$$

$$\theta_m(\xi,\eta) = \frac{1}{m!}\left.\frac{\partial^m \Theta(\xi,\eta;q)}{\partial \eta^m}\right|_{q=0},$$

$$\phi_m(\xi,\eta) = \frac{1}{m!}\left.\frac{\partial^m \Phi(\xi,\eta;q)}{\partial \eta^m}\right|_{q=0},$$

the final solutions are retrieved at $q=1$. The functions f_m, θ_m and ϕ_m are obtained from the deformation of Eqs. (7.27)–(7.32). Explicitly mth-order deformation problems corresponding to Eqs. (7.27)–(7.32) are

$$\mathcal{L}_f\left[f_m(\xi,\eta) - \chi_m f_{m-1}(\xi,\eta)\right] = \hbar_f\, R_m^f(\xi,\eta), \tag{6.37}$$

$$\mathcal{L}_\theta\left[\theta_m(\xi,\eta) - \chi_m \theta_{m-1}(\xi,\eta)\right] = \hbar_\theta\, R_m^\theta(\xi,\eta), \tag{6.38}$$

$$\mathcal{L}_\phi\left[\phi_m(\xi,\eta) - \chi_m \phi_{m-1}(\xi,\eta)\right] = \hbar_\phi\, R_m^\phi(\xi,\eta), \tag{6.39}$$

subject to the boundary conditions

$$\begin{cases} f_m(\xi,0)=0, \quad \left.\frac{\partial f_m(\xi,\eta)}{\partial\theta}\right|_{\eta=0}=0, \quad \left.\frac{\partial f_m(\xi,\eta)}{\partial\theta}\right|_{\eta\to+\infty}=0, \\ \theta_m(\xi,0)=0, \theta_m(\xi,\infty)=0, \phi_m(\xi,0)=0, \phi_m(\xi,\infty)=0, \end{cases} \tag{6.40}$$

where

$$R_m^f(\xi,\eta) = \frac{\partial^3 f_{m-1}}{\partial\eta^3} + \xi\sum_{k=0}^{m-1}\left[f_{m-1-k}\frac{\partial^2 f_k}{\partial\eta^2} - \frac{\partial f_{m-1-k}}{\partial\eta}\frac{\partial f_k}{\partial\eta}\right] + (1-\xi)\left(\frac{\eta}{2}\frac{\partial^2 f_{m-1}}{\partial\eta^2} - \xi\frac{\partial^2 f_{m-1}}{\partial\eta\partial\xi}\right), \tag{6.41}$$

$$R_m^\theta(\xi,\eta) = \xi\sum_{k=0}^{m-1}\left[f_{m-1-k}\frac{\partial\theta_k}{\partial\eta} + Nb\frac{\partial\theta_{m-1-k}}{\partial\eta}\frac{\partial\phi_k}{\partial\eta} + Nt\frac{\partial\theta_{m-1-k}}{\partial\eta}\frac{\partial\theta_k}{\partial\eta}\right] + \frac{1}{Pr}\frac{\partial^2\theta_{m-1}}{\partial\eta^2} + (1-\xi)\left(\frac{\eta}{2}\frac{\partial\theta_{m-1}}{\partial\eta} - \xi\frac{\partial\theta_{m-1}}{\partial\xi}\right), \tag{6.42}$$

$$R_m^\phi(\xi,\eta) = \frac{\partial^2\theta_{m-1}}{\partial\eta^2} + Sc\sum_{k=0}^{m-1} f_{m-1-k}\frac{\partial\phi_k}{\partial\eta} + \frac{Nt}{Nb}\frac{\partial^2\theta_{m-1}}{\partial\eta^2} + Sc(1-\xi)\left(\frac{\eta}{2}\frac{\partial\phi_{m-1}}{\partial\eta} - \xi\frac{\partial\phi_{m-1}}{\partial\xi}\right), \tag{6.43}$$

and

$$\chi_m = \begin{cases} 0, & m \le 1, \\ 1, & m > 1. \end{cases}$$

As pointed out by Liao [1] that convergence of the series solutions given in Eqs. (6.34)–(6.36) largely depends on the choice of auxiliary parameter $\hbar$. In order to obtain an appropriate value of $\hbar$, the $\hbar$-curves for the functions f, θ and ϕ at 15th-order of approximations are plotted in Fig. 6.1(a) and 6.1(b). Here the permissible range of $\hbar$ lies in the flat portion of $\hbar$-curves. Thus it is clear that interval of convergence for f, θ and ϕ is $[-0.7, -0.35]$ when $\xi = 1/2$. Our computations show that series solutions are uniformly convergent for all values of parameters when $\hbar = -0.4$. The initial unsteady and steady-state problems for temperature and concentration are solved numerically through command *NDSolve* of the software Mathematica. The 10th-order homotopy solution is compared with the numerical solution for the initial unsteady flow ($\xi = 0$). It is witnessed that both the solutions are in excellent agreement for different values of *Nb* and *Nt* (see Fig. 6.2). In addition the numerical values of reduced Nusselt and Sherwood numbers in the initial unsteady flow have been compared with those obtained via numerical solution (see Table 6.1). The 25th-order HAM solutions seem to be in a very good agreement with the numerical solutions for all the values of *Nb* and *Nt*.

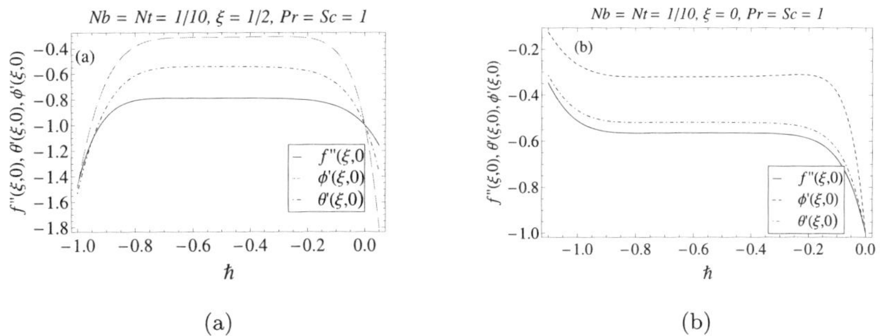

Fig. 6.1. $\hbar$-curves for the functions f, θ and ϕ when (a) ξ=1/2 and (b) ξ=0.

Fig. 6.3(a) plots the temperature profiles for different values of dimensionless time τ. We notice that profiles continuously develop from rest as τ increases until the steady-state situation is achieved. This observation leads to the conclusion that thermal boundary layer thickens as we move from initial unsteady flow ($\xi = 0$) to the final steady-state flow ($\xi = 1$). Further it is revealed that $\theta(0, +\infty)$ of the initial solution tends to zero much faster than $\theta(1, +\infty)$ of the steady-state solution. Nanoparticles concentration profiles ϕ for different (dimensionless) times are presented in

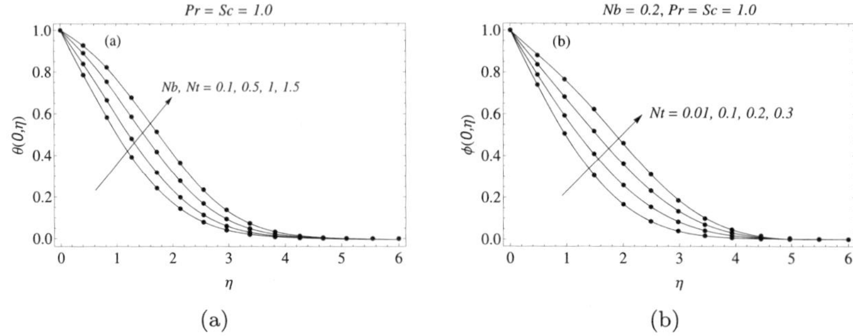

Fig. 6.2. Comparison of HAM and numerical solutions for the initial unsteady flow at $\xi = 0$. Lines: 10th-order HAM solutions at $\hbar = -0.4$; Points: Numerical solutions.

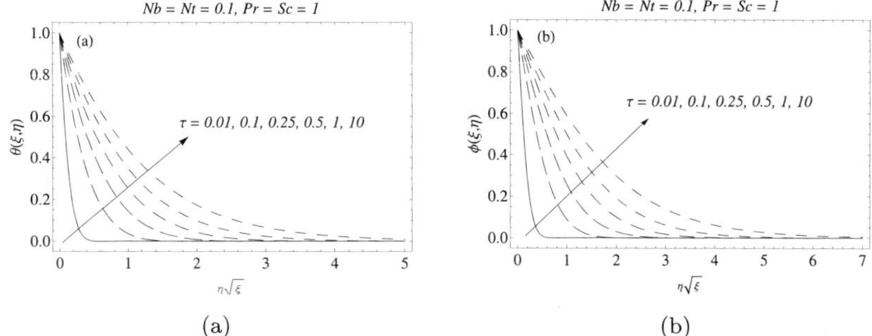

Fig. 6.3. Temperature and concentration profiles for various values of unsteadiness parameter τ.

Fig. 6.3(b). The behavior of dimensionless time τ on ϕ is quite similar to that accounted for temperature θ. Combined behavior of Nb and Nt on the reduced Nusselt number is sketched in Fig. 6.4(a). There is a decrease in the magnitude of Nur when Nb and Nt are increased. Interestingly, this reduction is uniform in both stronger and weaker Brownian motion and thermophoresis effects. The reduced Sherwood number Shr corresponding to the data given in Fig. 6.4(a) has been plotted in 6.4(b). We notice that for a weaker Brownian motion, the reduced Sherwood number drastically decreases with an increase in Nb. However the reduced Sherwood number is negligibly affected for sufficiently stronger Brownian motion effect. On the other hand, there is a slight increase in the mass transfer at the sheet with an increase in Brownian motion for a weaker thermophoretic effect. However when the strength of thermophoretic effect is increased i.e., Nt changes

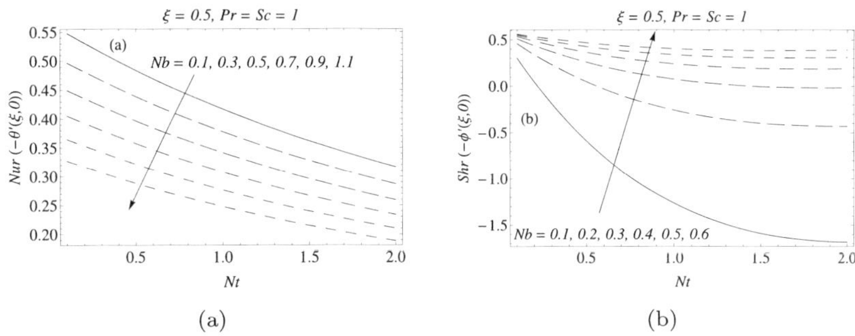

(a) (b)

Fig. 6.4. Dimensionless heat and mass transfer rates versus Nt for different values of Nb.

from 0.1 to 2, there is a massive increase in the nanoparticles concentration at the sheet with an increase in *Nb*. This is not surprising since the increasing values of *Nt* give rise to the effective movement of nanoparticles from the sheet to the fluid.

Table 6.1. Comparison of 25th-order HAM solutions with the numerical solutions for the steady-state flow ($\xi = 1$).

Nb	Nt	Nur HAM	Nur Numerical	Shr HAM	Shr Numerical
0.1	0.1	0.5372744	0.5372744	0.2244510	0.2244554
0.5		0.4351430	0.4351427	0.5275673	0.5275684
1.0		0.3282822	0.3282814	0.5634962	0.5634968
1.5		0.2428616	0.2428604	0.5741530	0.5741535
0.5	0.5	0.3861584	0.3861534	0.3471842	0.3471869
	1.0	0.3343331	0.3343109	0.1888924	0.1888942
	1.5	0.2913213	0.2912990	0.0842004	0.0842442
	2.0	0.2556406	0.2555893	0.0163227	0.0164096

6.3. Axisymmetric flow of nanofluid over a radially stretching sheet with convective boundary conditions

The flow of nanofluid (obeying the Buongiorno model [15]) past a radially stretching sheet has been investigated by Mustafa *et al.* [31]. The cylindrical coordinates system (r, θ, z) is adopted. The azimuthal component v of velocity $\mathrm{V} = [u, v, w]$ vanishes identically due to the rotational symmetry of the flow ($\partial/\partial\theta \equiv 0$). The sheet at $z = 0$ is stretched along the radial direction with the velocity $u_w = cr$. Here T_w denotes the convective temperature of the sheet while C_w is the nanoparticles concentration at the sheet. The ambient temperature and concentration are denoted by T_∞ and C_∞ respectively. Using the mathematical model formulated by Kuznetsov and Nield [21] the transport equations in cylindrical coordinates are as under

$$\frac{\partial u}{\partial r} + \frac{u}{r} + \frac{\partial w}{\partial z} = 0, \tag{6.44}$$

$$u\frac{\partial u}{\partial r} + w\frac{\partial u}{\partial z} = -\frac{1}{\rho_f}\frac{\partial \hat{p}}{\partial r} + \nu\left(\frac{\partial^2 u}{\partial r^2} + \frac{\partial^2 u}{\partial z^2} + \frac{1}{r}\frac{\partial u}{\partial r} - \frac{u}{r^2}\right), \tag{6.45}$$

$$u\frac{\partial w}{\partial r} + w\frac{\partial w}{\partial z} = -\frac{1}{\rho_f}\frac{\partial \hat{p}}{\partial z} + \nu\left(\frac{\partial^2 w}{\partial r^2} + \frac{\partial^2 w}{\partial z^2} + \frac{1}{r}\frac{\partial w}{\partial r}\right), \tag{6.46}$$

$$\begin{aligned} u\frac{\partial T}{\partial r} + w\frac{\partial T}{\partial z} &= \alpha\left(\frac{\partial^2 T}{\partial r^2} + \frac{1}{r}\frac{\partial T}{\partial r} + \frac{\partial^2 T}{\partial z^2}\right) \\ &\quad + \tau\left[D_B\left(\frac{\partial C}{\partial r}\frac{\partial T}{\partial r} + \frac{\partial C}{\partial z}\frac{\partial T}{\partial z}\right)\right] \\ &\quad + \tau\left[\frac{D_T}{T_\infty}\left[\left(\frac{\partial T}{\partial r}\right)^2 + \left(\frac{\partial T}{\partial z}\right)^2\right]\right], \end{aligned} \tag{6.47}$$

$$\begin{aligned} u\frac{\partial C}{\partial r} + w\frac{\partial C}{\partial z} &= D_B\left(\frac{\partial^2 C}{\partial r^2} + \frac{1}{r}\frac{\partial C}{\partial r} + \frac{\partial^2 C}{\partial z^2}\right) \\ &\quad + \frac{D_T}{T_\infty}\left(\frac{\partial^2 T}{\partial r^2} + \frac{1}{r}\frac{\partial T}{\partial r} + \frac{\partial^2 T}{\partial z^2}\right). \end{aligned} \tag{6.48}$$

The boundary conditions for the problem under consideration are

$$\begin{aligned} &u = u_w(r) = cr, -k\frac{\partial T}{\partial z} = h_s\left(T_w - T\right),\ C = C_w \quad \text{at} \quad y = 0, \\ &u \to 0, \quad T \to T_\infty, \quad C \to C_\infty \quad \text{as} \quad y \to \infty, \end{aligned} \tag{6.49}$$

in which u and w are the velocity components along r and z directions respectively, $\hat{p}$ is the pressure, ν is the kinematic viscosity, α is the thermal diffusivity, D_B is the Brownian motion coefficient, D_T is the thermophoretic diffusion coefficient, k is the thermal conductivity and $\tau = (\rho c)_p/(\rho c)_f$ is the ratio of effective heat capacity of the nanoparticle material to heat capacity of the fluid. Implementing the following transformations

$$\eta = \sqrt{\frac{c}{\nu}}z, u = crf^{'}(\eta), w = -2\sqrt{c\nu}f(\eta),$$
$$\theta(\eta) = \frac{T - T_\infty}{T_w - T_\infty}, \phi(\eta) = \frac{C - C_\infty}{C_w - C_\infty}, \quad (6.50)$$

into Eqs. (6.45) and (6.46) and eliminating the pressure gradient from the resulting equations we finally obtain (see Ariel [32])

$$f^{'''} + 2ff^{''} - f^{'2} = 0, \quad (6.51)$$

and Eqs. (6.47) and (6.48) in view of Eq. (6.50) become

$$\frac{1}{Pr}\theta^{''} + 2f\theta^{'} + Nb\theta^{'}\phi^{'} + Nt\theta^{'2} = 0, \quad (6.52)$$

$$\phi^{''} + 2Lef\phi^{'} + \frac{Nb}{Nt}\theta^{''} = 0, \quad (6.53)$$

subject to the boundary conditions

$$f(0) = 0,\ f^{'}(0) = 1,\ \theta^{'}(0) = -Bi(1 - \theta(0)),\ \phi(0) = 1,$$
$$f^{'}(\infty) \to 0,\ \theta(\infty) \to 0,\ \phi(\infty) \to 0. \quad (6.54)$$

where $Pr = \nu/\alpha$ is the Prandtl number, $Le = \nu/D_B$ is the Lewis number, Nb is the Brownian motion parameter, Nt is the thermophoresis parameter and $Bi = \frac{h_s}{k}\sqrt{\frac{\nu}{c}}$ is the Biot number. It is worth mentioning here that the boundary conditions (6.54) reduce to the the case of constant wall temperature ($\theta(0) = 1$) when $Bi \to \infty$. Further $Nb = Nt = 0$ corresponds to the situation in which Brownian motion and thermophoresis effects are absent. The skin friction coefficient C_f, the local Nusselt number Nu and the local Sherwood number Sh are given by

$$C_f = \frac{\mu(\frac{\partial u}{\partial z} + \frac{\partial w}{\partial r})_{z=0}}{\rho u_w^2}, Nu = -\frac{r(\frac{\partial T}{\partial z})_{z=0}}{k(T_w - T_\infty)}, Sh = -\frac{r(\frac{\partial C}{\partial z})_{z=0}}{D_B(C_w - C_\infty)},$$

Using (6.50) in the above formula gives

$$Re_r^{1/2}C_f = f^{''}(0), Nu/Re_r^{1/2} = -\theta^{'}(0) = Nur, Sh/Re_r^{1/2} = -\phi^{'}(0) = Shr,$$

where $Re_r = u_w r/\nu$ denotes the local Reynolds number.

The analytic solutions of Eqs. (6.51)–(6.53) subject to the boundary conditions (6.54) have been obtained by homotopy analysis method (HAM). Following initial guesses and linear operators for the functions f, θ and ϕ are selected

$$f_0(\eta) = 1 - \exp(-\eta), \theta_0(\eta) = \frac{Bi}{1 + Bi}\exp(-\eta), \phi_0 = \exp(-\eta), \quad (6.55)$$

$$\mathcal{L}_f(f) = f^{'''} - f^{'}, \mathcal{L}_\theta(\theta) = \theta^{''} - \theta, \mathcal{L}_\phi(\phi) = \phi^{''} - \phi. \tag{6.56}$$

We denote the non-zero auxiliary parameters for the functions f, θ and ϕ by $\hbar_f$, $\hbar_\theta$ and $\hbar_\phi$ respectively. In view of the Eqs. (6.55) and (6.56), the higher order deformation equations are formulated which are solved by symbolic computational software **Mathematica**. The admissible values of the auxiliary parameters can be determined from the line segment parallel to $\hbar$-axis in the Fig. 6.5. It can be seen that the permissible range of values of $\hbar_f$, $\hbar_\theta$ and $\hbar_\phi$ are $-1 \leq \hbar_f \leq -0.4$, $-1.4 \leq \hbar_\theta \leq -0.5$ and $-1.2 \leq \hbar_\phi \leq -0.7$ respectively. The series solutions are found to be convergent in the whole spatial domain when $\hbar_f = \hbar_\theta = \hbar_\phi = -1$ (see Table 6.2). Further the governing differential system has also been solved numerically through the software **Mathematica**. This software uses the shooting method as a default method for the numerical solution of nonlinear boundary value problems. Fig. 6.6 is displayed just to see the comparison of analytic and numerical solutions. It is clear that data obtained by both the solutions is virtually similar demonstrating the validation of current findings. The simultaneous behavior of Nb and Nt on the reduced Nusselt number is sketched in Fig. 6.7(a). There is a decrease in the magnitude of Nur when Nb and Nt are increased. Interestingly, this reduction is uniform in both stronger and weaker Brownian motion and thermophoresis effects. Fig. 6.7(b) plots the reduced sherwood number Shr against Nt for different values of Nb. It is seen that for a weaker thermophoretic effect, the change in the Brownian motion parameter slightly affects the reduced Sherwood number. However the nanoparticles concentration at the sheet effectively increases when Nb is increased from 0.1 to 0.5. Moreover the reduced Sherwood number is found to decrease upon increasing the strength of thermophoretic effect. This is not surprising since the increasing values of Nt give rise to the effective movement of nanoparticles from the sheet to the fluid. Table 6.3 shows the numerical values of Nur and Shr for different values of Pr, Le and Bi. As seen earlier in the graphical results that thermal boundary layer thickness decreases and profiles get closer to the boundary when Pr is increased . The reduced Nusselt and Sherwood numbers (which are proportional to the initial slopes) are therefore increasing functions of Pr and Le respectively. Further it is noticed that an increase in the Biot number Bi corresponds to the higher convective heating at the sheet which results in the increase of the rate of heat transfer at the sheet. Moreover it is observed that larger values of Bi have a little impact on the reduced Nusselt number Nur.

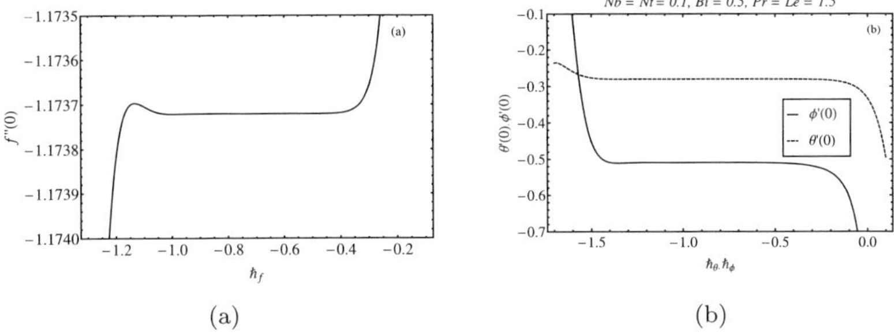

(a) (b)

Fig. 6.5. $\hbar$-curves for the functions f, θ and ϕ.

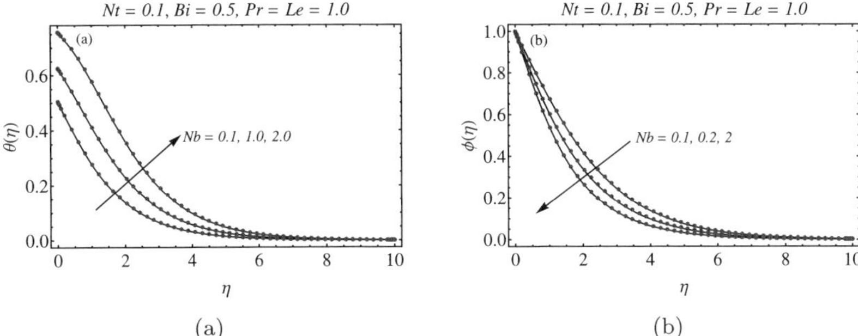

(a) (b)

Fig. 6.6. Comparison of 10th-order HAM solutions with the numerical solution: Lines: HAM solution; Points: Numerical solution.

6.4. Squeezing flow of nanofluid between parallel disks

The study of squeezing flows between parallel disks has particular relevance in many industrial applications which include polymer processing, compression, injection modeling, transient loading of mechanical components and the squeezed films in power transmission. The application of magnetic field in such flows allows us to prevent the unpredictable deviation of lubrication viscosity with temperature in certain extreme operating conditions. The seminal work on the topic under lubrication approximation has been studied by Stefan [33]. Mustafa *et al.* [25] recently considered the squeezing flow of nanofluid between parallel disks in which one of these is impermeable and the other is porous. The disks are separated by a distance $h(t) = H(1 - at)^{1/2}$. A magnetic field of strength $B(t) = B_0(1 - at)^{-1/2}$ is applied perpendicular to the disks. Here T_w and C_w denote the temperature and nanoparticles concentration at the lower disk while the tempera-

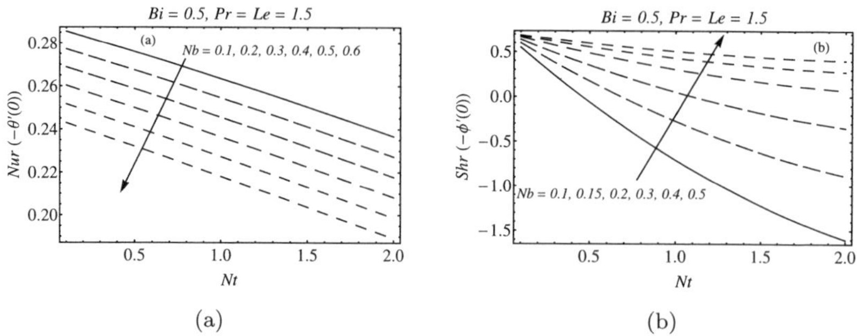

(a) (b)

Fig. 6.7. Dimensionless heat and mass transfer rates for different values of Nb and Nt.

Table 6.2. Convergence rate of HAM solutions when $Nb = Nt = 0.1$, $Pr = Le = 1$ and $\hbar_f = \hbar_\theta = \hbar_\phi = -0.8$. Parenthesis shows the numerical solutions.

Order of approximations	$Re_r^{1/2}C_f$	$Nu/Re_r^{1/2}$	$Sh/Re_r^{1/2}$
1	−1.166667	0.3145679	0.7333333
5	−1.173897	0.3078442	0.6540525
10	−1.173756	0.3078230	0.6501421
15	−1.173712	0.3078225	0.6501530
20	−1.173721	0.3078225	0.6501541
25	−1.173721	0.3078225	0.6501451
35	−1.173721	0.3078225	0.6501451
	(−1.17372)	(0.307825)	(0.650145)

ture and concentration at the upper disk are T_h and C_h respectively. The upper disk at $z = h(t)$ moves towards or away from the stationary lower disk with the velocity dh/dt. The equations governing the conservation of mass, momentum, energy and nanoparticles concentration in cylindrical coordinates are similar to those considered in the previous section with an addition of magnetic field term i.e., $-\sigma/\rho B(t)^2 u$ in the r-component of the momentum equation. The boundary conditions for this problem are (see Domairry and Aziz [34] and Joneidi *et al.* [35].)

$$
\begin{aligned}
&u = 0, \quad w = \frac{dh}{dt}, \quad T = T_h, \quad C = C_h \quad \text{at} \quad z = h(t),\\
&u = 0, \quad w = -\frac{w_0}{\sqrt{1 - at}}, \quad T = T_w, \quad C = C_w \quad \text{at} \quad z = 0. \quad (6.57)
\end{aligned}
$$

Table 6.3. Values of reduced Nusselt and Sherwood numbers $Nur = -\theta'(0)$ and $Shr = -\phi'(0)$ for different values of Pr, Le and Bi when $Nb = Nt = 0.1$.

Pr	Le	Bi	Nur	Shr
0.4	1	0.5	0.230078	0.710803
0.72			0.281948	0.671917
2			0.353934	0.605337
10			0.422650	0.51539
1	0.4		0.310557	0.193614
	0.7		0.308900	0.446367
	2		0.305905	1.160300
	10		0.303021	3.222890
	1.0	0.1	0.088980	0.793545
		0.4	0.266901	0.676928
		1.0	0.443452	0.561546
		2.0	0.567859	0.480446

Using the following transformations

$$u = \frac{ar}{2(1-at)}f'(\eta), \quad w = -\frac{aH}{\sqrt{1-at}}f(\eta), \quad \eta = \frac{z}{H\sqrt{1-at}},$$
$$B(t) = \frac{B_0}{\sqrt{1-at}}, \quad \theta = \frac{T-T_h}{T_w-T_h}, \quad \phi = \frac{C-C_h}{C_w-C_h}, \tag{6.58}$$

into the relevant equations and then eliminating the pressure gradient from the resulting equations we finally obtain

$$f'''' - S(\eta f''' + 3f'' - 2ff''') - M^2 f'' = 0. \tag{6.59}$$

Now dimensionless energy and equation of mass transfer become

$$\theta'' + PrS(2f\theta' - \eta\theta') + PrNb\theta'\phi' + PrNt\theta'^2 = 0, \tag{6.60}$$

$$\phi'' + LeS(2f\phi' - \eta\phi') + \frac{Nt}{Nb}\theta'' = 0, \tag{6.61}$$

with the boundary conditions

$$f(0) = A, \quad f'(0) = 0, \quad \theta(0) = \phi(0) = 1,$$
$$f(1) = \frac{1}{2}, \quad f'(1) = 0, \quad \theta(1) = \phi(1) = 0, \tag{6.62}$$

where S is the squeeze number, A is the suction/blowing parameter, M is the Hartman number, Nb is the Brownian motion parameter, Nt is the thermophoretic parameter, Pr is the Prandtl number and Le is the Lewis number which are defined as

$$A=\frac{w_0}{aH},\ S=\frac{aH^2}{2\nu},\ M=\sqrt{\frac{\sigma B_0^2H^2}{\nu}},\ Pr=\frac{\nu}{\alpha},\ Le=\frac{\nu}{D_e},$$
$$Nb=\frac{(\rho c)_pD_B(C_w-C_h)}{(\rho c)_f\nu},\ Nt=\frac{(\rho c)_pD_T(T_w-T_h)}{(\rho c)_fT_m\nu}.$$

The continuity equation is identically satisfied. It is worth mentioning here that $A>0$ indicates the suction of fluid from the lower disk while $A<0$ represents injection flow. For $Nb=Nt=0$, the problem reduces to the case of ordinary fluid (in which Brownian motion and thermophoretic effects are negligible). The physical quantities of interest are the skin friction coefficient C_{fr}, reduced Nusselt number Nur and reduced Sherwood number Shr which are defined by

$$C_{fr}=\frac{\tau_{rz}|_{z=h(t)}}{\rho\left(\frac{-aH}{2(1-at)^{1/2}}\right)^2},\quad Nu=\frac{Hq_w}{k(T_w-T_h)},\quad Sh=\frac{Hj_w}{D_B(C_w-C_h)},$$

where

$$\tau_{rz}=\mu\left(\frac{\partial u}{\partial z}+\frac{\partial w}{\partial r}\right)\bigg|_{z=h(t)},$$
$$q_w=-k\left(\frac{\partial T}{\partial z}\right)\bigg|_{z=h(t)},$$
$$j_w=-D_B\left(\frac{\partial C}{\partial z}\right)\bigg|_{z=h(t)}.$$

In terms of variables (6.58) we have

$$\frac{H^2}{r^2}Re_rC_{fr}=f''(1),\quad Nur=(1-at)^{1/2}Nu=-\theta'(1),$$
$$Shr=(1-at)^{1/2}Sh=-\phi'(1),\quad Re_r=\frac{raH(1-at)^{1/2}}{2\nu}.$$

The rule of solution expression and the involved boundary conditions direct us to select the initial guesses

$$f_0(\eta)=(-1+2A)\eta^3+\frac{1}{2}(3-6A)\eta^2+A,\theta_0(\eta)=\phi_0(\eta)=1-\eta,\quad (6.63)$$

and the auxiliary linear operators

$$L_f\equiv\frac{d^4}{d\eta^4}-\frac{d}{d\eta},\quad L_\theta\equiv\frac{d^2}{d\eta^2},\quad L_\phi\equiv\frac{d^2}{d\eta^2},\quad (6.64)$$

respectively. After solving the higher order deformation problems we seek the optimal values of the convergence control parameters. In this particular case $\hbar_1$ denotes the auxiliary parameter for the function f while $\hbar_2$ is the auxiliary parameter for the functions θ and ϕ. To determine the optimal values of these parameters (those will ensure rapid convergence) we define the averaged residuals errors for the functions f, θ and ϕ as (see Liao [13] for details)

$$E_{m,1}(\hbar_1) = \frac{1}{K}\sum_{i=0}^{K}\left[\mathcal{N}_f\left(\sum_{j=0}^{m} f_j(i\Delta x)\right)\right]^2, \tag{6.65}$$

$$E_{m,2}(\hbar_2) = \frac{1}{K}\sum_{i=0}^{K}\left[\mathcal{N}_\theta\left(\sum_{j=0}^{m} \theta_j(i\Delta x), \sum_{j=0}^{m} \phi_j(i\Delta x)\right)\right]^2, \tag{6.66}$$

$$E_{m,2}(\hbar_2) = \frac{1}{K}\sum_{i=0}^{K}\left[\mathcal{N}_\phi\left(\sum_{j=0}^{m} \theta_j(i\Delta x), \sum_{j=0}^{m} \phi_j(i\Delta x)\right)\right]^2, \tag{6.67}$$

where $\Delta x = 1/K$ and $K = 20$. Here $\mathcal{N}_f$, $\mathcal{N}_\theta$ and $\mathcal{N}_\phi$ denote the non-linear operators corresponding to the equations (6.59)–(6.61). The above averaged residual errors can be plotted versus the respective auxiliary parameters to determine the convergence region of the solutions. For a given order of approximations m, the optimal values of $\hbar_1$ and $\hbar_2$ can be determined by minimizing the averaged residual error given in Eqs. (6.65)–(6.67) using the command *Minimize* of the software **Mathematica 8.0**. In Tables 6.4 and 6.5 , the optimal values of $\hbar_1$ and $\hbar_2$ for the functions f, θ and ϕ corresponding to various values of the parameters are given. Here the corresponding averaged residuals are represented as $E^*_{m,1}$, $E^*_{m,2}$ and $E^*_{m,3}$ respectively. For a further check at the accuracy of our computations we compared HAM solutions with the numerical solutions obtained through the command `NDSolve` of the software **Mathematica 8.0**. The results are in an excellent agreement which can be seen from Tables 6.6 and 6.7 .Moreover the velocity profiles and dimensionless heat and mass transfer rates have been portrayed in the Figs. 6.8 and 6.9.

Table 6.4. Optimal values of $\hbar_1$ for different physical parameters.

		$A = 2$		$A = -2$	
M	S	Optimal value of of $\hbar_1$	$E^*_{m,1}$	Optimal value of $\hbar_1$	$E^*_{m,1}$
0	1	−0.868	3.06×10^{-14}	−0.889	8.32×10^{-12}
2		−0.855	2.23×10^{-15}	−0.833	1.00×10^{-12}
3		−0.856	1.21×10^{-15}	−0.772	9.18×10^{-14}
5		−0.791	1.04×10^{-11}	−0.720	1.64×10^{-11}
1	1/10	−0.904	8.97×10^{-24}	−0.868	6.16×10^{-24}
	1/2	−0.916	3.38×10^{-22}	−0.906	1.05×10^{-20}
	1	−0.862	1.68×10^{-14}	−0.874	4.92×10^{-12}
	2	−0.753	1.75×10^{-6}	−0.906	1.05×10^{-20}

Table 6.5. Optimal values of $\hbar_2$ for different physical parameters in case of $M = S = 1, A = 2$ and $\hbar_1 = -0.862$.

Nb	Nt	Optimal value of $\hbar_2$ for θ	$E^*_{m,2}$	Optimal value of $\hbar_2$ for ϕ	$E^*_{m,3}$
1/10	1/10	−0.908	1.47×10^{-11}	−0.936	2.51×10^{-10}
1/2		−0.915	1.62×10^{-13}	−0.941	1.18×10^{-12}
1		−0.931	9.10×10^{-13}	−0.894	4.11×10^{-13}
3/2		−0.921	4.51×10^{-12}	−0.886	2.86×10^{-13}
	1/2	−0.961	2.93×10^{-10}	−0.913	2.87×10^{-10}
	1	−1.021	1.21×10^{-8}	−0.953	2.97×10^{-8}
	3/2	−0.955	2.37×10^{-6}	−1.027	1.18×10^{-5}
	2	−0.820	1.00×10^{-4}	−0.916	4.12×10^{-4}

6.5. Boundary layer flow of non-Newtonian nanofluid over a stretching sheet

Many industrial fluids such as clay coating and suspensions, cosmetic products, oils and grease, drilling muds, coal water or coal-oil slurries and certain paints exhibit a nonlinear relationship between stress and shear rate and these are treated as non-Newtonian. The diverse characteristics of such fluids lead to the fact that these cannot be described by using a single constitutive relationship between stress and shear rate. Generally, the constitutive equations of non-Newtonian fluids yield the mathematical problems which are of the higher orders, more nonlinear and complicated than the corresponding problem for a Navier–Stokes fluid. The rheological model of second grade has been widely addressed by the researchers in view of its simplicity. This model can predict the effects of normal stresses in the flow.

Table 6.6. Results of skin friction coefficient $f''(1)$ for different values of M and S.

M	S	$f''(1)$ HAM	$f''(1)$ numerical
0	1	7.53316579	7.53316579
2		8.26387231	8.26387230
3		9.09732572	9.09732573
5		11.3492890	11.3492890
1	1/10	8.97552394	8.97552394
	1/2	8.3492478	8.34924578
	1	7.72194601	7.72194601
	2	6.94077326	6.94077334

Table 6.7. Results of reduced Nusselt number Nur and reduced Sherwood number Shr for different values of Nb and Nt in case of $A = 2, M = S = Pr = Le = 1$ and $\hbar_1 = -0.862$.

Nb	Nt	Nur HAM	Nur numerical	Shr HAM	Shr numerical
1/10	1/10	0.52628540	0.52628539	0.86604666	0.86604666
1/2		0.63433253	0.63433253	0.53012814	0.53012814
1		0.78636385	0.78636384	0.48603919	0.48603919
3/2		0.95569955	0.95569954	0.46986157	0.46986157
	1/2	1.17682119	1.17682119	0.40180718	0.40180718
	1	1.48581207	1.48581194	0.12619334	0.12619330
	3/2	1.82305276	1.82305354	0.39083080	0.39083988
	2	2.17915991	2.17922795	1.16777723	1.16800856

Mustafa *et al.* [36] considered the steady flow and heat transfer of second grade fluid when the nanoparticles are introduced into the base fluid. The fluid is electrically conducting in the presence of an applied magnetic field of strength B_0. The x- and y- axes are taken along and perpendicular to the sheet respectively. The velocity of the stretching sheet is $u_w = cx$ (where $c > 0$ is a positive constant). T_f denotes the convective surface temperature while T_∞ is the ambient temperature such that $T_f > T_\infty$. The nanoparticles concentration at the sheet is denoted by C_w. However C_∞ is the ambient concentration. Thus the boundary layer equations governing the conservation of mass, momentum, energy and nanoparticles volume

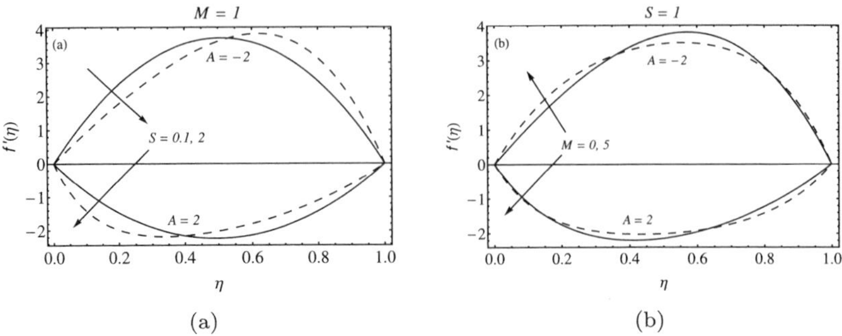

(a) (b)

Fig. 6.8. Velocity field f' for different values of S and M.

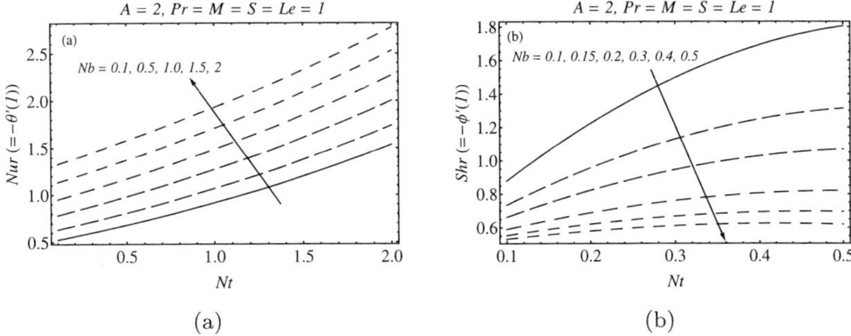

(a) (b)

Fig. 6.9. Dimensionless heat and mass transfer rates for different values of Nb and Nt.

fraction are (see Cortell [37] for more details)

$$\frac{\partial u}{\partial x}+\frac{\partial v}{\partial y}=0, \tag{6.68}$$

$$u\frac{\partial u}{\partial x}+v\frac{\partial u}{\partial y}=\nu\frac{\partial^2 u}{\partial y^2}+\frac{\alpha_1}{\rho}\left[\frac{\partial u}{\partial x}\frac{\partial^2 u}{\partial y^2}+u\frac{\partial^3 u}{\partial x\partial y^2}+\frac{\partial u}{\partial y}\frac{\partial^2 v}{\partial y^2}+v\frac{\partial^3 u}{\partial y^3}\right] -\frac{\sigma}{\rho}B_0^2 u, \tag{6.69}$$

$$u\frac{\partial T}{\partial x}+v\frac{\partial T}{\partial y}=\xi\frac{\partial^2 T}{\partial y^2}+\tau\left[D_B\frac{\partial C}{\partial y}\frac{\partial T}{\partial y}+\frac{D_T}{T_\infty}\left(\frac{\partial T}{\partial y}\right)^2\right], \tag{6.70}$$

$$u\frac{\partial C}{\partial x}+v\frac{\partial C}{\partial y}=D_B\frac{\partial^2 C}{\partial y^2}+\frac{D_T}{T_\infty}\frac{\partial^2 T}{\partial y^2}. \tag{6.71}$$

The relevant boundary conditions are

$$
\begin{aligned}
&u = u_w(x) = cx, \quad -k\frac{\partial T}{\partial y} = h_s(T_f - T), \ C = C_w \quad \text{at} \quad y = 0, \\
&u \to 0, \ \frac{\partial u}{\partial y} \to 0, \ T \to T_\infty, \ C \to C_\infty \quad \text{as} \quad y \to \infty.
\end{aligned} \tag{6.72}
$$

Here α_1 is the material fluid parameter, k is the thermal conductivity and h_s is the heat transfer coefficient and ξ is the thermal diffusivity. The rest of the quantities are already defined in the previous sections. Invoking the following dimensionless variables

$$
\begin{aligned}
&\eta = \sqrt{\frac{c}{\nu}}y, \ u = cxf^{'}(\eta), \ v = -\sqrt{c\nu}f(\eta), \\
&\theta(\eta) = \frac{T - T_\infty}{T_f - T_\infty}, \ \phi(\eta) = \frac{C - C_\infty}{C_w - C_\infty},
\end{aligned} \tag{6.73}
$$

equation (6.68) is identically satisfied and Eqs. (6.69)–(6.71) take the following forms:

$$
f^{'''} + ff^{''} - f^{'2} + \alpha\left(2f^{'}f^{'''} - f^{''2} - ff^{iv}\right) - M^2 f^{'} = 0, \tag{6.74}
$$

$$
\frac{1}{Pr}\theta^{''} + f\theta^{'} + Nb\theta^{'}\phi^{'} + Nt\theta^{'2} = 0, \tag{6.75}
$$

$$
\phi^{''} + Lef\phi^{'} + \frac{Nt}{Nb}\theta^{''} = 0, \tag{6.76}
$$

subject to the boundary conditions

$$
\begin{aligned}
&f(0) = 0, \ f^{'}(0) = 1, \ \theta^{'}(0) = -Bi(1 - \theta(0)), \ \phi(0) = 1, \\
&f^{'}(\infty) \to 0, \ \theta(\infty) \to 0, \ \phi(\infty) \to 0,
\end{aligned} \tag{6.77}
$$

where

$$
\begin{aligned}
&\alpha = \frac{c\alpha_1}{\mu}, \ Pr = \frac{\nu}{\xi}, \ Le = \frac{\nu}{D_B}, \ Bi = \frac{h_s}{k}\sqrt{\frac{\nu}{c}}, \\
&Nb = \frac{(\rho c)_p D_B (C_w - C_\infty)}{(\rho c)_f \nu}, \ Nt = \frac{(\rho c)_p D_T (T_f - T_\infty)}{(\rho c)_f T_\infty \nu}.
\end{aligned} \tag{6.78}
$$

Here α is the dimensionless second grade fluid parameter, Pr is the Prandtl number, Le is the Lewis number, Nb is the Brownian motion parameter, Nt is the thermophoresis parameter, and Bi is the Biot number, respectively. It is worth mentioning here that when $\alpha = 0$ the governing Eqs. (6.74)–(6.77) for viscous fluid case (originally considered by Makinde and Aziz [38] are deduced. Further $Nt = Nb = 0$ corresponds to the case when Brownian motion and thermophoretic effects are absent. Moreover the analysis for constant wall temperature ($\theta(0) = 1$) can be recovered by setting $Bi \to \infty$. Here the physical quantities of interest are the skin friction coefficient C_f, the local Nusselt number Nu and the local Sherwood number Sh defined by

$$C_f = \frac{\tau_w}{\rho u_w^2}, Nu = -\frac{x q_w}{k(T_w - T_\infty)}, Sh = -\frac{x j_w}{D_B(C_w - C_\infty)}, \tag{6.79}$$

where τ_w is the wall skin friction, q_w is the surface heat flux and j_w is the wall mass flux given by

$$\tau_w = \left[\mu\left(\frac{\partial u}{\partial y}\right) + \alpha_1\left(u\frac{\partial^2 u}{\partial x \partial y} + v\frac{\partial^2 u}{\partial y^2} - 2\frac{\partial v}{\partial y}\frac{\partial u}{\partial y}\right)\right]\Big|_{y=0},$$
$$q_w = k\left(\frac{\partial T}{\partial y}\right)_{y=0}, \quad j_w = D_B\left(\frac{\partial C}{\partial y}\right)_{y=0}.$$

Using Eq. (6.73) in Eq. (6.79) one obtains

$$Re_x^{1/2} C_f = (1 + 3\alpha) f''(0), \ Nu/Re_x^{1/2} = -\theta'(0) = Nur,$$
$$Sh/Re_x^{1/2} = -\phi'(0) = Shr.$$

It is worth mentioning here that the exact solution of Eq. (6.74) satisfying the boundary conditions given in Eq. (6.77) is given by (see Cortell [37])

$$f(\eta) = \left[\frac{1 - \exp(-K\eta)}{K}\right], K = \sqrt{\frac{1 + M^2}{1 + \alpha}}. \tag{6.80}$$

The analytic solutions of Eqs. (6.74)–(6.76) subject to the boundary conditions (6.77) have been computed by homotopy analysis method (HAM). The appropriate initial guesses satisfying the relevant boundary conditions for f, θ and ϕ are chosen as

$$f_0(\eta) = 1 - \exp(-\eta), \theta_0(\eta) = \frac{Bi}{1 + Bi}\exp(-\eta), \phi_0(\eta) = \exp(-\eta). \tag{6.81}$$

Further we have selected the following auxiliary linear operators

$$\mathcal{L}_f(f) = f''' - f', \ \mathcal{L}_\theta(\theta) = \theta'' - \theta, \ \mathcal{L}_\phi(\phi) = \phi'' - \phi. \tag{6.82}$$

The higher order deformation problems corresponding to Eqs. (6.74)–(6.76) have been solved using computational software Mathematica. We have denoted the so-called non-zero auxiliary parameters for the functions f, θ and ϕ by $\hbar_f, \hbar_\theta$ and $\hbar_\phi$ respectively. The appropriate values of these parameters have been determined by plotting the $\hbar$-curves in Fig. 6.10. It is found that admissible values of these parameters are $-0.9 \le \hbar_f \le -0.3$,

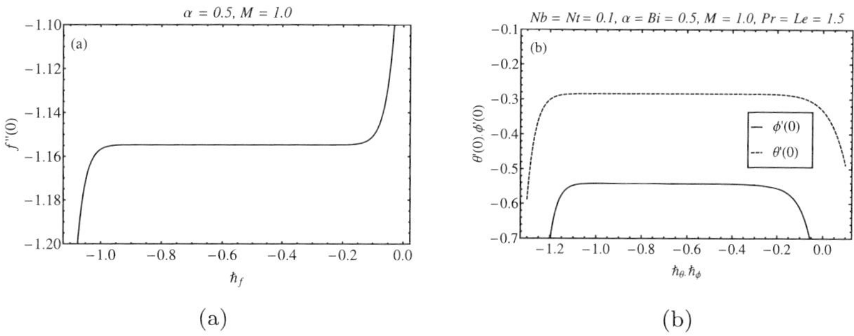

(a) (b)

Fig. 6.10. $\hbar$-curves for the functions f, θ and ϕ at 20th-order of approximations.

$-1.1 \leq \hbar_\theta \leq -0.3$ and $-1 \leq \hbar_\phi \leq -0.4$ respectively. After substituting the exact solution given in Eq. (6.80) into dimensionless equations for temperature and concentration given in (6.75) and (6.76) respectively, the resulting differential equations have been solved numerically by using **Mathematica**. The numerical solutions agree well with the homotopy solutions as can be seen from Fig. 6.11.

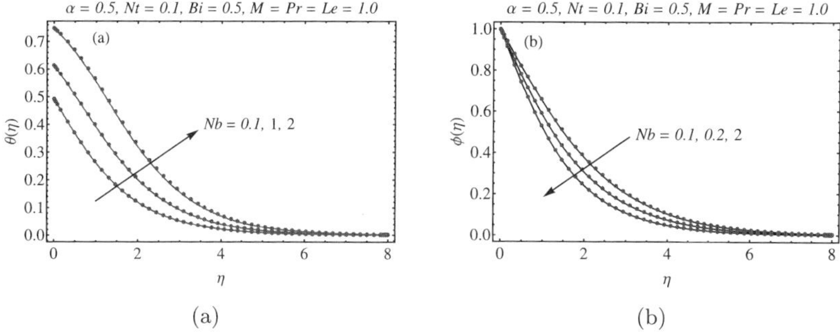

(a) (b)

Fig. 6.11. Comparison of numerical and series solutions. Points: Numerical solutions, Lines: 10th-order HAM solutions at $\hbar_f = -0.7, \hbar_\theta = \hbar_\phi = -1.2$.

Figure 6.2 displays the comparison of numerical and homotopy solutions for different values of parameters. A very good agreement is found between the solutions. Figure 6.12 plots the temperature and concentration profiles against the similarity variable η for different values of parameters. It is clear from this figure that an increase in the second grade fluid parameter or equivalently the normal stress differences shifts the profiles towards the boundary causing a reduction in the thermal boundary layer thickness. The temperature profiles monotonically decrease with an increase in η and

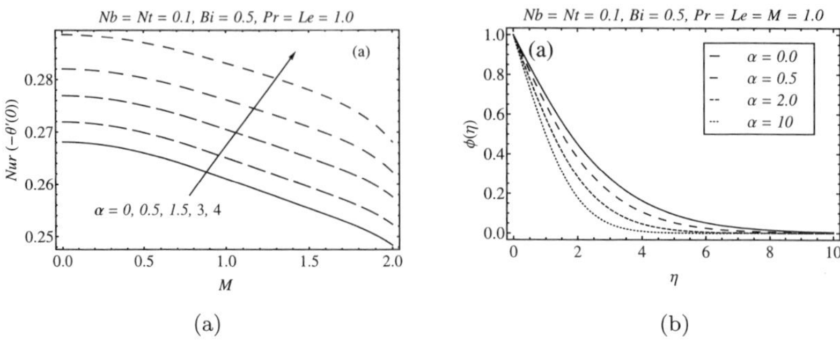

Fig. 6.12. Temperature and nanoparticles concentration for different values of α.

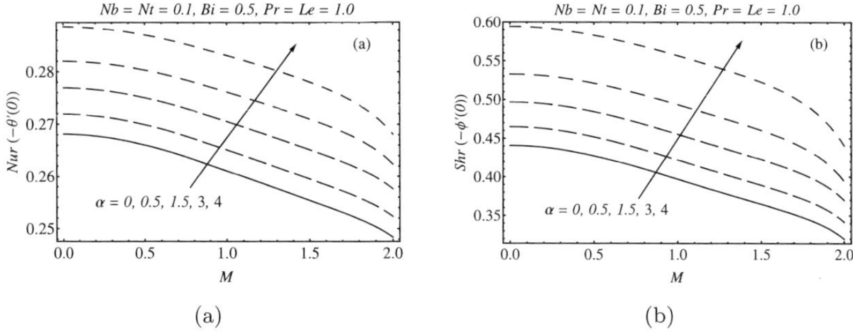

Fig. 6.13. Dimensionless heat and mass transfer rates for different values of α and M.

asymptotically reach the zero value as $\eta \to \infty$ representing the characteristic of boundary layer flow. Figure 6.13 is displayed to see the influence of second grade (viscoelastic) parameter on the reduced Nusselt and Sherwood numbers. We have seen earlier in the graphical results that thermal and concentration boundary layers reduce when the viscoelastic effects intensify. This reduction is compensated with the increase in the rate of heat and mass transfer at the bounding surface. Table 6.8 provides the numerical values of reduced Nusselt number for some values of Nb and Nt. In accordance with Makinde and Aziz [38] a slight decrease in the magnitude of Nur with an increase in Nt is observed for a weaker Brownian motion. Such reduction in Nur becomes prominent for larger Brownian motion parameter Nb. The values of reduced Sherwood number Shr corresponding to the data given in Table 6.1 have been provided in Table 6.9. There is a decrease in the magnitude of Shr with an increase in Nt. This decrease is more pronounced for a stronger Brownian motion. Further the reduced

Table 6.8. Comparison of numerical results (using shooting method) of reduced Nusselt number $Nur = -\theta'(0)$ for different values of Nb and Nt with the 25th-order HAM approximations (in parenthesis, using $\hbar_f = -0.7, \hbar_\theta = \hbar_\phi = -1.2$) in case of $\alpha = Bi = 0.5$ and $Pr = Le = M = 1.0$.

Nt	Nur $Nb = 0.2$	$Nb = 0.4$	$Nb = 0.6$
0.1	0.247493 (0.24749)	0.234266 (0.23427)	0.227170 (0.22717)
0.2	0.245638 (0.24564)	0.232295 (0.23230)	0.218641 (0.21864)
0.3	0.243761 (0.24376)	0.230303(0.23030)	0.216546 (0.21655)
0.4	0.241863 (0.24186)	0.228292 (0.22829)	0.214434 (0.21443)
0.5	0.239944 (0.23994)	0.226260 (0.22626)	0.212305 (0.21231)

Table 6.9. Comparison of numerical results (using shooting method) of reduced Nusselt number $Shr = -\phi'(0)$ for different values of Nb and Nt with the 25th-order HAM approximations (in parenthesis, using $\hbar_f = -0.7, \hbar_\theta = \hbar_\phi = -1.2$) in case of $\alpha = Bi = 0.5$ and $Pr = Le = M = 1.0$.

Nt	Shr $Nb = 0.2$	$Nb = 0.4$	$Nb = 0.6$
0.1	0.464939 (0.46494)	0.510999 (0.51010)	0.526370 (0.52637)
0.2	0.381010 (0.38101)	0.472329 (0.47233)	0.502799 (0.50280)
0.3	0.298766 (0.29877)	0.434532 (0.43453)	0.479824 (0.47982)
0.4	0.218223 (0.21822)	0.397613 (0.39761)	0.457448 (0.45745)
0.5	0.139396 (0.13940)	0.361579 (0.36158)	0.435673 (0.43567)

Sherwood number increases with an increase in Nb. Further the results are compared with those reported by Makinde and Aziz [38]. As already noticed in the graphical results that thermal and concentration boundary layers thin when Pr and Le are increased . The reduced Nusselt and Sherwood numbers (which are proportional to the initial slopes) are therefore increasing functions of Pr and Le respectively. It is also observed that magnitudes of Nur and Shr are increasing functions of Bi.

References

[1] S. J. Liao, An explicity, totally analytic approximate solution for Blasius' viscous flow problem, *International Journal of Nonlinear Mechanics.* **34**, 759–778 (1999).

[2] S. J. Liao and A. Campo, Analytic solutions of the temperature distribution in Blasius viscous flow problems, *Journal of Fluid Mechanics.* **453**, 411–425 (2002).

[3] S. J. Liao, On the analytic solution of magnetohydrodynamic flows of non-Newtonian fluids over a stretching sheet, *Journal of Fluid Mechanics.* **488**, 189–212 (2003).

[4] S. J. Liao, An analytic solution of unsteady boundary-layer flows caused by an impulsively stretching sheet, *Communications in Nonlinear Science and Numerical Simulations.* **11**, 326–339 (2006).

[5] M. Sajid and T. Hayat, Homotopy analysis of MHD boundary layer flow of an upper-convected Maxwell fluid, *International Journal of Engineering Science.* **45**, 393–401 (2007).

[6] M. Sajid and T. Hayat, On the analytic solution for thin film flow of a fourth grade fluid down a vertical cylinder, *Physics Letters A* **361**, 316–322 (2007).

[7] M. Sajid, I. Ahmad, T. Hayat and M. Ayub, Series solution for unsteady axisymmetric flow and heat transfer over a radially stretching sheet, *Communications in Nonlinear Science and Numerical Simulations.* **13**, 2193–2202 (2008).

[8] T. Hayat, M. Mustafa and M. Sajid, Influence of thermal radiation on Blasius flow of second grade fluid, *Zeitschrift fr Naturforschung A*, **64a**, 827–833 (2009).

[9] S. Abbasbandy and T. Hayat, Solution of the MHD Falkner-Skan flow by homotopy analysis method. *Communications in Nonlinear Science and Numerical Simulations.* **14**, 3591–3598 (2009).

[10] S. Abbasbandy and T. Hayat, On series solution for unsteady boundary layer equations in a special third grade fluid, *Communications in Nonlinear Science and Numerical Simulations.* **16**, 3140–3146 (2011).

[11] H. Xu, L. Z. Lin, S. J. Liao, Z. Wu and J. Majdalani, Homotopy based solutions of the Navier-Stokes equations for a porous channel with orthogonall moving walls, *Physics of Fluids.* **22**, 053601 (18 pages) (2010).

[12] M. M. Rashidi, S. A. M. Pour and S. Abbasbandy, Analytic approximate solutions for heat transfer of a micropolar fluid through a porous medium with radiation. *Communications in Nonlinear Science and Numerical Simulations.* **16**, 1874–1889 (2011).

[13] S. J. Liao, An optimal homotopy analysis approach for strongly nonlinear differential equations, *Communications in Nonlinear Science and Numerical Simulations.* **15**, 2003–2016 (2010).

[14] S. U. S. Choi, Enhancing thermal conductivity of fluids with nanoparticle, in: D.A. Siginer, H.P. Wang (eds.), Developments and Applications of Non-Newtonian Flows. ASME FED, **231**, 99–105 (1995).

[15] J. Buongiorno, Convective transport in nanofluids, *Journal of Heat Transfer-*

Transactions of the ASME. **128**, 240–250 (2006).

[16] W. Daungthongsuk and S. Wongwises, A critical review of convective heat transfer nanofluids, *Renewable and Sustainable Energy Reviews.* **11**, 797–817 (2007).

[17] X. Q. Wang and A. S. Mujumdar, A review on nanofluids - Part I: theoretical and numerical investigations, *Brazilian Journal of Chemical Engineering.* **25**, 613–630 (2008).

[18] X. Q. Wang and A. S. Mujumdar, A review on nanofluids - Part II: experiments and applications, *Brazilian Journal of Chemical Engineering.* **25**, 631–648 (2008).

[19] J. A. Eastman, S. U. S. Choi, S. Li, W. Yu and L. J. Thomson, Anomalously increased effective thermal conductivity of ethylene-glycol based nanofluids containing copper nanoparticles, *Applied Physics Letters.* **78**, 718–720 (2001).

[20] S. Kaka and A. Pramuanjaroenkij, Review of convective heat transfer enhancement with nanofluids, *International Journal of Heat and Mass Transfer.* **52**, 3187–3196 (2009).

[21] A. V. Kuznetsov and D. A. Nield, Natural convective boundary-layer flow of a nanofluid past a vertical plate, *International Journal of Thermal Sciences.* **49**, 243–247 (2010).

[22] M. Mustafa, T. Hayat, I. Pop S. Asghar and S. Obaidat, Stagnation-point flow of a nanofluid towards a stretching sheet, *International Journal of Heat and Mass Transfer.* **54**, 5588–5594 (2011).

[23] S. Dinarvand, A. Abbasi, R. Hosseini and I. Pop, Homotopy analysis method for mixed convective boundary layer flow of a nanofluid over a vertical circular cylinder. Thermal Science. 10.2298/TSCI120225165D (2012).

[24] S. Nadeem and C. Lee, Boundary layer flow of nanofluid over an exponentially stretching surface, *Nanoscale Research Letters.* **7**, 94, (2012).

[25] M. M. Hashmi, T. Hayat and A. Alsaedi, On the analytic solutions for squeezing flow of nanofluid between parallel disks. *Nonlinear Analysis: Modeling and Control.* **17**, 418–430 (2012).

[26] R. Ellahi, M. Raza and K. Vafai, Series solutions of non-Newtonian nanofluid with Reynolds' model and Vogel's model by means of the homotopy analysis method (HAM), *Mathematical and Computer Modelling.* **55**, 1876–1891 (2012).

[27] S. Nadeem, R. Mehmood and N. S. Akbar, Non-orthogonal stagnation-point flow of a nano non-Newtonian fluid towards a stretching surface with heat transfer, *International Journal of Heat and Mass Transfer* **57**, 679–689 (2013).

[28] L. Zheng, C. Zhang, X. Zhang and J. Zhang, Flow and radiation heat transfer of a nanofluid over a stretching sheet with velocity slip and temperature jump in porous medium, *Journal of the Franklin Institute.* DOI: http://dx.doi.org/10.1016/j. jfranklin. 2013.01.022.

[29] H. Xu, I. Pop and X. C. You, Flow and heat transfer in a nano-liquid film over an unsteady stretching surface, *International Journal of Heat and Mass Transfer.* **60**, 646–652 (2013).

[30] M. Mustafa, T. Hayat and A. Alsaedi, Unsteady boundary layer flow of nanofluid past an impulsively stretching sheet, Journal of Mechanics. DOI: 10.1017/jmech.2013.9 (2013).
[31] M. Mustafa, T. Hayat and A. Alsaedi, Axisymmetric flow of a nanofluid over a radially stretching sheet with convective boundary conditions, *Current Nanoscience.* **8**, 328-334 (2012a).
[32] P. D. Ariel, Axisymmetric flow of a second grade fluid past a stretching sheet, *International Journal of Engineering Science.* **39**, 529–553 (2001).
[33] M. J. Stefan, Versuch Uber die scheinbare adhesion. Sitzungsberichte der Akademie der Wissenschaften in Wien, *Mathematic-Naturwissen.* **69**, 713–721 (1874).
[34] G. Domairry and A. Aziz, Approximate analysis of MHD Squeeze flow between two parallel disks with suction or injection by homotopy perturbation method, Mathematical Problems in Engineering. doi:10.1155/2009/603916.
[35] A. Joneidi, G. Domairry and M. Babaelahi, Effect of mass transfer on the flow in the magnetohydrodynamic squeeze film between two parallel disks with one porous disk, *Chemical Engineering Communications.* **198**, 299–311 (2011).
[36] M. Mustafa, M. Nawaz, T. Hayat and A. Alsaedi, MHD boundary layer flow of second grade nanofluid over a stretching sheet with convective boundary conditions. *Journal of Aerospace Engineering.* doi:10.1061/(ASCE)AS.1943-5525.0000314 (2012b).
[37] R. Cortell, MHD flow and mass transfer of an electrically conducting fluid of second grade in a porous medium over a stretching sheet with chemically reactive species. *Chemical Engineering Processing.* **46**, 721–728 (2007).
[38] O. D. Makinde and A. Aziz, Boundary layer flow of a nanofluid past a stretching sheet with a convective boundary condition. *International Journal of Thermal Science.* **50**, 1326–1332 (2011).

Chapter 7

Homotopy Analysis Method for Fractional Swift–Hohenberg Equation

S. Das* and K. Vishal

Department of Applied Mathematics,
Indian Institute of Technology (BHU), Varanasi-221005, India
**subir_das08@hotmail.com*

In this chapter, homotopy analysis method is used to obtain approximate analytic solution of the time-fractional Swift–Hohenberg equation with a given initial condition. The fractional derivatives are considered in the Caputo sense. Effects of parameters on the convergence of the approximate series solution by minimizing the averaged residual error are calculated numerically and presented through graphs and tables for different particular cases.

Contents

7.1. Introduction 292
7.2. Basic ideas of fractional calculus 295
7.3. Basic ideas of the homotopy analysis method 296
7.4. Solution given by the homotopy analysis method 298
7.5. Numerical results and discussion 301
7.6. Conclusion 304
References 306

7.1. Introduction

The Swift–Hohenberg (S–H) equation was derived by Jack Swift and Pierre Hohenberg [1] in the year 1977 as

$$\frac{\partial u}{\partial t} = \mu u - \left(1 + \frac{\partial^2}{\partial x^2}\right)^2 u + \gamma u^2 - u^3, \tag{7.1}$$

where μ and γ are parameters, which is a model pattern-forming equation used as a model for a fluid which is thermally convecting. This evolution equation is a parabolic equation involving a fourth order spatial derivative. It is an effective model equation for a variety of systems in Physics and Mechanics. The mathematical model for the Rayleigh–Benard convection involves the Navier–Stokes equations coupled with the transport equation for temperature. The Swift–Hohenberg equation has important role in different branches of physics, ranging from hydrodynamics to nonlinear optics, such as Taylor–Couette flow [2] and in the study of lasers [3]. The equation also plays an important role in the study of pattern formation [4]. Large-time behaviour of solutions of the Swift–Hohenberg equation on a one-dimensional domain has been studied by Peletier and Rottschafer [5]. In the last few decades, fractional differential equations (FDEs) have been focused due to their frequent appearances in various applications in fluid mechanics, viscoelasticity, biology, physics, electrical network, control theory of dynamical systems, chemical physics, optics and signal processing, which are successfully modelled by linear and non-linear fractional order differential equations [6–8]. Due to its important applications in engineering and physics, the authors are motivated to solve the following Swift–Hohenberg equation with fractional order time derivative $\alpha (0 < \alpha \leq 1)$:

$$\begin{aligned}&\frac{\partial^\alpha u(x,t)}{\partial t^\alpha} + \frac{\partial^4 u(x,t)}{\partial x^4} + 2\frac{\partial^2 u(x,t)}{\partial x^2} + (1-\mu)u(x,t) - 2u^2(x,t)\\ &+u^3(x,t) = 0, (x,t) \in \Omega,\end{aligned} \tag{7.2}$$

subject to the initial condition

$$u(x,0) = \frac{1}{10}\sin\left(\frac{\pi x}{l}\right), \tag{7.3}$$

where μ is the real bifurcation parameter and $\frac{\partial^\alpha}{\partial t^\alpha}$ is the Caputo derivative of order α.

In [5] it is shown that if $\mu \leq 0$, then the steady case of said equation has only the trivial solution and hence, for every $x \in (0, l)$, $u(x,t) \to 0$ as $t \to \infty$, So we have taken the case $\mu > 0$. In 2010, Akyildiz *et al.* [9] have solved the Swift–Hohenberg equation for the standard motion. In 2011, the S–H equation with fractional order time derivative is studied by Vishal *et al.* [10]. Recently, the S–H equation in the presence of dispersion with fractional order time derivative is studied by Vishal *et al.* [11].

Previously, modeling was mainly restricted to linear systems for which analytical treatment is tractable. But due to the advent of powerful computers and with improved computational techniques, nowadays it is possible to tackle even nonlinear problems to some extent. Nonlinearity is a phenomenon that is exhibited by most of the systems in nature and has gained increasing popularity during last few decades. Most of the nonlinear problems do not have a precise analytical solution; especially it is hard to obtain it for the fractional order nonlinear equations. These types of equations should be solved by various approximate analytical methods or by using different numerical methods. Owing to the difficulty of obtaining solutions in closed form for fractional order systems, numerical simulations are quite often used to investigate the behavior of the systems. Obviously, the integer-order model can be viewed as a special case from the more general fractional order model. Namely, the integer-order model can be retrieved by putting all fractional orders of the derivatives equal to unity. In other words, the ultimate behavior of the fractional order system response must converge to the response of the integer-order version of the model. Recently, the fractional differential equations (FDEs) have gained much attention due to the fact that fractional order system response ultimately converges to the integer order system response (Podlubny [6], Miller and Ross [12], Oldham and Spanier [13]). Analysis of fractional partial differential equations, which are obtained from the classical equations in Mathematical Physics by replacing the first order time derivative by a fractional derivative of order α satisfying $0 < \alpha \leq 1$, has been a field of growing interest as evident from literature survey. This shows that fractional calculus is the extension of classical mathematics where derivatives are taken as rational, irrational and complex orders. In the last two decades, fractional differential equations have been widely used by the researchers not only in science and engineering but also in economics and finance. It is also a powerful tool in modeling multi scale problems, characterized by wide time or length scale. The attribute of fractional order systems for which they have gained popularity in the investigation of dynamical systems is that they allow greater flexibility in the model. An integer order differential operator is a local operator. Whereas the fractional order differential operator is non-local in the sense that it takes into account the fact that the future state not only depends upon the present state but also upon all of the history of its previous states. An important characteristic of these evolution equations is that they generate the fractional Brownian motion (FBM) which is a generalization of Brownian motion (BM).

For physical systems, one should have to keep in mind two things for application of fractional order in the system for making a decisive step for the penetration of mathematics of fractional calculus into a body of natural sciences. Firstly to analyze, the importance and physical influence of the memory effects on time or space or both. Secondly, to give proper interpretation is of general meaning of non integer operators. However, fractional calculus has scarcely been applied to ecological problems due to their nonlinear nature and difficulties faced while confronting the problems with fractional derivatives which are addressed by Das *et al.* [14–16]. Researchers are working hard to overcome it. The main advantage of the fractional calculus is that fractional derivatives provide an excellent instrument for the description of memory and hereditary properties of various materials and processes.

Homotopy Analysis Method (HAM) proposed by S. J. Liao [17] is based on homotopy, a fundamental concept in topology and differential geometry. It is an analytical approach to get the series solutions of linear and nonlinear differential equations. The difference with the other perturbation methods is that this method is independent of small/large physical parameters. Another important advantage of this method as compared to the other existing perturbation and non-perturbation method lies in the flexibility to choose proper base function to get better approximate solution of the problems. This method offers certain advantages over routine numerical methods. Numerical methods use discretization which gives rise to rounding off errors causing loss of accuracy, and requires large computer memory and time. This computational method is better since it does not involve discretization of the variables and hence is free from rounding off errors and does not require large computer memory or time. This method has been successfully applied by many researchers for solving linear and non-linear partial differential equations [18–25]. Das *et al.* [26] have successfully applied the method to investigate the influences of auxiliary parameter to find the region of convergence through h-curve analysis for solving the fractional diffusion equation. In 2010, Liao [27] has given the optimal homotopy analysis method which is generalization of HAM, where one can ensure the convergence of series solution by means of optimal value of convergence-control parameter. Recently, Vishal and Das [28] have successfully used optimal homotopy analysis method to obtain approximate analytical solution of time fractional nonlinear diffusion equation in the presence of an external force and an absorbant term. In another article, Das *et al.* [29] have solved approximate solution of telegraph equation with fractional order time derivative for both fractional Brownian motions and standard motion. The method is very effective and has been proven to be a very efficient tool when applied to solve various linear and nonlinear fractional order differential equations having physical relevance in physical models as well as biological models ([30–39]).

The beauty of this chapter can be attributed to the successful application of more accurate, flexible and very powerful analytical method HAM for the solution of the nonlinear fractional order S–H equation and the convergence of the series solution through minimization of averaged residual error with proper choices of optimal values of auxiliary homotopy parameter confirming the validity and potential of the method. The salient features of the chapter are the numerical and graphical presentations of the effects of bifurcation parameter and the length of the domain on the solution with time and also the periodic nature of the solutions for different values of parameters.

7.2. Basic ideas of fractional calculus

In this section, we give some definitions and properties of the fractional calculus [6] which are used in this chapter.

Definition 7.1. A real function $f(t), t > 0$, is said to be in the space C_μ, $\mu \in \Re$, if there exists a real number $p > \mu$, such that $f(t) = t^p f_1(t)$, where $f_1(t) \in C(0,\infty)$, and it is said to be in the space C_μ^n, if and only if $f^{(n)} \in C_\mu$, $n \in N$.

Definition 7.2. The Riemann–Liouville fractional integral operator J_t^α of order $\alpha \geq 0$, of a function $f \in C_\mu$, $\mu \geq -1$, is defined as

$$J_t^\alpha f(t) = \frac{1}{\Gamma(\alpha)} \int_0^t (t-\xi)^{\alpha-1} f(\xi) d\xi, \alpha > 0, t > 0, \tag{7.4}$$
$$J_t^0 f(t) = f(t),$$

where $\Gamma(\alpha)$ is the well-known gamma function. Some of the properties of the operator $J_t^\alpha f(t)$, which are needed here, are as follows: For $f \in C_\mu$, $\mu \geq -1$, $\alpha, \beta \geq 0$ and $\gamma \geq -1$,

(i) $J_t^\alpha J_t^\beta f(t) = J_t^{\alpha+\beta} f(t)$,
(ii) $J_t^\alpha J_t^\beta f(t) = J_t^\beta J_t^\alpha f(t)$,
(iii) $J_t^\alpha t^\gamma = \frac{\Gamma(\gamma+1)}{\Gamma(\alpha+\gamma+1)} t^{\alpha+\gamma}$.

Definition 7.3. The fractional derivative D_t^α of $f(t)$, in the Caputo sense is defined as

$$D_t^\alpha f(t) = \frac{1}{\Gamma(n-\alpha)} \int_0^t (t-\xi)^{n-\alpha-1} f^n(\xi) d\xi, \tag{7.5}$$

for $n-1 < \alpha < n$, $n \in N$, $t > 0$, $f \in C_{-1}^n$. The following are two basic properties of the Caputo fractional derivative [7]:

(i) Let $f \in C_{-1}^n$, $n \in N$, and $D_t^\alpha f$, $0 \leq \alpha \leq n$ is well defined and $D_t^\alpha f \in C_{-1}$.
(ii) Let $n - 1 \leq \alpha \leq n$, $n \in N$, and $f \in C_\mu^n$, $\mu \geq -1$.

Then

$$(J_t^\alpha D_t^\alpha) f(t) = f(t) - \sum_{k=0}^{n-1} f^{(k)}(0^+) \frac{t^k}{k!}. \tag{7.6}$$

7.3. Basic ideas of the homotopy analysis method

To describe the basic ideas of HAM, consider the following fractional order differential equation

$$\frac{\partial^\alpha u(x,t)}{\partial t^\alpha} = \mathcal{N}[u(x,t)], t > 0, 0 < \alpha \leq 1, \tag{7.7}$$

where $u(x,t)$ is an unknown function and $\mathcal{N}$ is a nonlinear operator. By means of HAM, we first construct the so-called zeroth-order deformation equation

$$\begin{aligned}(1-q)&\mathcal{L}[\phi(x,t;q) - u_0(x,t)] \\ &= q\, \hbar\, H(x,t) \left[\frac{\partial^\alpha}{\partial t^\alpha} \phi(x,t;q) - \mathcal{N}[\phi(x,t;q)] \right],\end{aligned} \tag{7.8}$$

where $q \in [0,1]$ is the embedding parameter; $\hbar \neq 0$, is the so-called convergence-control parameter, $H(x,t) \neq 0$ is an auxiliary function, $u_0(x,t)$ is the initial guess of $u(x,t)$ and

$$\mathcal{L}[\phi(x,t;q)] = \frac{\partial^\alpha \phi(x,t;q)}{\partial t^\alpha}$$

is an auxiliary linear operator with property $\mathcal{L}[c] = 0$, where c is an integral constant.

It is obvious that when the embedding parameter $q = 0$ and $q = 1$, the zeroth-order deformation equation (7.8) has the solution $\phi(x,t;0) = u_0(x,t)$ and $\phi(x,t;1) = u(x,t)$ respectively. Thus as q increases from zero to unity, the solution $\phi(x,t;q)$ varies from the initial guess $u_0(x,t)$ to the exact solution $u(x,t)$. Expanding $\phi(x,t;q)$ in Taylor series with respect to q, one has

$$\phi(x,t;q) = u_0(x,t) + \sum_{k=1}^{\infty} u_k(x,t)q^k, \tag{7.9}$$

where

$$u_k(x,t) = \frac{1}{k!}\,\frac{\partial^k \phi(x,t;q)}{\partial q^k}\bigg|_{q=0}.$$

The convergence of the series given in Eq. (7.9) depends on the convergence-control parameter $\hbar$. If it is convergent at $q = 1$, one has

$$u(x,t) = \phi(x,t;1) = u_0(x,t) + \sum_{k=1}^{\infty} u_k(x,t), \tag{7.10}$$

which must be one of the solutions of the original fractional order differential equation, as proven by Liao [17].

Now defining the vector

$$\tilde{u}_n(x,t) = (u_0(x,t), u_1(x,t), u_2(x,t), ..., u_n(x,t)), \tag{7.11}$$

the mth-order deformation equation reads

$$\mathcal{L}[u_m(x,t) - \chi_m u_{m-1}(x,t)] = \hbar\; H(x,t)\; R_m(\tilde{u}_{m-1}(x,t), \tag{7.12}$$

with the initial condition

$$u_m(x,0) = 0, \tag{7.13}$$

where

$$\begin{aligned} &R_m(\tilde{u}_{m-1}(x,t)) \\ &= \frac{1}{(m-1)!}\frac{\partial^{m-1}}{\partial q^{m-1}}\left[\frac{\partial^\alpha \phi(x,t;q)}{\partial t^\alpha} - N[\phi(x,t;q)]\right]\bigg|_{q=0} \end{aligned} \tag{7.14}$$

and

$$\chi_m = \begin{cases} 0, \; m \le 1, \\ 1, \; m > 1. \end{cases} \tag{7.15}$$

Now the solution of mth-order deformation equation (7.12) for $m \ge 1$ reads

$$u_m(x,t) = \chi_m u_{m-1}(x,t) + \hbar\mathcal{L}^{-1}\left[H(x,t)R_m(\tilde{u}_{m-1}(x,t))\right] + c, \tag{7.16}$$

where c is the integration constant which is determined by the initial condition given in Eq. (7.13).

7.4. Solution given by the homotopy analysis method

To solve Eq. (7.2) by homotopy analysis method, we choose the initial approximation

$$u_0(x,t) = \frac{1}{10}\sin\left(\frac{\pi x}{l}\right), \tag{7.17}$$

and the linear auxiliary operator

$$\mathcal{L}[\phi(x,t;q)] = \frac{\partial^\alpha \phi(x,t;q)}{\partial t^\alpha}, \tag{7.18}$$

with the property

$$\mathcal{L}[c] = 0, \tag{7.19}$$

where c is integral constant, $\phi(x,t;q)$ is an unknown function. Furthermore, in the view of Eq. (7.2), we have defined the nonlinear operator as

$$\begin{aligned} \mathcal{N}[\phi(x,t;q)] = {} & \frac{\partial^\alpha \phi(x,t;q)}{\partial t^\alpha} + \frac{\partial^4 \phi(x,t;q)}{\partial x^4} + 2\frac{\partial^2 \phi(x,t;q)}{\partial x^2} \\ & + (1-\mu)\phi(x,t;q) - 2\phi^2(x,t;q) + \phi^3(x,t;q). \end{aligned} \tag{7.20}$$

By means of the homotopy analysis method, Liao [17] has constructed the so-called zeroth-order deformation equation as

$$(1-q)\mathcal{L}[\phi(x,t;q) - u_0(x,t)] = q\,\hbar\,\mathcal{N}[\phi(x,t;q)], \tag{7.21}$$

where $q \in [0,1]$ is the embedding parameter, $\hbar \neq 0$ is the convergence-control parameter, $u_0(x,t)$ is the initial guess of $u(x,t)$. It is obvious that when the embedding parameter $q = 0$ and $q = 1$, Eq. (7.21) becomes

$$\begin{aligned} \phi(x,t;0) &= u_0(x,t), \\ \phi(x,t;1) &= u(x,t), \end{aligned}$$

respectively. Next we expand $\phi(x,t;q)$ as in Eq. (7.9). If the auxiliary linear operator $\mathcal{L}$, the initial guess $u_0(x,t)$ and the convergence-control parameter $\hbar$ are properly chosen so that the series described in Eq. (7.9) converges at $q = 1$ as discussed in previous section, then $u(x,t)$ defined as in Eq. (7.10) will be one of the solutions of our considered problem given in Eq. (7.2).

Differentiating the zero-order deformation Eq. (7.21) m times with respect to q and then dividing it by $m!$ and finally setting $q = 0$, one has the so called mth-order deformation equation as

$$\mathcal{L}[u_m(x,t) - \chi_m u_{m-1}(x,t)] = \hbar\, R_m[u_{m-1}(x,t)], \tag{7.22}$$

with the initial condition

$$u_m(x,0) = 0, \tag{7.23}$$

where

$$\begin{aligned} R_m[\tilde{u}_{m-1}(x,t)] = {} & \frac{\partial^\alpha u_{m-1}(x,t)}{\partial t^\alpha} + \frac{\partial^4 u_{m-1}(x,t)}{\partial x^4} \\ & + 2\frac{\partial^2 u_{m-1}(x,t)}{\partial x^2} + (1-\mu)u_{m-1}(x,t) \\ & + \sum_{i=0}^{m-1} u_{m-1-i}\left(\sum_{j=0}^{i} u_j u_{i-j}\right) - 2\sum_{i=0}^{m-1} u_i u_{m-1-i}, \end{aligned} \tag{7.24}$$

and χ_m is defined in Eq. (7.15). Then, we have

$$u_m(x,t) = \chi_m u_{m-1}(x,t) + \hbar J_t^\alpha\, R_m[u_{m-1}(x,t)] + c, \tag{7.25}$$

where

$$J_t^\alpha[f(t)] = \frac{1}{\Gamma(\alpha)}\int_0^t (t-\xi)^{\alpha-1} f(\xi)d\xi,$$

and the integration constant c is determined by the initial condition Eq. (7.23).

Now from Eq. (7.25), the values $u_m(x,t)$ for $m = 1, 2, 3, ...$ can be obtained and the series solutions are thus gained. Finally the approximate solution is obtained by truncating the series as

$$\tilde{u}_m(x,t) = \sum_{i=0}^{m} u_i(x,t). \tag{7.26}$$

It is clear from Eq. (7.26) that $\tilde{u}_m(x,t)$ contains one convergence-control parameter $\hbar$, which determines the convergence region and rate of the homotopy-series solution.

As given by Liao [27], at the mth-order of approximation, one can define the exact square residual as

$$\Delta_m = \int\int_\Omega \left(N\left[\sum_{i=0}^{m} u_i(x,t)\right]\right)^2 dx\ dt \tag{7.27}$$

However, it was illustrated by Liao [27] that the exact residual error Δ_m defined by Eq. (7.27) needs too much CPU time to calculate even if the order of approximation is not very high.

Thus, to overcome this difficulty i.e., to decrease the CPU time, we use here the so-called averaged residual error defined by

$$E_m = \frac{1}{(M_x+1)(M_t+1)} \sum_{j=1}^{M_x} \sum_{k=1}^{M_t} \left(N \left[\sum_{i=0}^{m} u_i(j\Delta x, k\Delta t) \right] \right)^2, \quad (7.28)$$

where $M_x = M_t = 14$ and $\Delta x = \Delta t = 0.5$ for the considered problem. The optimal value of $\hbar$ can be obtained by means of minimizing the so-called averaged residual error Eq. (7.28), corresponding to the nonlinear algebraic equation

$$\frac{dE_m}{d\hbar} = 0. \quad (7.29)$$

An asymptotic expansion or a Taylor series expansion can often be accelerated quite dramatically (or even turned from divergent to convergent) by being re-arranged into a ratio of two such expansions.

A Padé approximation

$$P_n^m(x) = \frac{\sum_{k=0}^{m} a_k x^k}{\sum_{k=0}^{n} b_k x^k} \quad (7.30)$$

(normalized by $b_0 = 1$) generalizes the Taylor series expansion with the same total number of coefficients

$$T_{m+n}(x) = \sum_{k=0}^{m+n} c_k x^k. \quad (7.31)$$

From a truncated Taylor series expansion given by Eq. (7.31), one determines the corresponding Padé coefficients by requiring that if Eq. (7.30) is expanded as Taylor series and the result shall match all the terms given in Eq. (7.31). One of the main applications of Padé approximations is to extract as much information as we can form a power series expansion that is known only to a few terms. Conversion from Taylor to Padé form usually accelerates convergence, and often allows good accuracy even outside the radius of convergence of a power series (which in case of divergent asymptotic expansions may be zero).The so-called homotopy-Padé technique [22] was proposed by combining the above-mentioned traditional Padé technique with the homotopy analysis method. Hence, in order to calculate the $[m, n]$

homotopy-Padé approximant of $\phi(x,t;q)$, first the traditional Padé technique is used with expansion about the homotopy embedding parameter q as

$$\phi(x,t;q) = \frac{\sum_{k=0}^{m} A_{m,k}(x,t)q^k}{\sum_{k=0}^{n} B_{m,k}(x,t)q^k} \tag{7.32}$$

where the coefficients $A_{m,k}$ and $B_{m,k}$ are determined by the first several approximations

$$u_0(x,t), u_1(x,t), u_2(x,t), \ldots\ldots\ldots u_{m+n}(x,t). \tag{7.33}$$

Then, setting $q = 1$ in Eq. (7.32) and using Eq. (7.10), we have the so-called $[m, n]$ homotopy-Padé approximant

$$u(x,t) \approx \frac{\sum_{k=0}^{m} A_{m,k}(x,t)}{\sum_{k=0}^{n} B_{m,k}(x,t)} = \frac{P_m(x,t)}{Q_m(x,t)}. \tag{7.34}$$

The [1,1] and [2,2] homotopy-Padé approximants in terms of the basis functions u_k read

$$u \approx \frac{u_0u_1 + u_1^2 - u_2u_0}{u_1 - u_2}$$

and

$$u \approx \frac{P_2(x,t)}{Q_2(x,t)}, \tag{7.35}$$

where

$$\begin{aligned} P_2(x,t) = {} & u_0u_2^2 + u_1u_2^2 + u_2^3 - u_0u_1u_3 - u_1^2u_3 - u_0u_2u_3 - 2u_1u_2u_3 \\ & + u_0u_3^2 + u_0u_1u_4 + u_1^2u_4 - u_0u_2u_4, \\ Q_2(x,t) = {} & u_2^2 + u_3^2 - u_1u_3 + u_1u_4 - u_2u_3 - u_2u_4. \end{aligned}$$

Likewise, the $[i, j]$ homotopy-Padé approximant can be obtained from the terms in the $(i + j)$th-order series approximation.

7.5. Numerical results and discussion

In this section, the numerical results of probability density function $u(x,t)$ for the non-linear time fractional Swift–Hohenberg equation have been obtained for the one-dimensional domain. The optimal values of $\hbar$ and minimum averaged residual error are provided through Tables 7.1 to 7.6 and are displayed through Fig. 7.1. The effects of the parameter μ and the length of the domain l on solution profile of $u(x,t)$ are presented graphically through

Table 7.1. Optimal value of $\hbar$ in case of $l = 10, \mu = 2$ and $\alpha = 1$.

Order of Approx.	$\hbar$	E_m
1	0.189270	2.79031×10^{-3}
3	0.082567	2.36788×10^{-3}
5	0.053414	2.26236×10^{-3}

Table 7.2. Optimal value of $\hbar$ in case of $l = 8, \mu = 2$ and $\alpha = 1$.

Order of Approx.	$\hbar$	E_m
1	0.174622	2.84695×10^{-3}
3	0.076513	2.40528×10^{-3}
5	0.049551	2.29546×10^{-3}

Table 7.3. Optimal value of $\hbar$ in case of $l = 10, \mu = 2$ and $\alpha = 0.75$.

Order of Approx.	$\hbar$	E_m
1	0.300154	2.23036×10^{-3}
3	0.196166	1.83711×10^{-3}
5	0.119303	1.59495×10^{-3}

Table 7.4. Optimal value of $\hbar$ in case of $l = 8, \mu = 2$ and $\alpha = 0.75$.

Order of Approx.	$\hbar$	E_m
1	0.273515	2.26256×10^{-3}
3	0.179471	1.84955×10^{-3}
5	0.109371	1.59816×10^{-3}

Figs. 7.2 to 7.9. It is seen from Figs. 7.2 to 7.9, which are drawn for $\mu = 2$, $\alpha = 1, 0.5$ and $l = 10, 8, 6, 2$ that the behaviours of solutions are periodic in nature for different time.

It is clear from Tables 7.1 and 7.2 that optimal values of $\hbar$ are 0.053414 and 0.049551 for $l = 10$ and $l = 8$ respectively for $\mu = 2, \alpha = 1$ in the case of 5th-order approximation. Tables 7.3 and 7.4 exhibit that optimal values of $\hbar$ are 0.119303 and 0.109371 for $l = 10$ and $l = 8$ respectively for $\mu = 2, \alpha = 0.75$. It is also seen from the Table 7.5 and Table 7.6 that optimal values of $\hbar$ are 0.293335 and 0.272750 for $l = 10$ and $l = 8$

Table 7.5. Optimal value of $\hbar$ in case of $l = 10, \mu = 2$ and $\alpha = 0.5$.

Order of Approx.	$\hbar$	E_m
1	0.565759	1.55442×10^{-3}
3	0.418052	1.09422×10^{-3}
5	0.293335	8.21554×10^{-4}

Table 7.6. Optimal value of $\hbar$ in case of $l = 8, \mu = 2$ and $\alpha = 0.5$.

Order of Approx.	$\hbar$	E_m
1	0.488307	1.56634×10^{-3}
3	0.364019	1.07741×10^{-3}
5	0.272750	7.79888×10^{-4}

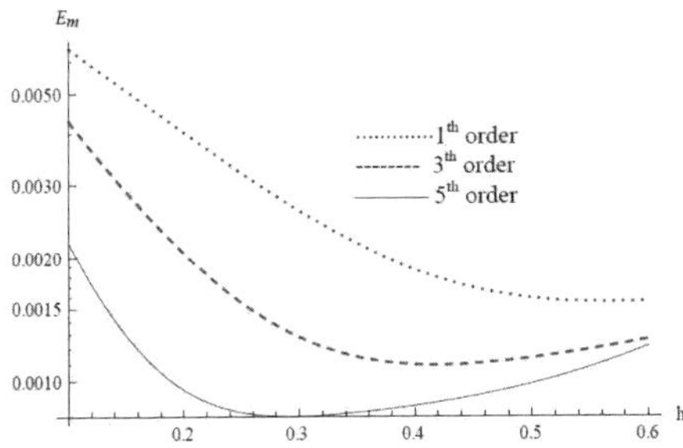

Fig. 7.1. Plots of average residual error E_m vs. $\hbar$ for $\mu = 2, \alpha = 0.5$ and $l = 10$.

respectively when $\mu = 2, \alpha = 0.5$.

It is also observed from Tables 7.1, 7.3 and 7.5, which are calculated for $\mu = 2, l = 10$ that as the values of α decrease the optimal value of $\hbar$ increases and goes away from $\hbar = -1$, the case of usual HAM. Similarly, Tables 7.2, 7.4 and 7.6, which are calculated for $\mu = 2, l = 8$ show that as the values of α decreases the optimal value of $\hbar$ increases and goes away from $\hbar = -1$. We see that the optimal value of the convergence control parameter depends not only on the values of α and l, but also on the number of terms employed in the series approximation of the solution. We have found that the value of the residual error decreases while we increase the number of terms of the series. This shows that HAM is a good alternative to a numerical method. Further HAM provides analytical solutions which can be readily available

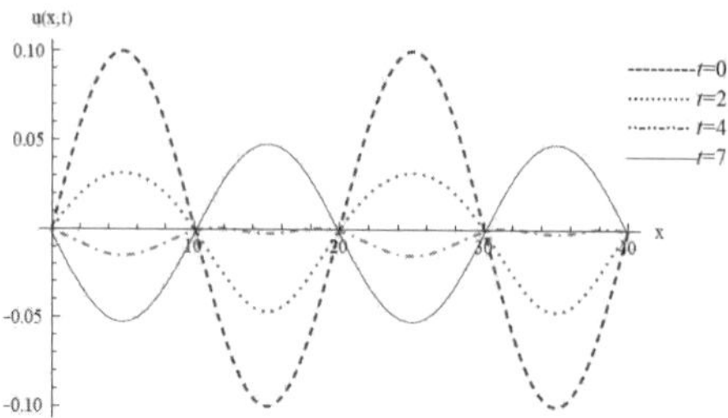

Fig. 7.2. Plots of profile of $u(x,t)$ vs. x at $\mu = 2, \alpha = 1, l = 10$ for different values of t at 5th order of approximation.

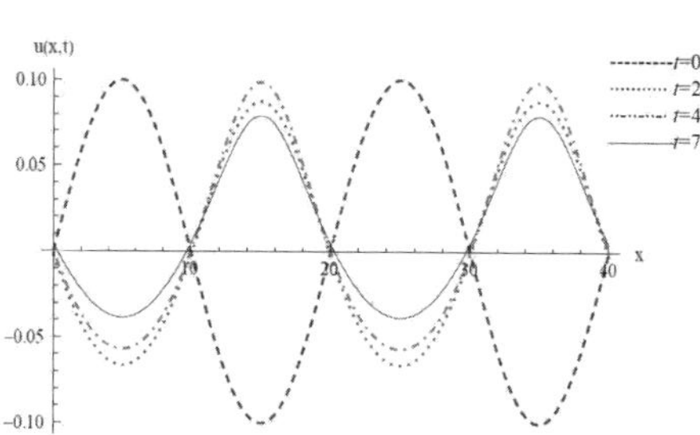

Fig. 7.3. Plots of profile of $u(x,t)$ vs. x at $\mu = 2, \alpha = 0.5, l = 10$ for different values of t at 5th order of approximation.

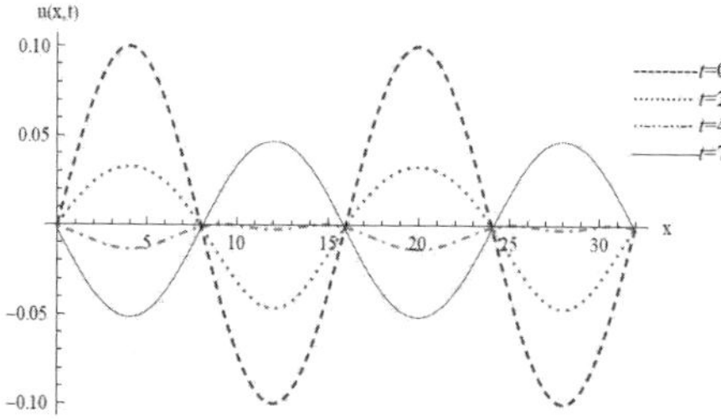

Fig. 7.4. Plots of profile of $u(x,t)$ vs. x at $\mu = 2, \alpha = 1, l = 8$ for different values of t at 5th order of approximation.

for other purposes.

7.6. Conclusion

In this chapter three important goals are achieved. First one is the successful application of the homotopy analysis method for finding the solution

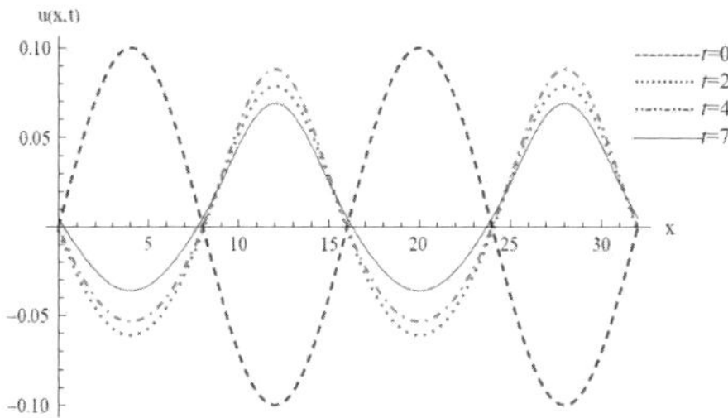

Fig. 7.5. Plots of profile of $u(x,t)$ vs. x at $\mu = 2, \alpha = 0.5, l = 8$ for different values of t at 5th order of approximation.

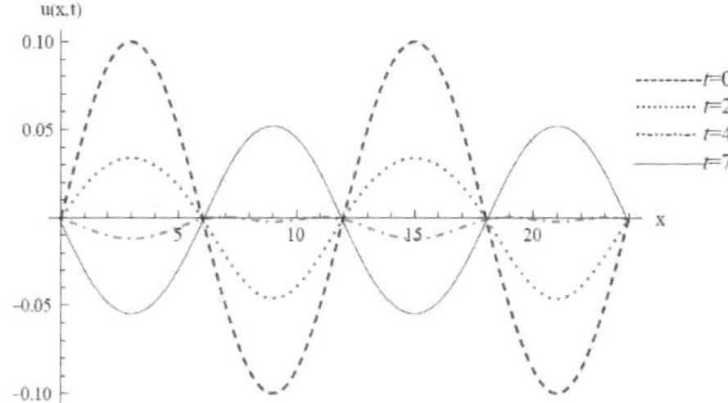

Fig. 7.6. Plots of profile of $u(x,t)$ vs. x at $\mu = 2, \alpha = 1, l = 6$ for different values of t at 5th order of approximation.

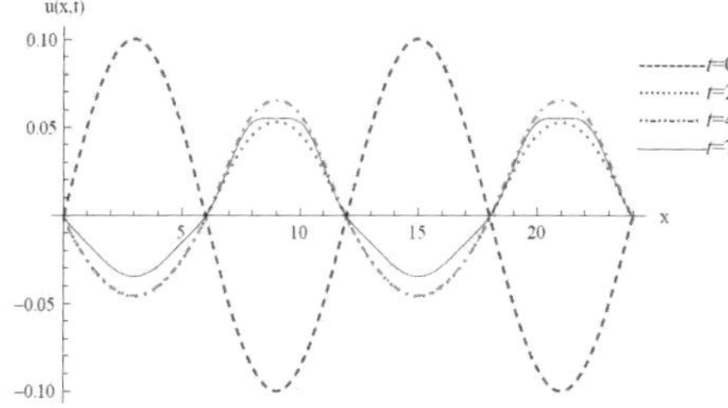

Fig. 7.7. Plots of profile of $u(x,t)$ vs. x at $\mu = 2, \alpha = 0.5, l = 6$ for different values of t at 5th order of approximation.

of the nonlinear S–H equation with fractional order time derivative. Secondly, the optimal value of the convergence-control parameter that gives rise to a convergent series solution. Third one is the successful numerical and graphical presentations of the effects of bifurcation parameter on the probability density function. The effects of bifurcation parameter in spatial domain are also described during the study. The Swift–Hohenberg equation

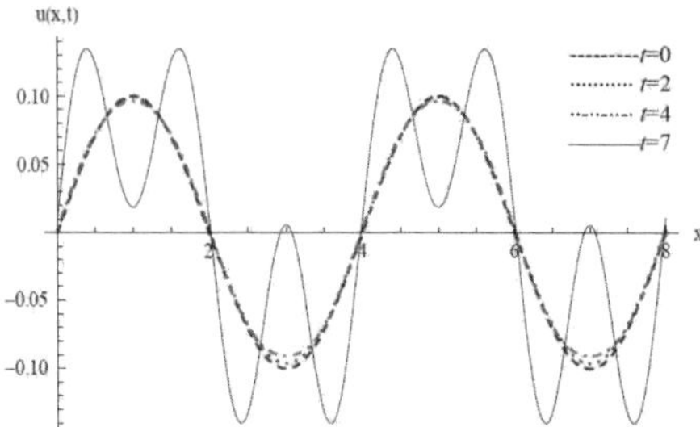

Fig. 7.8. Plots of profile of $u(x,t)$ vs. x at $\mu = 2, \alpha = 1, l = 2$ for different values of t at 5th order of approximation.

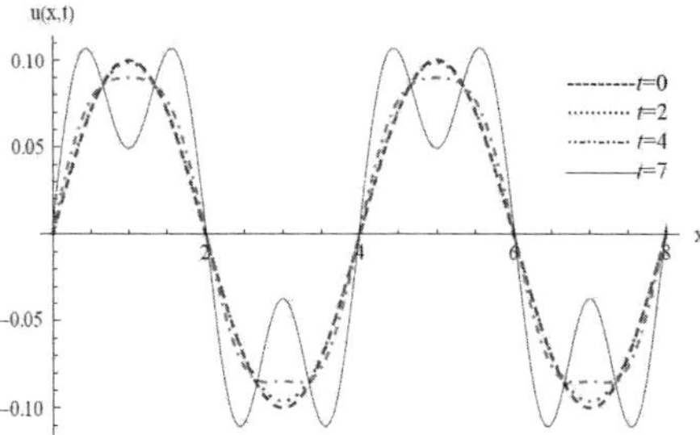

Fig. 7.9. Plots of profile of $u(x,t)$ vs. x at $\mu = 2, \alpha = 0.5, l = 2$ for different values of t at 5th order of approximation.

describes the convective heat current in a Rayleeigh–Benard cell and the nature of hydrodynamic stability. The equation further allows the study of convective pattern that results as a function of the forcing mecanism. A clear conclusion from the numerical results is that the HAM provides highly accurate numerical solutions for the nonlinear partial differential equation with fractional time derivative. It has been observed that the proper choice of the convergence control parameter greatly improves the convergence rate of the approximate series solution.

References

[1] J.B. Swift and P.C. Hohenberg, Hydrodynamics fluctuations at the convective Instability, *Phys. Rev. A.* **15**, 319–328 (1977).

[2] P.C. Hohenberg and J.B. Swift, Effects of additive noise at the onset of Rayleigh-Benard Convection, *Phys. Rev. A.* **46**, 4773–4785 (1992).

[3] L. Lega, J.V. Moloney and A.C. Newell, Swift- Hohenberg Equation for lasers, *Phys. Rev. Lett.* **73**, 2978–2981 (1994).

[4] M.C. Cross and P.C. Hohenberg, Pattern formulation outside of equiblirium, *Rev. Mod. Phys.* **65**, 851–1112 (1993).

[5] L.A. Peletier and V. Rottschafer, Pattern selection of solutions of the Swift–Hohenberg equation, *Physica D* **194**, 95–126 (2004).

[6] I. Podlubny, *Fractional Differential Equations,* Academic Press, New York, (1999).

[7] R. Gorenflo and F. Mainardi, Fractional calculus: Integral and differential equations of fractional order, in: A. Carpinteri, F. Mainardi (Eds.), Fractals and Fractional Calculus, New York, (1997).

[8] A.A. Kilbas, H.M. Srivastava and J.J. Trujillo, *Theory and Applications of Fractional Differential Equations,* Elsevier, Amsterdam, (2006).

[9] F.T. Akyildiz, D.A. Siginer, K. Vajravelu and R.A.V. Gorder, Analytical and numerical results for the Swift–Hohenberg equation, *Appl. Math. Comput.* **216**, 221–226 (2010).

[10] K. Vishal, S. Kumar and S. Das, Application of homotopy analysis method for fractional Swift–Hohenberg equation-Revisited, *Appl. Math. Model.* **36**, 3630–3637 (2011).

[11] K. Vishal, S. Das, S.H. Ong and P. Ghosh, On the solutions of fractional Swift–Hohenberg equation with dispersion, *Appl. Math. Comput.* **219**, 5792–5801 (2013).

[12] K.S. Miller and B. Ross, *An introduction to the fractional calculus and fractional differential equations,* John Wiley and Sons, Inc., New York, (2003).

[13] K.B. Oldham and J. Spanier, *The fractional calculus,* Academic Press, New York/London, 1974.

[14] S. Das, K. Vishal and P.K. Gupta, Approximate approach to the Das Model of Fractional Logistic Population Growth, *Appl. and Appl. Math.* **5**, 1702–1708 (2010).

[15] S. Das and P.K. Gupta, A Mathematical Model on fractional Lotka–Volterra equations, *J. Theoret. Biol.* **277**, 1–6 (2011).

[16] S.P. Ansari, S.K. Agarwal and S. Das, Stability Analysis of fractional order Generalized Chaotic SIR epidemic model and its Synchronization using Active Control Method, *Cent. Eur. J. Math.* Accepted (2012).

[17] S.J. Liao, *The proposed homotopy analysis technique for the solution of nonlinear problems,* Ph.D. Thesis, Shanghai Jiao Tong University, (1992).

[18] S.J. Liao, Notes on the homotopy analysis method: some definition and theorems, *Comm. in Nonlin. Sci. Numer. Simul.* **14**, 983–997 (2009).

[19] S.J. Liao, An approximate solution technique which does not depend upon small parameters: a special example, *Int. J. Nonlinear Mech.* **30**, 371–380 (1995).

[20] S.J. Liao, An approximate solution technique which does not depend upon small parameters (Part 2): an application in fluid mechenics, *Int. J. Nonlinear Mech.* **32**, 815–822 (1997).

[21] S.J. Liao, An explicit totally analytic approximation of Blasius viscous flow problems, *Int. J. Nonlinear Mech.* **34**, 759–778 (1999).

[22] S.J. Liao, *Beyond perturbation: introduction to the homotopy analysis method,* CRC Press, Boca Raton, Chapman and Hall, (2003).

[23] S.J. Liao, On the homotopy analysis method for nonlinear problems, *Appl. Math. Comput.* **147**, 499–513 (2004).

[24] S.J. Liao, Homotopy analysis method: a new analytical technique for nonlinear problems, *Commun. Nonlinear Sci. Numer. Simulat.* **2**, 95–100 (1997).

[25] S.J. Liao, A new branch of solutions of boundary-layer flows over an impermeable stretched plate, *Int. J. Heat Mass Transfer.* **48**, 2529–2539 (2005).

[26] S. Das, K. Vishal, P.K. Gupta and S.S. Ray, Homotopy Analysis Method for solving Fractional Diffusion equation, *Int. J. Appld. Math. & Mech.* **7**, 28–37 (2011).

[27] S.J. Liao, An optimal homotopy-analysis approach for strongly nonlinear differential equations, *Comm. in Nonlin. Sci. Numer. Simul.* **15**, 2003–2016 (2010).

[28] K. Vishal and S. Das, Solution of nonlinear fractional diffusion equation with absorbent term and external force using optimal homotopy analysis method, *Z. Naturforsch.* **67a**, 203–209 (2012).

[29] S. Das, K. Vishal, P.K. Gupta and A. Yildirim, An approximate analytical solution time fractional telegraph equation, *Appl. Math. Comput.* **217**, 7405–7411 (2011).

[30] L. Song and H. Zhang, Application of homotopy analysis method to fractional KdV Burgers–Kuramoto equation, *Phys. Lett. A.* **367**, 1–2 (2007).

[31] M. Zurigat, S. Momani, Z. Odibat and A. Alawneh, The homotopy analysis method for handling systems of fractional differential equations, *Appl. Math. Model.* **34**, 24–35 (2010).

[32] Z. Odibat, S. Momani and H. Xu, A reliable algorithm of homotopy analysis method for solving nonlinear fractional differential equations, *Appl. Math. Model.* **34**, 593–600 (2010).

[33] H. Saberi Nik and R. Buzhabadi, Solving fractional Fornberg–Whitham equation by homotopy analysis method, *J. Appl. Math. Statistics and Informatics.* **7**, No. 2. (2011).

[34] A. Neamaty, R. Darzi and A. Dabbaghian, Homotopy analysis method for solving fractional Sturm–Liouville problems, *Austral. J. Pure Appl. Sci.* **4**, 5108–5027 (2010).

[35] A.A.M. Arafa, S. Z. Rida and H. Mohamed, Homotopy analysis method for solving biological population model, *Commun. Theretical Phys.* **56**, 797–800 (2011).

[36] T. Hayat, C. Fetecau and S. Asghar, Series solution for the upper-convected Maxwell fluid over a porous stretching plate, *Phys. Lett. A* **358**, 396–403 (2006).

[37] F.M. Allan, Construction of analytic solution to chaotic dynamical systems using the homotopy analysis method, *Chaos Solitons Fract.* **39**, 1744–1752 (2009).

[38] T. Hayat and M. Khan, Homotopy solution for a generalized second grade fluid past a porous plate, *Nonlinear Dynam.* **42**, 395–405 (2005).

[39] H. Xu, Analytical approximations for a population growth model with fractional order, *Commun. Nonlinear Sci. Numer. Simulat.* **14**, 1978–1983 (2009).

Chapter 8

HAM-Based Package NOPH for Periodic Oscillations of Nonlinear Dynamic Systems

Yinping Liu

Department of Computer Science
East China Normal University, Shanghai 200241, China
ypliu@cs.ecnu.edu.cn

Based on Wu's elimination method and the homotopy analysis method (HAM), the Maple package `NOPH` (version 1.0.2) is developed for periodically oscillating systems of center and limit cycle types, which delivers accurate approximations of frequency, mean of motion and amplitudes of oscillation automatically. Since the HAM is valid for highly nonlinear problems, the package can be used to find accurate approximate solutions of nonlinear oscillation systems with strong nonlinearity. For systems with physical parameters, it can provide possible constraint conditions on parameters. In this chapter, we briefly describe the discussed problem, an efficient algorithm, a Maple package `NOPH` and so on. A simple users guide is given in the appendix of this chapter. The Maple package `NOPH` is free available (Accessed 20 Jan 2013, will be updated in the future) at `http://numericaltank.sjtu.edu.cn/NOPH.htm`.

Contents

8.1. Introduction . . . 310
8.2. Problem description . . . 312
8.3. Algorithm . . . 313
8.3.1. Rule of solution expression . . . 314
8.3.2. The zero-order deformation equation . . . 314
8.3.3. The higher-order deformation equation . . . 316
8.3.4. Solving nonlinear algebraic equations with parameters . . . 318
8.3.5. The choice of the convergence-control parameter $\hbar$ . . . 320
8.3.6. The homotopy-Padé technique . . . 321
8.4. The package NOPH . . . 322
8.5. Applications of NOPH . . . 324
8.5.1. Example 1 . . . 325
8.5.2. Example 2 . . . 327
8.5.3. Example 3 . . . 331
8.5.4. Example 4 . . . 335
8.5.5. Example 5 . . . 338
8.6. Discussions and conclusions . . . 342
Appendix A. The procedures contained in the NOPH . . . 342
Appendix B. A simple user's guide . . . 343
B.1. Example 1 . . . 344
B.1.1. All physical parameters are unknown . . . 344
B.1.2. In the case of $\alpha = 5$, $a = 1/2$ . . . 345
B.2. Example 2 . . . 347
B.2.1. All physical parameters are unknown . . . 348
B.2.2. In the case of $\epsilon = 2$ . . . 349
B.3. Example 3 . . . 352
B.3.1. All physical parameters are unknown . . . 352
B.3.2. In the case of $\varepsilon = 1/2$ . . . 354
B.3.3. In the case of $\varepsilon = 1/2$, $\mu = 8$ . . . 355
References . . . 357

8.1. Introduction

Nonlinear oscillation is a type of common nonlinear initial-value problems, which can be seen anywhere anytime in science and engineering. Considerable interests have been focused on the study of analytic solutions of nonlinear oscillations, for these solutions may give more insights into internal aspects of nonlinear problems. Foremost among the analytic techniques are perturbation methods in terms of a small/large physical parameter [1–4]. The basic premise of perturbation methods is to transfer a nonlinear problem to a set of infinite number of linear sub-problems by means of a small/large *physical* parameter. Unfortunately, many nonlinear equations do not contain such types of small/large physical parameters.

Besides, perturbation approximations are generally only accurate enough for weakly nonlinear problems, but often break down when the physical parameters become large. In addition, restricted by such types of small/large physical parameters, the perturbation techniques cannot provide freedom to choose the equation type of the linear sub-problems. This can regularly result in major difficulties when solving nonlinear oscillation problems. To overcome the restrictions of perturbation methods, some non-perturbation methods, such as the artificial small parameter method, the δ-expansion method and Adomian's decomposition method, have been developed [5–7]. However, it is a pity that these perturbation and non-perturbation methods cannot provide a convenient way to adjust and control the convergence of approximation series.

HAM [8, 9] is an analytic approximation method for strongly nonlinear problems, which has been successfully applied to solve many nonlinear problems in science and engineering [8–23]. This is especially true for several solutions that were not identified using analytic or even numerical techniques, as presented by Liao [14].

Compared with other analytic or numerical techniques, HAM has the following advantages.

- HAM does *not* need any small/large *physical* parameters. Therefore, its use is valid in a greater number of nonlinear problems.
- HAM introduces an auxiliary parameter, called convergence-control parameter, which provides a simple way to control and adjust the convergence of approximation solutions [8, 9, 15, 16].
- HAM provides great freedom to choose the proper equation-type and base functions to approximate solution of nonlinear problems efficiently.
- Using HAM, accurate approximations of the required solution can be obtained, as well as possible dependent relations of the physical parameters from the system under consideration. These relations are helpful for better understandings the inner aspect of nonlinear problems.
- Multiple solutions of nonlinear problems can be gained by means of HAM.

These advantages of HAM imply that it is possible to develop some general symbolic computation packages for some types of nonlinear problems. Following a HAM-based Mathematica package `BVPh` for coupled, highly nonlinear boundary-value/eigenvalue problems[a], this Maple package NOPH 1.0.2 has been successfully developed for strongly periodic nonlinear oscillations, the old version of the package has been reported [24], the current version of the package can work for some single parametric or forced oscillation. However, as there are external and internal at least two different frequencies in these two types of oscillations. Therefore, we will continue to develop new algorithm to solve parametric or forced oscillations, especially with vector frequency. These packages are freely available on the website http:/numericaltank.sjtu.edu.cn/.

8.2. Problem description

Consider periodic solutions of nonlinear oscillations governed by

$$\ddot{U}_i(t) = f_i[\mathbf{U}(t), \dot{\mathbf{U}}(t), \ddot{\mathbf{U}}(t)], \qquad 1 \leq i \leq \kappa, \tag{8.1}$$

where the dependent variable $\mathbf{U}(t)$ has κ components $U_1(t), \ldots, U_\kappa(t)$, the independent variable t denotes time, the dot denotes differentiation with respect to t, and $f_i[\mathbf{U}(t), \dot{\mathbf{U}}(t), \ddot{\mathbf{U}}(t)]$ is either an algebraic or rational function of $\mathbf{U}(t), \dot{\mathbf{U}}(t)$ and $\ddot{\mathbf{U}}(t)$. Note that Eq. (8.1) is rather general which can be used to describe lots of periodic oscillations of nonlinear problems in science and engineering. In this chapter, the focus is mainly on single conservative oscillations and self-excited oscillations. The former corresponds to periodic oscillations of the center type, and the latter to periodic oscillations of the limit cycle type. By the way we should mention that self-excited oscillators can only appear as relaxation oscillation when the coefficient of the main nonlinear term is large enough. As the intermediate expressions "swells" so fast, the current version of the package NOPH does not work for a relaxation oscillator. It should be emphasized that Eq. (8.1) does not necessarily contain small/large physical parameters.

Let $\Omega = \{\omega_1, \ldots, \omega_\kappa\}$ denote the set of frequencies of the periodic oscillations. For simplicity, we just consider the case that Ω contains single ω in what follows. For multidimensional systems, if $\omega_i\,(1 \leq i \leq \kappa)$ are different, we could reduce Ω to $\{\omega\}$ by the pretreatments: $\omega_2 = p_2\omega, \ldots, \omega_\kappa = p_\kappa\omega$. In this case, $p_i\,(i = 2, \ldots, \kappa)$ can be looked upon as positive parameters, and they will be determined by using the second main strategy in §8.3.4. It should be pointed out that according to the property of oscillations, the algorithm can only deal with integer $p_i\,(2 \leq i \leq \kappa)$, and it may not work well for rational or irrational ratios $p_i\,(2 \leq i \leq \kappa)$.

[a]The package is free available online http://numericaltank.sjtu.edu.cn/BVPh.htm.

Define the mean of motions

$$\delta_i = \frac{1}{T}\int_0^T U_i(t)dt,\ i = 1, \ldots, \kappa, \tag{8.2}$$

where $T = 2\pi/\omega$ is the period of oscillation. Note that if f_i in Eq. (8.1) contains only odd nonlinearity about $U_i(t)$, the mean of motion δ_i is equal to zero.

Without loss of generality, let us consider nonlinear periodic oscillations with initial conditions

$$\begin{aligned} &U_i(0) = a_i + \delta_i,\quad \dot{U}_i(0) = \omega\, b_i,\qquad i = 1, \ldots, \kappa - 1,\\ &U_\kappa(0) = a_\kappa + \delta_\kappa,\ \dot{U}_\kappa(0) = 0, \end{aligned} \tag{8.3}$$

where a_i, b_i and δ_i are unknowns to be determined later. Physically, the frequency ω can be regarded as a time-scale. Hence, introducing the transformations

$$\tau = \omega\, t,\quad U_i(t) = \delta_i + u_i(\tau),\qquad\qquad 1 \le i \le \kappa,$$

Eqs. (8.1) and (8.3) become

$$\omega^2\, \ddot{u}_i(\tau) = f_i[\delta + \mathbf{u}(\tau), \omega\, \dot{\mathbf{u}}(\tau), \omega^2\, \ddot{\mathbf{u}}(\tau)], \tag{8.4}$$

subject to initial conditions

$$u_\kappa(0) = a_\kappa,\quad \dot{u}_\kappa(0) = 0,\quad u_i(0) = a_i,\quad \dot{u}_i(0) = b_i,\quad 1 \le i \le \kappa - 1, \tag{8.5}$$

in which $\mathbf{u}$ and δ have κ components $u_1(\tau), \ldots, u_\kappa(\tau)$ and $\delta_1, \ldots, \delta_\kappa$, respectively. In Eqs. (8.4) and (8.5), apart from the unknown functions $u_i(\tau)$ and $i = 1, \ldots, \kappa$, there are 3κ unknown physical quantities, including the frequency ω, the mean of motion $\delta_i\,(1 \le i \le \kappa)$, the amplitudes $a_j,\ b_j\ (1 \le j \le \kappa - 1)$ and a_κ. All of these unknowns can be determined by the so-called "rule of solution expression" in the frame of HAM, as described later.

8.3. Algorithm

Although HAM has been widely applied while solving nonlinear problems in science and engineering, it is not very easy to apply to a system of coupled nonlinear oscillations. The reason lies in that one has to solve κ coupled nonlinear differential equations together with a set of 3κ highly nonlinear algebraic equations related to the unknown physical quantities $\omega,\ \delta_i\,(1 \le i \le \kappa),\ a_j,\ b_j\,(1 \le j \le \kappa - 1)$ and a_κ. In this section, HAM is combined with Wu's elimination method to develop an efficient algorithm.

8.3.1. *Rule of solution expression*

In general, a system of coupled nonlinear oscillators is more complicated. Here, the focus is only on periodic solutions, which can be expressed in terms of the base functions

$$\{\cos(m\,\tau),\ \ \sin(m\,\tau)\,|\,m=1,\,2,\,3,\,\ldots\}. \tag{8.6}$$

This is the so-called "rule of solution expression".

To search for periodic solutions, initial guesses were chosen

$$\begin{cases} u_{\kappa,0}(\tau)=a_{\kappa,0}\cos(\tau), \\ u_{i,0}(\tau)=a_{i,0}\cos(\tau)+b_{i,0}\sin(\tau), \quad i=1,\ldots,\kappa-1, \end{cases} \tag{8.7}$$

where $a_{i,0}$ and $b_{i,0}$ are unknowns to be determined later, and an auxiliary linear operator

$$\mathcal{L}[\Phi(\tau,q)]={\omega_0}^2\left[\frac{\partial^2\Phi(\tau,q)}{\partial\tau^2}+\Phi(\tau,q)\right] \tag{8.8}$$

with property $\mathcal{L}[C_1\sin(\tau)+C_2\cos(\tau)]=0$, in which $q\in[0,1]$ is the embedding-parameter, ω_0 is the initial guess of the frequency ω, $\Phi(\tau,q)$ is a function of τ and q, while C_1 and C_2 are integral constants.

8.3.2. *The zero-order deformation equation*

Based on Eq. (8.4), the following nonlinear operators are defined

$$\begin{aligned}\mathcal{N}_i[\mathbf{\Phi}(\tau,q),\Omega(q),\mathbf{\Delta}(q)]=\Omega^2(q)\frac{\partial^2\mathbf{\Phi}(\tau,q)}{\partial\tau^2}-f_i\Bigg[\mathbf{\Delta}(q)+\mathbf{\Phi}(\tau,q),\\ \Omega(q)\frac{\partial\mathbf{\Phi}(\tau,q)}{\partial\tau},\Omega^2(q)\frac{\partial^2\mathbf{\Phi}(\tau,q)}{\partial\tau^2}\Bigg],\end{aligned} \tag{8.9}$$

where $\Omega(q)$ and $\mathbf{\Delta}(q)$ are functions of q, $\mathbf{\Phi}(\tau,q)$ and $\mathbf{\Delta}(q)$ have κ components, corresponding to the function $\mathbf{u}(\tau)$ and the mean of motion δ, respectively.

HAM is based on continuous variations $\mathbf{\Phi}(\tau,q)$, $\Omega(q)$ and $\mathbf{\Delta}(q)$, as the embedding-parameter q varies from 0 to 1, $\mathbf{\Phi}(\tau,q)$ varies from the initial guess $\mathbf{u}_0(\tau)$ to the exact solution $\mathbf{u}(\tau)$. So does $\Omega(q)$ from the initial guess ω_0 to the exact frequency ω, and $\mathbf{\Delta}(q)$ from the initial guess $\delta_{\mathbf{0}}$ to the exact mean of motion δ. Then, with the aid of homotopy in topology, the following equation is constructed (called the zeroth-order deformation equation)

$$(1-q)\mathcal{L}\left[\mathbf{\Phi}(\tau,q)-\mathbf{u}_0(\tau)\right]=q\,\hbar\mathcal{N}_i\left[\mathbf{\Phi}(\tau,q),\Omega(q),\mathbf{\Delta}(q)\right], \tag{8.10}$$

subject to initial conditions

$$\Phi_i(0,q) = \bar{a}_i(q), \qquad \left.\frac{\partial \Phi_j(\tau,q)}{\partial \tau}\right|_{\tau=0} = \bar{b}_j(q),$$
$$\left.\frac{\partial \Phi_\kappa(\tau,q)}{\partial \tau}\right|_{\tau=0} = 0, \qquad 1 \le i \le \kappa, \qquad 1 \le j \le \kappa - 1, \tag{8.11}$$

where $\bar{a}_i(q)$ and $\bar{b}_j(q)$ are functions of q, corresponding to a_i and b_j, respectively.

When $q = 0$, it follows from (8.7) and (8.10) that

$$\mathbf{\Phi}(\tau,0) = \mathbf{u}_0(\tau), \quad \Omega(0) = \omega_0, \quad \mathbf{\Delta}(0) = \delta_{\mathbf{0}}. \tag{8.12}$$

When $q = 1$, since $\hbar \neq 0$, the zeroth-order deformation equation (8.10) is equivalent to

$$\mathcal{N}_i[\mathbf{\Phi}(\tau,1), \Omega(1), \mathbf{\Delta}(1)] = 0, \qquad 1 \le i \le \kappa, \tag{8.13}$$

which is exactly the same as the original equation (8.4), provided

$$\mathbf{\Phi}(\tau,1) = \mathbf{u}(\tau), \quad \Omega(1) = \omega, \quad \mathbf{\Delta}(1) = \delta. \tag{8.14}$$

Thus, according to (8.12) and (8.14), as the embedding-parameter q increases from 0 to 1, $\mathbf{\Phi}(\tau,q)$ varies continuously from the initial guess $\mathbf{u}_0(\tau)$ to the exact solution $\mathbf{u}(\tau)$. So does $\Omega(q)$ from the initial guess ω_0 to the exact frequency ω, and $\mathbf{\Delta}(q)$ from the initial guess $\delta_{\mathbf{0}}$ to the exact mean of motion δ.

According to Taylor's theorem and using (8.12), $\Phi_i(\tau,q)$, $\Delta_i(q)$, $\Omega(q)$, $\bar{a}_i(q)$ and $\bar{b}_i(q)$ are expanded in the power series of q as follows

$$\Phi_i(\tau,q) = u_{i,0}(\tau) + \sum_{n=1}^{+\infty} u_{i,n}(\tau)\, q^n,$$
$$\Omega(q) = \omega_0 + \sum_{n=1}^{+\infty} \omega_n\, q^n, \qquad \Delta_i(q) = \delta_{i,0} + \sum_{n=1}^{+\infty} \delta_{i,n}\, q^n,$$
$$\bar{a}_i(q) = a_{i,0} + \sum_{n=1}^{+\infty} a_{i,n}\, q^n, \qquad \bar{b}_j(q) = b_{j,0} + \sum_{n=1}^{+\infty} b_{j,n}\, q^n, \tag{8.15}$$

where $1 \le i \le \kappa$, $1 \le j \le \kappa - 1$, and

$$u_{i,n}(\tau) = \frac{1}{n!}\left.\frac{\partial^n \Phi_i(\tau,q)}{\partial q^n}\right|_{q=0}, \quad \delta_{i,n} = \frac{1}{n!}\left.\frac{\partial^n \Delta_i(q)}{\partial q^n}\right|_{q=0},$$

$$\omega_n = \frac{1}{n!}\left.\frac{\partial^n \Omega(q)}{\partial q^n}\right|_{q=0}, \quad a_{i,n} = \frac{1}{n!}\left.\frac{\partial^n \bar{a}_i(q)}{\partial q^n}\right|_{q=0}, \quad b_{j,n} = \frac{1}{n!}\left.\frac{\partial^n \bar{b}_j(q)}{\partial q^n}\right|_{q=0}.$$

Assuming that the parameter $\hbar$ is properly chosen so that the above series are convergent at $q = 1$, and due to (8.12) there is

$$u_i(\tau) = u_{i,0}(\tau) + \sum_{n=1}^{+\infty} u_{i,n}(\tau), \qquad \omega = \omega_0 + \sum_{n=1}^{+\infty} \omega_n,$$

$$\delta_i = \delta_{i,0} + \sum_{n=1}^{+\infty} \delta_{i,n}, \quad a_i = a_{i,0} + \sum_{n=1}^{+\infty} a_{i,n}, \quad b_j = b_{j,0} + \sum_{n=1}^{+\infty} b_{j,n}, \tag{8.16}$$

where $1 \le i \le \kappa$ and $1 \le j \le \kappa - 1$. At the Mth-order approximations, there is

$$\tilde{u}_i(\tau) \approx \sum_{n=0}^{M} u_{i,n}(\tau), \quad \tilde{\omega} \approx \sum_{n=0}^{M-1} \omega_n, \quad \tilde{\delta}_i \approx \sum_{n=0}^{M-1} \delta_{i,n},$$

$$\tilde{a}_i \approx \sum_{n=0}^{M-1} a_{i,n}, \quad \tilde{b}_j \approx \sum_{n=0}^{M-1} b_{j,n}, \quad 1 \le i \le \kappa, \ 1 \le j \le \kappa - 1. \tag{8.17}$$

8.3.3. *The higher-order deformation equation*

Differentiating the zero-order deformation equation (8.10) n times with respect to q, then dividing them by $n!$, and finally setting $q = 0$, the nth-order deformation equation can be obtained

$$\mathcal{L}[u_{i,n}(\tau) - \chi_n \, u_{i,n-1}(\tau)] = \hbar \, R_{i,n}, \ 1 \le i \le \kappa, \tag{8.18}$$

subject to initial conditions

$$u_{\kappa,n}(0) = a_{\kappa,n}, \quad \dot{u}_{\kappa,n}(0) = 0, \quad u_{i,n}(0) = a_{i,n},$$
$$\dot{u}_{i,n}(0) = b_{i,n}, \quad 1 \le i \le \kappa - 1, \tag{8.19}$$

where

$$R_{i,n} = \frac{1}{(n-1)!} \frac{\partial^{n-1} \mathcal{N}_i \left[\mathbf{\Phi}(\tau, q), \Omega(q), \mathbf{\Delta}(\mathbf{q})\right]}{\partial q^{n-1}}\bigg|_{q=0}, \quad 1 \le i \le \kappa, \tag{8.20}$$

and

$$\chi_n = \begin{cases} 0, \ n \le 1, \\ 1, \ n > 1. \end{cases} \tag{8.21}$$

The above nth-order deformation equation (8.18) is a set of κ linear differential equations about $u_{i,n}(\tau)$ and $1 \le i \le \kappa$. However, apart from κ unknown functions $u_{i,n}(\tau)$, there are 3κ unknown parameters

$\omega_{n-1}, a_{i,n-1}, \delta_{i,n-1}\,(1 \le i \le \kappa)$ and $b_{j,n-1}\,(1 \le j \le \kappa-1)$. According to the properties of trigonometric functions, $R_{i,n}$ can be expressed by

$$R_{i,n} = A_0^{i,n} + \sum_{j=1}^{\phi(i,n)} A_j^{i,n}\cos(j\tau) + \sum_{j=1}^{\psi(i,n)} B_j^{i,n}\sin(j\tau), \quad 1 \le i \le \kappa,$$

where the integers $\phi(i,n)$ and $\psi(i,n)$ depend on Eq. (8.4) and the order n. To search for a periodic solution and avoid the so-called secular terms, $A_0^{i,n}$, $A_1^{i,n}$ and $B_1^{i,n}$, where $1 \le i \le \kappa$, must vanish, i.e.,

$$A_0^{i,n} = 0, \quad A_1^{i,n} = 0, \quad B_1^{i,n} = 0, \qquad 1 \le i \le \kappa. \tag{8.22}$$

Equations (8.22) provide us with a set of 3κ algebraic equations to determine the above mentioned 3κ unknowns, i.e., ω_{n-1}, $a_{i,n-1}$, $b_{j,n-1}$ and $\delta_{i,n-1}$, where $1 \le i \le \kappa$ and $1 \le j \le \kappa-1$. Then, it is easy to obtain the solutions for the nth-order deformation equation

$$u_{i,n}(\tau) = C_1^{i,n}\sin(\tau) + C_2^{i,n}\cos(\tau) + \sum_{j=2}^{\phi(i,n)} \frac{A_j^{i,n}}{1-j^2}\cos(j\tau) + \sum_{j=2}^{\psi(i,n)} \frac{B_j^{i,n}}{1-j^2}\sin(j\tau),$$

where $1 \le i \le \kappa$, the 2κ integral constants $C_1^{i,n}$ and $C_2^{i,n}$ are determined by the 2κ initial conditions given in (8.19).

It is to be stressed that Eq. (8.4) is more general, which include single conservative oscillations and self-excited oscillations, etc. For single conservative oscillation the base functions (8.6) can be simplified as $\{\cos(m\,\tau)\,|\,m = 1, 2, 3, \ldots\}$, since in this case, all coefficients of sine functions become zero. Thus Eqs. (8.22) are reduced to

$$A_0^{1,n} = 0, \quad A_1^{1,n} = 0,$$

which are used to determine ω_{n-1} and $\delta_{1,n-1}$. Note that different types of oscillations have different characteristics, for example, the amplitude for a single conservative oscillation is fixed, or it is even known sometimes, but the amplitude for a self-excited oscillation system is unknown and needs to be determined. The problem can be removed when the conservative oscillation amplitude is unknown, as this means it can be considered to be a parameter and treated as such.

8.3.4. *Solving nonlinear algebraic equations with parameters*

It should be pointed out that, Eqs. (8.22) are linear when $n > 1$, but are 3κ coupled nonlinear algebraic equations when $n = 1$. For a single nonlinear oscillation, it is easy to solve Eqs. (8.22) for ω_0, $\delta_{1,0}$ and $a_{1,0}$. However, for coupled nonlinear oscillation, it is difficult to solve the corresponding nonlinear algebraic Eqs. (8.22) for ω_0, $\delta_{i,0}$, $a_{i,0}$ and $b_{j,0}$, where $1 \le i \le \kappa$ and $1 \le j \le \kappa - 1$, which may become a bottleneck for the whole algorithm. To overcome this difficulty and to improve the efficiency of this program, various techniques were applied, apart from the divide-and-conquer technique and freeze technique. Here, the three main strategies are listed below:

- Firstly, Wu's elimination method [25] is a powerful tool for solving nonlinear algebraic equations. Wang *et al.* [26, 27] completely implemented Wu's elimination method and developed a Maple package `CharSets`, in which the `Csolve` solver can be directly used to solve the nonlinear algebraic equations. In this Maple code, Eqs. (8.22) are solved by means of the package `CharSets`, whose greatest advantage is that it can avoid missing solutions. Besides, since the orders of variables can be specified when solving nonlinear algebraic equations by using the package, the orders of variables can be optimized so as to improve the computational efficiency. In addition, the `CharSets` package is very efficient, due to some excellent techniques having been employed in it, such as decomposing expression, extracting subexpression as well as sorting and specifying the orders of equations to be solved, and so on.
- Secondly, in Eqs. (8.22), usually the degrees of parameters involved in the governing equations are lower than those of unknowns, so treating these parameters as unknowns might greatly weaken the difficulty of the problem. Doing this it is possible to obtain compact special solutions with parameter constraints.
- Thirdly, even with the above two strategies, for some coupled nonlinear oscillating systems, the obtained equations (8.22) are still very difficult to deal with. In this case, some extra conditions to simplify Eqs. (8.22) might be imposed. With this strategy, the complicated algebraic equations (8.22) could be greatly simplified, and then some special solutions may be obtained.

Here, an example is given to illustrate how to apply these techniques to simplify and solve a nonlinear algebraic equations. Consider a three coupled self-excited oscillator

$$\begin{cases} \ddot{U} - \varepsilon\,(1 - U^2)\,\dot{U} + U = \varepsilon\mu(W - U), \\ \ddot{V} - \varepsilon\,(1 - V^2)\,\dot{V} + V = \varepsilon\mu(W - V), \\ \ddot{W} - \varepsilon\,(1 - W^2)\,\dot{W} + p^2\,W = \varepsilon\mu(U + V - 2W), \end{cases} \tag{8.23}$$

with initial conditions

$$\begin{cases} U(0) = a_1, \;\; \dot{U}(0) = b_1, \;\; V(0) = a_2, \;\; \dot{V}(0) = b_2, \\ W(0) = a_3, \;\; \dot{W}(0) = 0, \end{cases} \tag{8.24}$$

where ε, μ and p are unknown parameters, and a_1, a_2, a_3, b_1 and b_2 are unknowns to be determined.

As this system only contains odd nonlinearity, $\delta_i = 0$ and $i = 1, 2, 3$. Direct calculations, following the procedure of HAM, yield

$$\begin{cases} \varepsilon\omega_0 b_{1,0}(b_{1,0}^2 + a_{1,0}^2 - 4) - 4[\varepsilon\mu a_{3,0} + a_{1,0}(\omega_0^2 - 1 - \varepsilon\mu)] = 0, \\ \varepsilon\omega_0(a_{1,0}^3 + b_{1,0}^2 a_{1,0}) - 4(b_{1,0} + \varepsilon\omega_0 a_{1,0} + \varepsilon\mu b_{1,0} - \omega_0^2 b_{1,0}) = 0, \\ \varepsilon\omega_0 b_{2,0}(b_{2,0}^2 + a_{2,0}^2 - 4) - 4[\varepsilon\mu a_{3,0} + a_{2,0}({\omega_0}^2 - 1 - \varepsilon\mu)] = 0, \\ \varepsilon\omega_0(a_{2,0}^3 + b_{2,0}^2 a_{2,0}) - 4(b_{2,0} + \varepsilon\omega_0 a_{2,0} + \varepsilon\mu b_{2,0} - b_{2,0}{\omega_0}^2) = 0, \\ \varepsilon\,\mu\,a_{1,0} + {\omega_0}^2\,a_{3,0} - p^2\,a_{3,0} - 2\,\varepsilon\,\mu\,a_{3,0} + \varepsilon\,\mu\,a_{2,0} = 0, \\ 4\,a_{3,0}\omega_0 - 4\mu\,b_{2,0} - \omega_0\,a_{3,0}^3 - 4\,\mu\,b_{1,0} = 0. \end{cases} \tag{8.25}$$

It is difficult to solve Eqs. (8.25) for the unknowns ω_0, $a_{1,0}$, $b_{1,0}$, $a_{2,0}$, $b_{2,0}$ and $a_{3,0}$, because the highest degrees of $a_{1,0}$, $b_{1,0}$, $a_{2,0}$, $b_{2,0}$ and $a_{3,0}$ are all 3, and the highest degree of ω_0 is 2. Due to the known fact that the solution expression of a cubic equation is complicated, it is natural that the solution expressions of Eqs. (8.25) would be huge. It becomes even worse when the intermediate expressions "swell" so fast that the memory will overflow finally. Therefore, treating the parameters ε, μ and p as unknowns may reduce such types of difficulties. If solving the complicated equations (8.25) with some or all of ε, μ, p, ω_0, $a_{1,0}$, $b_{1,0}$, $a_{2,0}$, $b_{2,0}$ and $a_{3,0}$ as unknowns, unfortunately, a trivial solution may be obtained with the constraint $\varepsilon = 0$, or it may take too long to obtain the result, or even in some cases the memory could overflow. In order to obtain nontrivial solutions, some extra conditions could be imposed, such as $\{b_{1,0}=0\}$, to reduce Eqs. (8.25) as

$$\begin{cases} {\omega_0}^2\, a_{1,0} - a_{1,0} - \varepsilon\,\mu\, a_{1,0} + \varepsilon\,\mu\, a_{3,0} = 0, \\ \varepsilon\omega_0 b_{2,0}(b_{2,0}^2 - 4 + a_{2,0}^2) - 4a_{2,0}({\omega_0}^2 - 1 - \varepsilon\mu) - 4\varepsilon\mu a_{3,0} = 0, \\ \varepsilon\omega_0 a_{2,0}(4 - b_{2,0}^2 - a_{2,0}^2) + 4b_{2,0}(1 + \varepsilon\mu - {\omega_0}^2) = 0, \\ \varepsilon\,\mu\, a_{1,0} + {\omega_0}^2\, a_{3,0} - p^2\, a_{3,0} - 2\,\varepsilon\,\mu\, a_{3,0} + \varepsilon\,\mu\, a_{2,0} = 0, \\ 4\, a_{3,0}\,\omega_0 - 4\,\mu\, b_{2,0} - \omega_0\, a_{3,0}^3 = 0, \\ a_{1,0}^2 - 4 = 0. \end{cases} \tag{8.26}$$

Clearly, the equations in (8.26) are much simpler than those in (8.25). Treating the parameters ε, μ and p as unknowns and solving Eqs. (8.26) with the Maple package `CharSets`, the following solutions are obtained

$$\begin{aligned} &\{\omega_0 = \sqrt{1+\varepsilon\,\mu},\, a_{1,0} = \mp 2,\, a_{2,0} = \pm 2,\, b_{1,0} = 0,\, b_{2,0} = 0,\, a_{3,0} = 0\}, \\ &\{\mu = \frac{1-p^2}{2\varepsilon},\, \omega_0 = \sqrt{2-p^2},\, a_{1,0} = a_{2,0} = \mp 2,\, b_{1,0} = b_{2,0} = 0,\, a_{3,0} = \pm 2\}, \\ &\{p = \mp 1,\, \omega_0 = 1,\, a_{1,0} = -2,\, a_{2,0} = -2,\, b_{1,0} = 0,\, b_{2,0} = 0,\, a_{3,0} = -2\}, \\ &\{p = \mp 1,\, \omega_0 = 1,\, a_{1,0} = 2,\, a_{2,0} = 2,\, b_{1,0} = 0,\, b_{2,0} = 0,\, a_{3,0} = 2\}. \end{aligned}$$

As p appears in the form of p^2 in Eqs. (8.23), it is easy to see that if $U(t), V(t)$ and $W(t)$ are solutions of the systems (8.23) and (8.24), then $-U(t), -V(t)$ and $-W(t)$ are also solutions for the system. Therefore, there are three essentially different solutions in (8.27), given by

$$\begin{aligned} &\{\omega_0 = \sqrt{1+\varepsilon\,\mu},\, a_{1,0} = 2,\, a_{2,0} = 2,\, b_{1,0} = 0,\, b_{2,0} = 0,\, a_{3,0} = 0\}, \\ &\{\mu = \frac{1-p^2}{2\varepsilon},\, \omega_0 = \sqrt{2-p^2},\, a_{1,0} = a_{2,0} = a_{3,0} = 2,\, b_{1,0} = b_{2,0} = 0\}, \\ &\{p = 1,\, \omega_0 = 1,\, a_{1,0} = 2,\, a_{2,0} = 2,\, b_{1,0} = 0,\, b_{2,0} = 0,\, a_{3,0} = 2\}. \end{aligned}$$

Obviously, there are parameter constraints for the last two solutions. It is easy to see that under the different conditions imposed, the obtained solutions for the systems (8.23) and (8.24) would be different.

8.3.5. *The choice of the convergence-control parameter $\hbar$*

Following the above steps, a family of solution series can be obtained in the convergence-control parameter $\hbar$. How can a proper value of $\hbar$ be chosen to guarantee a fast enough convergence of the solution series for any given physical parameters? The optimal value of $\hbar$ is determined by minimizing the squared residual error of Eq. (8.4), i.e.,

$$\Lambda(\hbar) = \sum_{i=1}^{\kappa} \int_0^{2\pi} \epsilon_i(\hbar, \tau)^2 d\tau,$$

where $\epsilon_i(\hbar, \tau) = \mathcal{N}_i[\tilde{\mathbf{u}}(\tau), \tilde{\omega}, \tilde{\delta}]$, $\tilde{\mathbf{u}}(\tau) = \{\tilde{u}_1(\tau), \ldots, \tilde{u}_\kappa(\tau)\}$, $\tilde{\delta} = \{\tilde{\delta}_1, \ldots, \tilde{\delta}_\kappa\}$ and $1 \le i \le \kappa$.

8.3.6. *The homotopy-Padé technique*

Padé approximant expands a function as a ratio of two power series. When a function is given in the form of $R(x) = (\sum_{k=0}^{M} a_k\, x^k)/(1 + \sum_{k=1}^{N} b_k\, x^k)$, then $R(x)$ is said to be a Padé approximant to the series $f(x) = \sum_{k=0}^{\infty} c_k\, x^k$, if

$$R(0) = f(0), \quad \frac{d^k}{dx^k} R(x)\Big|_{x=0} = \frac{d^k}{dx^k} f(x)\Big|_{x=0}, \quad k = 1, 2, \ldots, M+N. \tag{8.27}$$

The conditions (8.27) provide $M + N + 1$ equations for the unknowns $a_0, \ldots, a_M$ and $b_1, \ldots, b_N$. The Padé technique can be combined with HAM to generate the so-called homotopy-Padé method, as shown by Liao [8]. The general procedure is as follows. First, employ the traditional Padé technique to the series (8.15) about the embedding-parameter $q \in [0, 1]$ to obtain the $[m, n]$ Padé approximant

$$TP_{m,n} = \left(\sum_{k=0}^{m} W_k(\tau) q^k\right) \Big/ \left(1 + \sum_{k=1}^{n} W_{m+k}(\tau) q^k\right),$$

where $W_k(\tau), k = 0, 1, \ldots, m+n$, are functions determined by the first $m + n + 1$ components $u_{i,j}(\tau)$, $j = 0, 1, \ldots, m+n$. Then, setting $q = 1$, the so-called $[m, n]$ homotopy-Padé approximant was obtained

$$HP_{m,n} = \left(\sum_{k=0}^{m} W_k(\tau)\right) \Big/ \left(1 + \sum_{k=1}^{n} W_{m+k}(\tau)\right).$$

More details can be found in [8].

It has been found that the homotopy-Padé approximant usually converges faster than the corresponding traditional $[m, n]$ Padé approximant. In many cases, the $[m, m]$ homotopy-Padé approximant does not depend upon the convergence-control parameter $\hbar$. In general, the homotopy-Padé technique can greatly enlarge the convergence region of solution series. Hence, it is employed in the code.

8.4. The package NOPH

Although the algorithm described in §8.3 is relatively simple in principle, the calculations would be very complicated if they had to be performed by hand. In this section, based on the package `CharSets`, a Maple package `NOPH` is presented, which can be used to automatically derive homotopy analysis solutions for nonlinear oscillation equations in the forms of (8.1) and (8.3).

In `NOPH`, the main interface is *main*(*eqns*, *ini_con*, [*h_order*, *sol_order*]), in which *eqns* represents an oscillation equations to be solved and *ini_con* represents the corresponding initial conditions. Note that *eqns* should be put in a list, as should *ini_con*. The other two parameters *h_order* (the default value is 8) and *sol_order* (the default value is 10) are optional, which are used to control the process flow of the program. In order to meet different needs, the computations in the program are divided into two parts: the first part is dependent on the convergence-control parameter $\hbar$, and the second part is free of the convergence-control parameter $\hbar$. They are described as follows:

(1) When there are unknown parameters in *eqns* or *ini_con*, just perform the first part computation, namely, calculate the first *h_order* components of the solution series to serve as an approximation of the required solution. In this case, the optional parameter *sol_order* can be omitted, and the program will output the [*h_order*/2, *h_order*/2] homotopy-Padé approximate expression of the frequency ω as well as the mean of motions δ_i, $i = 1, \ldots, \kappa$.
(2) If there is no unknown parameter in *eqns* or *ini_con*, first obtain the first *h_order* components of the solution series by the first part computation, and further calculate the squared residual error of the input equations to determine the optimal value of $\hbar$. Once the value of $\hbar$ is determined, a more accurate approximate solution (i.e., continue to calculate from *h_order* + 1 to *sol_order* order) may be obtained by the second part

computation. In this case, the program will output different order approximations and the corresponding errors of the frequency ω, as well as the mean of motions δ_i, $i = 1, \ldots, \kappa$, etc. In addition, comparison graphs of the different order approximations will be produced for the required solutions.

As described in §8.3, there are different characteristics for different types of oscillations. Therefore, the output of the program is flexible and diverse, not only depending on the number of unknown parameters, but also depending upon the type of oscillations. The output of the program can be summerized as follows.

- If there are more than one unknown parameters in input system, the program delivers homotopy Padé approximants of ω and δ. For oscillator with just odd nonlinearity, δ will be degenerated to zero. The below is same.
- If there is just one unknown parameter in input system, the program outputs not only homotopy Padé approximants of ω and δ, but also comparison graphs for different order approximants of ω and δ.
- If there is no unknown parameter in input system, for conservative oscillator, the program outputs approximants of ω and δ, as well as comparison graph for different order approximations of the required solution; for self-excited oscillator or coupled oscillator, the program delivers approximants of ω, δ, and the amplitudes a_i, b_j, meanwhile, the program also outputs limit cycles of the required solutions.

Similarly, the process flow of the program mainly depends on types of nonlinear oscillations. There are three different cases.

- For conservative oscillator, the solving process is completely automatic.
- For self-excited oscillator, the obtained nonlinear algebraic equations for $a_{i,j}$ and $b_{i,j}$ may exist multiple solutions. In this case, users have to select one solution and go on further calculation direction.
- For coupled nonlinear oscillator, apart from multiple solutions, the obtained nonlinear algebraic equations for $a_{i,j}$ and $b_{i,j}$ may become too complicated to be dealt with. In this case, the program provides several input interfaces so as to perform manual intervention.

Apart from all the output results listed in the next section, if necessary, users could also get components of the required solution series. The components of obtained solution series are stored in a set that is named as 'export'. All these sets are listed as follows:

- **Know_u:** the components set of $U_i(t)$ and $i = 1, \ldots, \kappa$.
- **Know_omega:** the components set of the frequency ω; and **Pade_omega:** the homotopy-Padé approximations set of ω.
- **Know_delta:** the components set of the mean of motion δ_i and $i = 1, \ldots, \kappa$, in accordance; and **Pade_delta:** the homotopy-Padé approximations set of δ_i and $i = 1, \ldots, \kappa$.
- **Know_abc:** the components set of the amplitudes a_i, b_i, $i = 1, \ldots, \kappa-1$ and a_κ. It should be noted that **Know_abc** is set exclusively for self-excited oscillating systems.

It is very simple to list the contents of the above sets. For example, to display the set **Know_u**, one just needs to input the set name **Know_u** after the command prompt">", that is,

> *Know_u*;

The procedures of the package are outlined in Appendix A. A simple user guide is given in Appendix B, three different types of examples are given to show how to use the package `NOPH`. The Maple package `NOPH` is free available online via the website `http://numericaltank.sjtu.edu.cn/NOPH.htm`.

8.5. Applications of NOPH

In this section, some examples are given to show the effectiveness of the package. More than 30 oscillation equations have been solved using the package `NOPH`. It should be mentioned that all the results shown in this section were obtained on a laptop with Intel Core i5-2410M CPU as the processor and a frequency of 2.30GHz. Due to space limitations, only five different types of nonlinear oscillation systems are considered here.

To demonstrate the convergence of the obtained homotopy analysis solutions, these solutions are compared with exact solutions or numerical solutions obtained using a Fehlberg fourth-fifth order Runge–Kutta method (`rkf45`) with degree four interpolant. As `rkf45` does not work for a system with unknown parameters, each unknown parameter was assigned the value delivered by the package NOPH.

8.5.1. *Example 1*

Consider the free oscillation of a conservative system with odd nonlinearity [8]

$$\ddot{U} + U + \epsilon\, U^3 = 0,$$

subject to initial conditions $U(0) = a, \quad \dot{U}(0) = 0$.

For this system, introducing the transformation $V(t) = U(t)\sqrt{\epsilon}$, yields

$$\ddot{V} + V + V^3 = 0, \tag{8.28}$$

subject to initial conditions

$$V(0) = b, \quad \dot{V}(0) = 0, \tag{8.29}$$

where $b = a\sqrt{\epsilon}$ is an unknown parameter.

There is only one parameter b in the system (8.28) and (8.29), the package automatically delivers (by $h\_order = 5$) the [1,1] homotopy-Padé approximation

$$\omega_{P_{1,\,1}} = \frac{(279\, b^4 + 768\, b^2 + 512)\,\sqrt{3\, b^2 + 4}}{2\,(285\, b^4 + 768\, b^2 + 512)},$$

and the [2,2] homotopy-Padé approximation

$$\begin{aligned}\omega_{P_{2,\,2}} = {} & [574118847\, b^{14} + 5452491348\, b^{12} + 22121717760\, b^{10} \\ & + 49711251456\, b^8 + 66831974400\, b^6 + 53758656512\, b^4 \\ & + 23957864448\, b^2 + 4563402752] \Big/ [2\sqrt{3\, b^2 + 4}(195618321\, b^{12} \\ & + 1584935424\, b^{10} + 5335944192\, b^8 + 9556131840\, b^6 \\ & + 9602662400\, b^4 + 5133828096\, b^2 + 1140850688)].\end{aligned}$$

Moreover, the package automatically delivers a figure to compare approximations of ω at different orders. As shown in Fig. 8.1, the above [1,1] and [2,2] homotopy-Padé approximations of ω agree quite well even for $0 \le b < +\infty$.

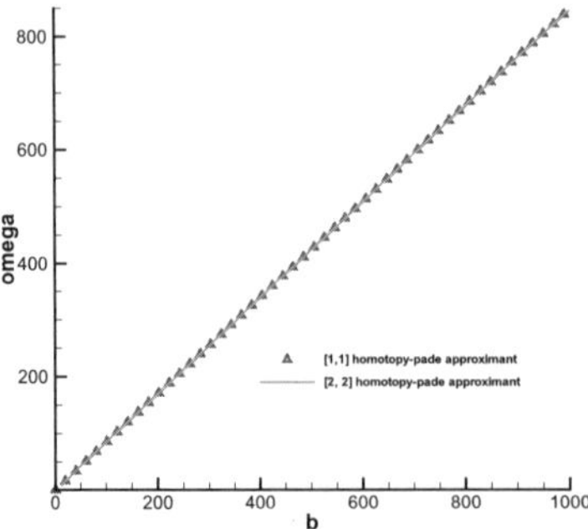

Fig. 8.1. The homotopy-Padé approximants of ω in Eq. (8.28) as a function of b.

To further show the convergence of obtained solution series, the case of $b = 1$ is considered. In this case, according to (13) and (14) given in [28], an exact solution for the system (8.28) and (8.29) is in the form of $V(t) = \mathrm{cn}(\mathrm{t}\sqrt{2}; \frac{1}{4})$, the frequency $\omega = \sqrt{2}\pi/K(1/4)$, where $K(1/4) = \int_0^{2\pi}(1 - \frac{1}{4}\sin^2\theta)^{-\frac{1}{2}}d\theta$ is the complete elliptic integral of the first kind. In this case, all physical parameters are known, when taking $h\_order = 5$ and $sol\_order = 10$, the package automatically delivers $\hbar = -1$. At the same time, the package also outputs approximants of ω, and a comparison graph for different order approximations of $V(t)$ within 10 seconds. The approximate solutions are compared with the above exact one in Fig. 8.2.

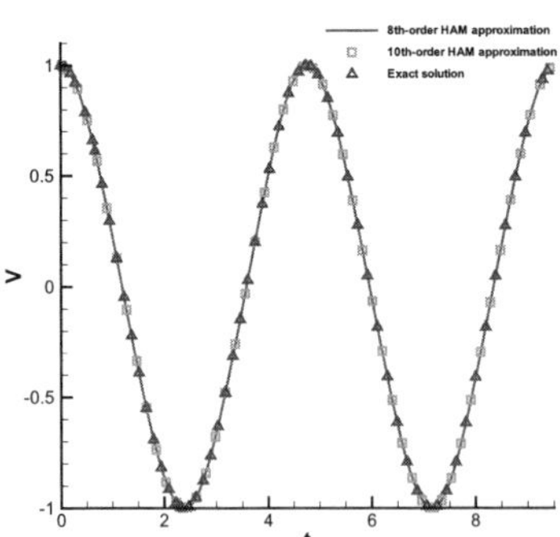

Fig. 8.2. The comparison of $V(t)$ for Eq. (8.28) when $b = 1$ and taking $\hbar = -1$.

As shown in Fig. 8.2, the 8th-order and 10th-order HAM approximations of $V(t)$ agree very well with the exact solution shown above. Besides, the ith-order approximation of ω and the corresponding $[i, i]$ homotopy-Padé approximants are given in Tables 8.1 and 8.2, respectively. The frequency delivered by the package is in good agreement with the exact one, the homotopy-Padé approximants of ω converge faster than the corresponding homotopy approximants.

Table 8.1. The ith-order approximations of ω in Eq. (8.28) when $b = 1$ and taking $\hbar = -1$.

i	ω
1	1.3229
2	1.3178
3	1.3178

Table 8.2. The $[i, i]$ homotopy-Padé approximations of ω in Eq. (8.28) when $b = 1$.

i	ω
1	1.3178
2	1.3178

8.5.2. *Example 2*

For free oscillation of a conservative system with quadratic nonlinearity, the package automatically delivers approximations of both the frequency ω and the mean of motion δ_i ($i = 1, \ldots, \kappa$). For example, consider a free oscillation with quadratic nonlinearity [8]

$$\ddot{U}(t) + U(t) + \varepsilon U^2(t) = 0, \tag{8.30}$$

subject to initial conditions $U(0) = 1/2 + \delta$ and $\dot{U}(0) = 0$, where ε is a parameter and δ is the mean of motion.

When ε is unknown, the package automatically delivers (by h_order = 7) within 52 seconds the [1,1] homotopy-Padé approximants

$$\omega_{P_{[1,1]}} = [\sqrt{2}((50\epsilon^5 - 212\epsilon^3 + 224\epsilon)\sqrt{4-2\epsilon^2} - 37\epsilon^6 + 224\epsilon^4 - 492\epsilon^2 + 384)]$$
$$/[(4-2\epsilon^2)^{3/4}((54\epsilon^3 - 112\epsilon)\sqrt{4-2\epsilon^2} - 35\epsilon^4 + 154\epsilon^2 - 192)],$$

$$\delta_{P_{[1,1]}} = \frac{(3\epsilon^3 - 48\epsilon^2 - 18\epsilon + 96)\sqrt{4-2\epsilon^2} - 14\epsilon^3 + 36\epsilon - 48(\epsilon^2-2)^2}{4\epsilon((24\epsilon^2 - 48)\sqrt{4-2\epsilon^2} + 7\epsilon^3 - 18\epsilon)},$$

the [2,2] homotopy-Padé approximants

$$\begin{aligned}
\omega_{P_{[2,2]}} = -[&(2072291\,\epsilon^{12} - 21339966\,\epsilon^{10} + 88093860\,\epsilon^8 - 178279336\,\epsilon^6 \\
&+ 165696000\,\epsilon^4 - 33134592\,\epsilon^2 - 29491200)\sqrt{4-2\,\epsilon^2} \\
&- 106700\,\epsilon^{13} + 4451720\,\epsilon^{11} - 43305488\,\epsilon^9 + 186308576\,\epsilon^7 \\
&- 408762112\,\epsilon^5 + 449421312\,\epsilon^3 - 197001216\,\epsilon]\,/\,[\sqrt[4]{16-8\,\epsilon^2} \\
&((42994\,\epsilon^{11} + 1674424\,\epsilon^9 - 17106744\,\epsilon^7 + 59363648\,\epsilon^5 \\
&- 89012224\,\epsilon^3 + 49250304\,\epsilon)\sqrt{4-2\,\epsilon^2} - 2184891\,\epsilon^{12} \\
&+ 22445294\,\epsilon^{10} - 91960132\,\epsilon^8 + 183929320\,\epsilon^6 \\
&- 168259584\,\epsilon^4 + 32520192\epsilon^2 + 29491200)],
\end{aligned}$$

$$\begin{aligned}
\delta_{P_{[2,2]}} = -[&(1935360 - 5483520\,\epsilon^5 + 5807232\,\epsilon^3 - 2460672\,\epsilon - 27113\,\epsilon^{10} \\
&+ 302154\,\epsilon^8 - 1405356\,\epsilon^6 + 3347640\,\epsilon^4 - 4025088\,\epsilon^2 + 58248\,\epsilon^{11} \\
&- 613440\,\epsilon^9 + 2591424\,\epsilon^7)\sqrt{4-2\,\epsilon^2} - 3870720 + 9985536\,\epsilon^2 \\
&+ 4921344\,\epsilon + 6507984\,\epsilon^6 + 11261952\,\epsilon^5 - 10895184\,\epsilon^4 \\
&- 11775744\,\epsilon^3 - 37999\,\epsilon^{12} - 123504\,\epsilon^{11} + 443564\,\epsilon^{10} + 1288320\,\epsilon^9 \\
&- 2271600\,\epsilon^8 - 5384832\,\epsilon^7]\,/\,[2((27113\,\epsilon^{10} - 302154\,\epsilon^8 \\
&+ 1405356\,\epsilon^6 - 3347640\,\epsilon^4 + 4025088\,\epsilon^2 - 1935360)\sqrt{4-2\,\epsilon^2} \\
&+ 123504\,\epsilon^{11} - 1288320\,\epsilon^9 + 5384832\,\epsilon^7 - 11261952\,\epsilon^5 \\
&+ 11775744\,\epsilon^3 - 4921344\,\epsilon)],
\end{aligned}$$

and the [3,3] homotopy-Padé approximants are omitted here due to space limitations.

It was found that the [2,2] and [3,3] homotopy-Padé approximants of the frequency ω and the mean of motion δ agree well, as shown in Figs. 8.3 and 8.4.

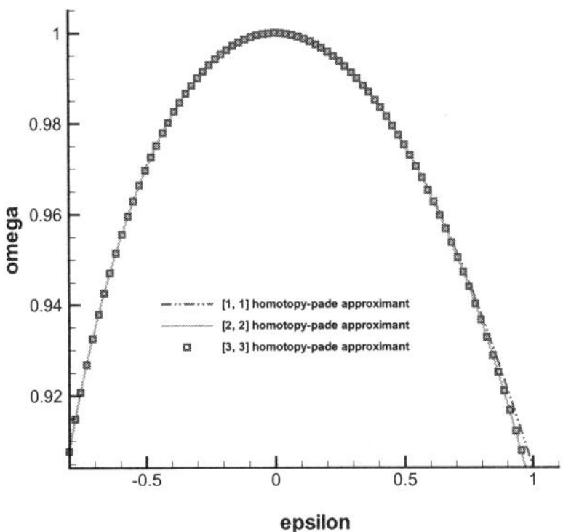

Fig. 8.3. The HAM-Padé approximants of ω in Eq. (8.30) as a function of ε.

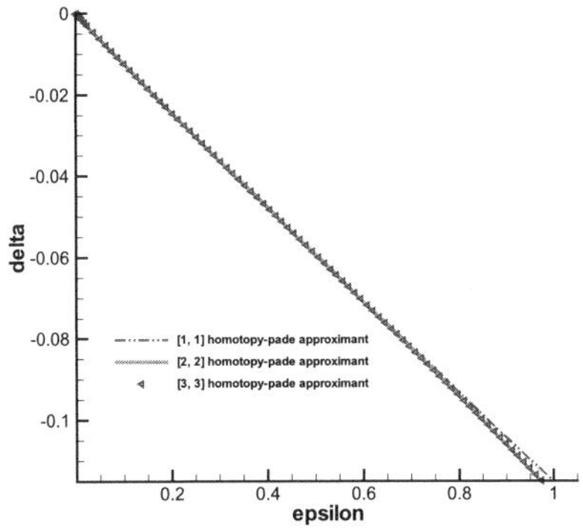

Fig. 8.4. The HAM-Padé approximants of δ in Eq. (8.30) as a function of ε.

To show the accuracy of these solutions, the case of $\varepsilon = 1$, for Rand [29] is considered and presented as an exact solution to Eq. (8.30) for $\varepsilon =$

1, which has the form $U(t) = A + a\mathrm{sn}^2(\mathrm{wt};\mathrm{m})$, as shown in [30]. For convenience of comparison, δ is assigned the value delivered by the package, namely $\delta = -0.1181$, then $A = 1/2 + \delta = 1/2 - 0.1181 = 0.3819$. Following the steps shown in [30], $\{a = -0.9147, m = 0.7269, w = 0.5371\}$ is obtained. For the case of $\varepsilon = 1$, there is no unknown parameter and the package automatically derives (by *h_order* = 12 and *sol_order* = 15) $\hbar = -0.4402$, and approximants of ω and δ as well as a comparison graph for different order approximations of $U(t)$ within 820 seconds.

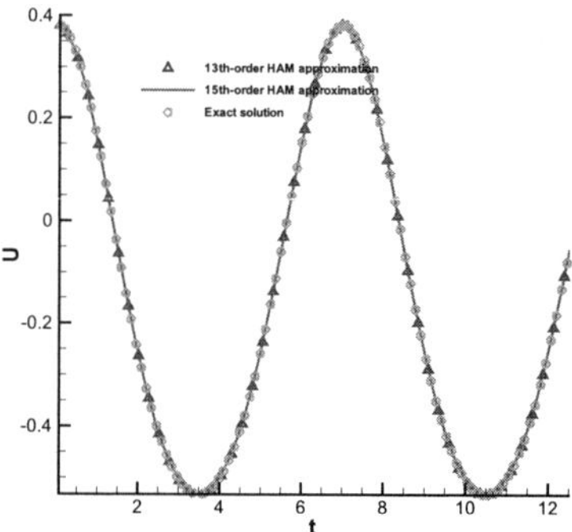

Fig. 8.5. The comparison of $U(t)$ for Eq. (8.30) when $\varepsilon = 1$, and taking $\hbar = -0.4402$.

Table 8.3. The approximations of ω and δ in Eq. (8.30) when $\varepsilon = 1$ and taking $\hbar = -0.4402$.

i	ω	δ
1	0.8409	−0.1281
2	0.8781	−0.1205
3	0.8934	−0.1180
4	0.8983	−0.1176
5	0.8991	−0.1178
6	0.8987	−0.1179
7	0.8983	−0.1181
8	0.8981	−0.1181

Table 8.4. The homotopy-Padé approximations of ω and δ in Eq. (8.30) when $\varepsilon = 1$.

i	ω	δ
1	0.9041	−0.1149
2	0.8975	−0.1184
3	0.8981	−0.1180
4	0.8981	−0.1181

As shown in Fig. 8.5, the 13th-order and 15th-order HAM approximations of $U(t)$ agree very well with the above exact solution. Table 8.3 indicates that the frequency ω as well as the mean of motion δ converge quickly, and their homotopy-Padé approximants converge even faster, as shown in Table 8.4.

8.5.3. *Example 3*

The package also works for rational function f with respect to $U, \dot{U}$ and $\ddot{U}$. For example, consider a free periodic oscillator governed by

$$\ddot{U} = -\frac{\lambda U + \varepsilon_1 U \dot{U}^2 + 2\varepsilon_2 U^3 \dot{U}^2 + \varepsilon_3 U^3 + \varepsilon_4 U^5}{1 + \varepsilon_1 U^2 + \varepsilon_2 U^4}, \tag{8.31}$$

subject to initial conditions $U(0) = a$ and $\dot{U}(0) = 0$, where λ is an integer that may take values -1, 0 or 1, a is the amplitude and $\varepsilon_1 > 0, \varepsilon_2 > 0, \varepsilon_3 > 0$ and $\varepsilon_4 > 0$ are physical parameters. Obviously, Eq. (8.31) only contains odd nonlinearity, so the mean of motion δ equals zero.

In Chen's work [31], the approximate solution was constructed by a new differential transformation method. To show the effectiveness of their solutions, the authors compared their solutions in four different parameter modes with numerical solutions obtained using the fourth-order Runge–Kutta method; their solutions agreed well with numerical results.

For this example, it is solved using the package in the following three cases.

8.5.3.1. *All physical parameters are unknown*

When setting *h_order* = 3, the program automatically delivers the $[1,1]$ homotopy-Padé approximate expression of ω within 30 seconds. However, the obtained approximate expression is so complicated that it is omitted here due to space limitations.

8.5.3.2. *In the case of* $\lambda = 1$, $\varepsilon_1 = \frac{1}{2}$, $\varepsilon_2 = \frac{1}{2}$, $\varepsilon_3 = 1$ *and* $a = 1$

In this case, there is only one unknown parameter ϵ_4. Setting *h_order* = 5, it takes about 910 seconds for the package to deliver the [1,1] homotopy-Padé approximant

$$\begin{aligned}\omega_{P_{1,1}} = {}& \sqrt{644 + 230\epsilon_4}(282621400{\epsilon_4}^4 + 1040639690{\epsilon_4}^3 - 2101840661{\epsilon_4}^2 \\ & - 10283156607\epsilon_4 - 7846822521)/[23(289246200{\epsilon_4}^4 + 1013871130{\epsilon_4}^3 \\ & - 2120380083{\epsilon_4}^2 - 10191069819\epsilon_4 - 7768422279)],\end{aligned}$$

and the [2,2] homotopy-Padé approximant

$$\begin{aligned}\omega_{P_{2,2}} = {}& 2(72160080229783054982000000000{\epsilon_4}^{13} \\ & + 1238302188977775732096400000000{\epsilon_4}^{12} \\ & + 7705851616838919062770725000000{\epsilon_4}^{11} \\ & + 13400710210748113794843474000000\,{\epsilon_4}^{10} \\ & - 76768592799546957341281400850 0\,{\epsilon_4}^{9} \\ & - 474626306272512684685625296215 0\,{\epsilon_4}^{8} \\ & - 898250996680388129498943631061 5\,{\epsilon_4}^{7} \\ & + 53889675635936948591262486916 41\,{\epsilon_4}^{6} \\ & + 603350459682705442334194042527 63\,{\epsilon_4}^{5} \\ & + 134137324911540129238896999420609\,{\epsilon_4}^{4} \\ & + 160727905086757562229362014143834\,{\epsilon_4}^{3}\end{aligned}$$

$$
\begin{aligned}
&+ 11468688337000976819551360862795l\,{\epsilon_4}^2 \\
&+ 47071103220848811048576739633845\,\epsilon_4 \\
&+ 88597224652446656451883532l7170)/ \\
&[\sqrt{644 + 230\,\epsilon_4}(147948325632742461644000000\,{\epsilon_4}^{12} \\
&+ 209549537893151903060560000\,{\epsilon_4}^{11} \\
&+ 956529410574667548376097000\,{\epsilon_4}^{10} \\
&- 69098271018605460645706800\,{\epsilon_4}^{9} \\
&- 15399002893815058537846929290\,{\epsilon_4}^{8} \\
&- 5161574431434277164919l9938930\,{\epsilon_4}^{7} \\
&- 32765815090991422404506573l355\,{\epsilon_4}^{6} \\
&+ 20333115019598877496028312279O1\,{\epsilon_4}^{5} \\
&+ 63465833605044811052996486966 25\,{\epsilon_4}^{4} \\
&+ 8882680801800394612648600646085\,{\epsilon_4}^{3} \\
&+ 7006874832650635013575763874036\,{\epsilon_4}^{2} \\
&+ 3105495703111427773085478899457\,\epsilon_4 \\
&+ 626473417036090662183962398527)].
\end{aligned}
$$

The above two approximations agree very well for $\epsilon_4 > 0$, as shown in Fig. 8.6.

8.5.3.3. *In the case of* $a = \lambda = \varepsilon_3 = \varepsilon_4 = 1$ *and* $\varepsilon_1 = \varepsilon_2 = \frac{1}{2}$

In this case, there is no unknown parameter and the program automatically gives (by setting *h_order* = 6, *sol_order* = 8) $\hbar = -0.3384$, approximants of ω as well as a comparison graph for different order approximations of $U(t)$ within 2000 seconds. As shown in Fig. 8.7, the 6th-order and 8th-order

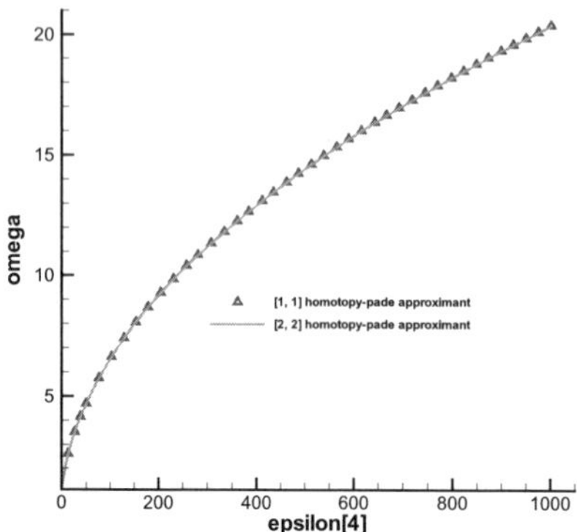

Fig. 8.6. The homotopy-Padé approximants of ω in Eq. (8.31) when $a = \lambda = \varepsilon_3 = 1, \varepsilon_1 = \varepsilon_2 = \frac{1}{2}$.

HAM approximations of $U(t)$ agree very well. Meanwhile these curves agree well with the numerical imitating curve of the numerical solution for the given values $a = \lambda = \varepsilon_3 = \varepsilon_4 = 1, \varepsilon_1 = \varepsilon_2 = \frac{1}{2}$ and the absolute error 10^{-7}. Moreover, the frequency ω also converges quickly, as shown in Tables 8.5 and 8.6. It needs to be emphasized that for the four different modes of the parameters provided in [31], the obtained homotopy analysis solutions also agree well with the numerical solution obtained using the `rkf45`, while the comparison graph is omitted due to space limitations.

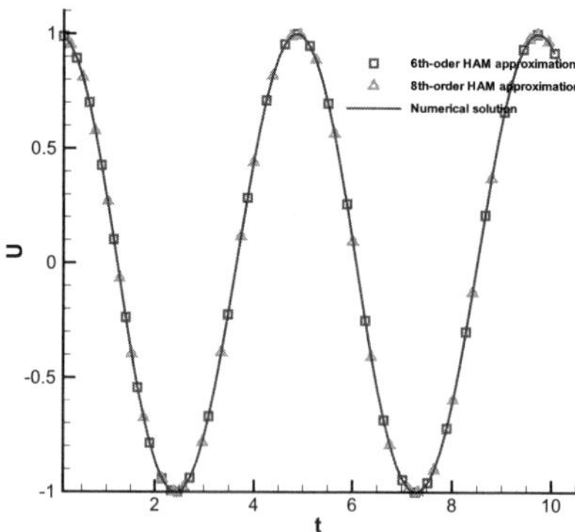

Fig. 8.7. The HAM approximations of $U(t)$ for Eq. (8.31) when $a = \lambda = \varepsilon_3 = \varepsilon_4 = 1$, $\varepsilon_1 = \varepsilon_2 = \frac{1}{2}$ and taking $\hbar = -0.3384$.

Table 8.5. The ith-order HAM approximations of ω in Eq. (8.31) when $a = \lambda = \varepsilon_3 = \varepsilon_4 = 1, \varepsilon_1 = \varepsilon_2 = \frac{1}{2}$ and taking $\hbar = -0.3384$.

i	ω
1	1.2854
2	1.2942
3	1.2944
4	1.2945
5	1.2945

Table 8.6. The $[i, i]$ homotopy-Padé approximants of ω in Eq. (8.31) when $a = \lambda = \varepsilon_3 = \varepsilon_4 = 1$ and $\varepsilon_1 = \varepsilon_2 = \frac{1}{2}$.

i	ω
1	1.2944
2	1.2945

8.5.4. *Example 4*

Consider the J. W. S. equation

$$\ddot{U}(t) + U(t) = \mu(1 - \beta\dot{U}(t)^2)\dot{U}(t), \tag{8.32}$$

with initial conditions $U(0) = c_1$ and $\dot{U}(0) = 0$, where μ, β are unknown parameters.

There are two unknown parameters in this system, it takes about 114 seconds for the program to deliver (by *h_order* = 7)

$$\{\{\omega_0 = 1, c_{1,0} = 2/\sqrt{3\beta}\}, \quad \{\omega_0 = 1, c_{1,0} = -2/\sqrt{3\beta}\}\}.$$

By selecting the first solution, it further outputs:
the [1,1] homotopy-Padé approximation

$$\omega_{P_{1,1}} = \frac{\mu^2 + 32}{3\,\mu^2 + 32},$$

the [2,2] homotopy-Padé approximation

$$\omega_{P_{2,2}} = \frac{9\,\mu^6 + 16640\,\mu^2 + 960\,\mu^4 + 49152}{45\,\mu^6 + 19712\,\mu^2 + 1920\,\mu^4 + 49152},$$

and the [3,3] homotopy-Padé approximation

$$\begin{aligned}\omega_{P_{3,3}} = (&-109231931392\,\mu^4 - 2382205288448\,\mu^2 - 3020544\,\mu^8 \\ &- 1970749440\,\mu^6 - 1175344644096 + 178848\,\mu^{1}0 + 729\,\mu^{1}2)/ \\ (&-256206766080\,\mu^4 - 2455664328704\,\mu^2 - 73979136\,\mu^8 \\ &- 4347887616\,\mu^6 - 1175344644096 + 598752\,\mu^{1}0 + 5103\,\mu^{1}2).\end{aligned}$$

As the above homotopy-Padé approximants of ω are but functions of μ, the package also automatically outputs a figure to compare approximations of ω at different orders, as shown in Fig. 8.8. It can be seen that the curves of [2,2] and [3,3] homotopy-Padé approximants of ω agree well.

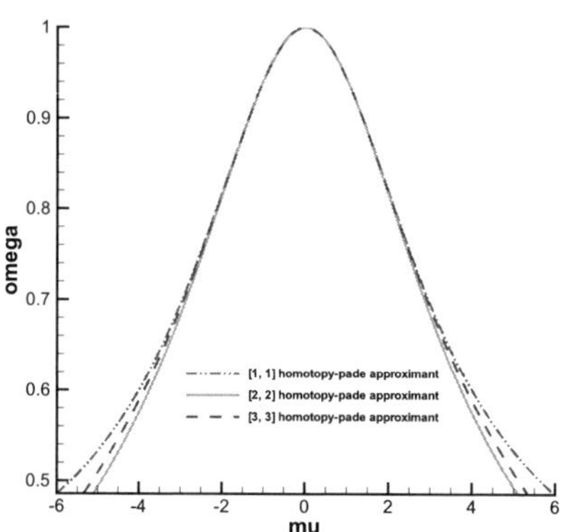

Fig. 8.8. The homotopy-Padé approximants of ω in Eq. (8.32) as a function of μ.

In addition, in the case of $\mu = 1, \beta = 3$, there is no unknown parameter in the input system and the package takes about 530 seconds to automatically output (by *h_order* $= 8$ and *sol_order* $= 11$) $\hbar = -0.9453$, the approximants of ω, c_1, and the comparison graph for different order approximations of $U(t)$. As shown in Fig. 8.9, the 9th-order and 11th-order HAM approximations of $U(t)$ agree very well. Meanwhile, these curves agree well with the numerical imitating curve of the numerical solution for the given values $\mu = 1$, $\beta = 3$, $c_1 = 0.7193$ and the absolute error 10^{-7}. The frequency ω and the amplitude c_1 converge quickly for the given value $\hbar = -0.9453$, as shown in Tables 8.7 and 8.8.

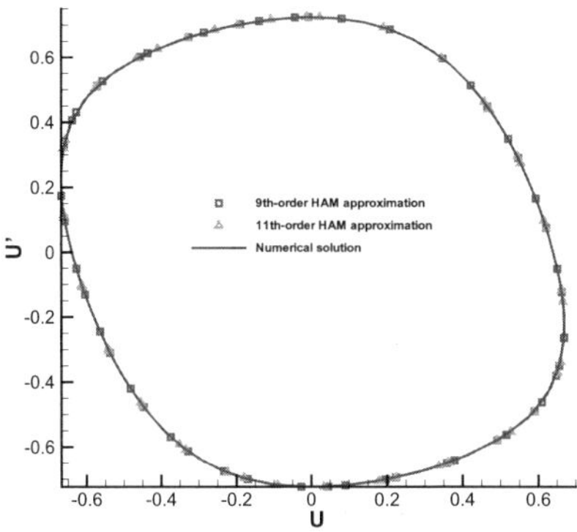

Fig. 8.9. The approximations of $U(t)$ for Eq. (8.32) when $\mu = 1$, $\beta = 3$ and taking $\hbar = -0.9453$.

Table 8.7. The homotopy approximations of ω and c_1 in Eq. (8.32) when $\mu = 1$, $\beta = 3$ by means of $\hbar = -0.9453$.

i	ω	c_1
1	1.0000	0.6667
2	0.9409	0.7061
3	0.9429	0.7179
4	0.9425	0.7187
5	0.9426	0.7196
6	0.9430	0.7194
7	0.9429	0.7193
8	0.9430	0.7193

Table 8.8. The homotopy-Padé approximations of ω and c_1 in Eq. (8.32) when $\mu = 1$, $\beta = 3$.

i	ω	c_1
1	0.9429	0.7230
2	0.9426	0.7196
3	0.9427	0.7196
4	0.9430	0.7192
5	0.9430	0.7193

8.5.5. *Example 5*

Consider two directly coupled van der Pol oscillators

$$\begin{cases} \ddot{U}(t) - \varepsilon(1 - U^2(t))\dot{U}(t) + U(t) = \varepsilon\mu(V(t) - U(t)), \\ \ddot{V}(t) - \varepsilon(1 - V^2(t))\dot{V}(t) + p^2 V(t) = \varepsilon\mu(U(t) - V(t)), \end{cases} \tag{8.33}$$

with initial conditions

$$U(0) = a_1, \quad \dot{U}(0) = b_1, \quad V(0) = c_1, \quad \dot{V}(0) = 0, \tag{8.34}$$

where ε, μ and p are physical parameters.

Similar to the above examples, this system is solved by considering three different cases.

8.5.5.1. *All physical parameters are unknown*

In this case there are three unknown parameters ε, μ and p, and the package takes about 20 seconds to deliver 8 group solutions for ω_0, $a_{1,0}$, $b_{1,0}$ and $c_{1,0}$:

$$\begin{aligned} &\{p = \mp 1,\ \omega_0 = 1,\ a_{1,0} = 2,\ b_{1,0} = 0,\ c_{1,0} = 2\}, \\ &\{p = \mp 1,\ \omega_0 = 1,\ a_{1,0} = -2,\ b_{1,0} = 0,\ c_{1,0} = -2\}, \\ &\{p = \mp 1,\ \omega_0 = \sqrt{1 + 2\,\varepsilon\,\mu},\ a_{1,0} = 2,\ b_{1,0} = 0,\ c_{1,0} = -2\}, \\ &\{p = \mp 1,\ \omega_0 = \sqrt{1 + 2\,\varepsilon\,\mu},\ a_{1,0} = -2,\ b_{1,0} = 0,\ c_{1,0} = 2\}. \end{aligned} \tag{8.35}$$

As p appears in the form of p^2 in Eq. (8.33), it is easy to verify that if $U(t)$ and $V(t)$ are solutions of the systems (8.33) and (8.34), then $-U(t)$ and $-V(t)$ are also solutions for the system. Therefore, there are two essentially different solutions in (8.35) as follows:

$$\begin{aligned} &\{p = 1,\ \omega_0 = 1,\ a_{1,0} = 2,\ b_{1,0} = 0,\ c_{1,0} = 2\}, \\ &\{p = 1,\ \omega_0 = \sqrt{1 + 2\,\varepsilon\,\mu},\ a_{1,0} = -2,\ b_{1,0} = 0,\ c_{1,0} = 2\}. \end{aligned} \tag{8.36}$$

Select a solution and input its order, such as 2, then the program further delivers:
the parameter constraint condition: $\{p = 1\}$,
the [1,1] homotopy-Padé approximant

$$\omega_{P_{1,1}} = \frac{(\varepsilon^2 + 64\,\varepsilon\,\mu + 32)\,\sqrt{2\,\varepsilon\,\mu + 1}}{3\,\varepsilon^2 + 64\,\varepsilon\,\mu + 32}, \tag{8.37}$$

and the [2,2] homotopy-Padé approximant

$$\begin{aligned}\omega_{P_{2,2}} &= (16640\,\varepsilon^2 + 66560\,\varepsilon^3\,\mu + 960\,\varepsilon^4 + 294912\,\varepsilon\,\mu + 66560\,\varepsilon^4\,\mu^2 \\ &\quad + 589824\,\varepsilon^2\,\mu^2 + 49152 + 393216\,\varepsilon^3\,\mu^3 + 1920\,\varepsilon^5\,\mu \\ &\quad + 9\,\varepsilon^6)\sqrt{2\,\varepsilon\,\mu + 1}\,/(45\,\varepsilon^6 + 393216\,\varepsilon^3\,\mu^3 + 589824\,\varepsilon^2\,\mu^2 \\ &\quad + 78848\,\varepsilon^4\,\mu^2 + 78848\,\varepsilon^3\,\mu + 294912\,\varepsilon\,\mu + 3840\,\varepsilon^5\,\mu \\ &\quad + 19712\,\varepsilon^2 + 49152 + 1920\,\varepsilon^4). \end{aligned} \tag{8.38}$$

8.5.5.2. *In the case of* $\varepsilon = 1/2$

From (8.36), it is known that there is a parameter constraint $p = 1$ in each solution. Therefore, letting $\varepsilon = 1/2$, there will be only one unknown parameter μ contained in the input system, then the program automatically delivers (by selecting the 2nd solution in (8.36) and setting *h_order* $= 5$) the following results within 590 seconds.
The parameter constraint condition: $\{p = 1\}$,
the [1,1] homotopy-Padé approximant

$$\omega_{P_{1,1}} = \frac{(128\,\mu + 129)\,\sqrt{\mu + 1}}{128\,\mu + 131}, \tag{8.39}$$

and the [2,2] homotopy-Padé approximant

$$\omega_{P_{2,2}} = \frac{(9973504\,\mu + 9703424\,\mu^2 + 3415817 + 3145728\,\mu^3)\,\sqrt{\mu + 1}}{3468845 + 3145728\,\mu^3 + 9752576\,\mu^2 + 10075648\,\mu}. \tag{8.40}$$

Moreover, the program also provides a figure to compare homotopy-Padé approximants of ω at different orders, as shown in Fig. 8.10. It indicates that $[1, 1]$ and $[2, 2]$ homotopy-Padé approximants of ω agree very well.

8.5.5.3. *In the case of* $\varepsilon = 1/2, \ \ \mu = 8$

In this case, there are no unknown parameters in the input system, for there is the parameter constraint $p = 1$ involved in each solution. The program automatically outputs (by setting *h_order* $= 7$ and *sol_order* $= 9$) $\hbar = -0.498$, approximants of ω, a_1, b_1 and c_1, as well as comparison graphs for different order approximations of $U(t)$ and $V(t)$ within 755 seconds.

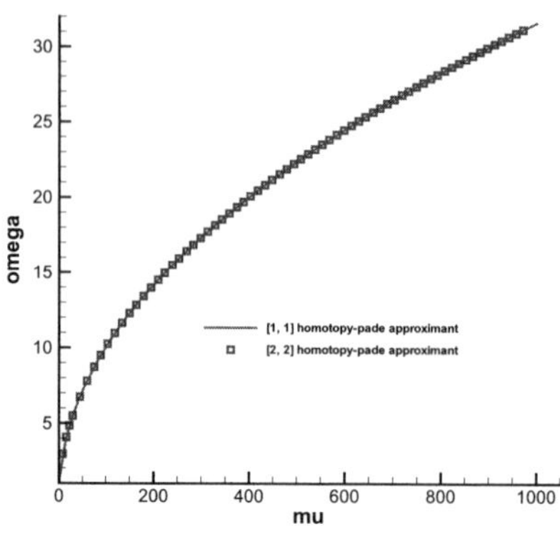

Fig. 8.10. The homotopy-Padé approximants of ω in Eq. (8.33) as a function of μ.

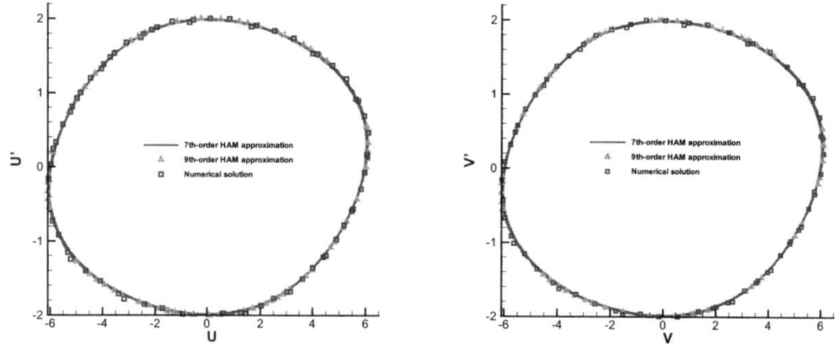

Fig. 8.11. Approximations of $U(t)$, $V(t)$ for Eq. (8.33) when $\varepsilon = 1/2$, $\mu = 8$ by means of $\hbar = -0.498$.

As shown in Fig. 8.11, the 7th-order and 9th-order HAM approximations of $U(t)$ and $V(t)$ agree very well. Meanwhile, both agree very well with the corresponding imitating curves of the numerical solutions for the given values $a_1 = 1.9964$, $b_1 = -0.1234, c_1 = -1.9964$ and the absolute error 10^{-7}.

Table 8.9 indicates that the homotopy approximants of the frequency ω and the amplitudes a_1, b_1 and c_1 converge very quickly when setting $\hbar = -0.498$. It is shown from Table 8.10 that the homotopy-Padé approximants converge even faster than the corresponding homotopy approximants.

Table 8.9. The HAM approximants of ω, a_1, b_1, c_1 in Eq. (8.33) when $\varepsilon = 1/2$, $\mu = 8$ and taking $\hbar = -0.498$.

i	ω	a_1	b_1	c_1
1	3.0000	2.0000	−0.1245	−2.0000
2	2.9964	2.0000	−0.1248	−2.0000
3	2.9964	1.9964	−0.1234	−1.9964
4	2.9964	1.9964	−0.1234	−1.9964

Table 8.10. The HAM-Padé approximants of ω, a_1, b_1, c_1 in Eq. (8.33) when $\varepsilon = 1/2$, $\mu = 8$.

i	ω	a_1	b_1	c_1
1	2.9964	2.0000	−0.1248	−2.0000
2	2.9964	1.9964	−0.1234	−1.9964

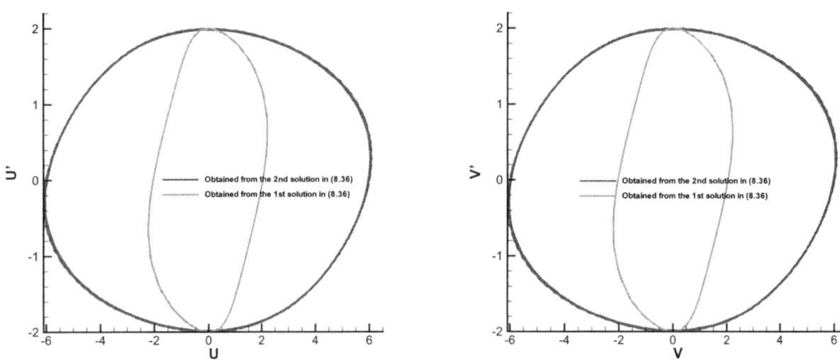

Fig. 8.12. Different approximations obtained from different solutions in (8.36) when $\varepsilon = 1/2$, $\mu = 8$.

To indicate that the solutions obtained from the different solutions in (8.36) are different, $U(t)$ and $V(t)$ obtained from the 1st and 2nd solutions in (8.36), respectively, were compared. As shown in Fig. 8.12, it can be seen that the solutions obtained from different solutions in (8.36) are indeed different.

8.6. Discussions and conclusions

In this chapter, an algorithm and a Maple package are presented for automated reasoning of analytic approximate solutions of nonlinear oscillating systems. The package is user-friendly as one need only input a set of governing equations and initial conditions, and then periodic analytic approximations are returned in a short time. The package NOPH has the following three merits: Firstly, it can work for highly nonlinear oscillating systems; Secondly, if there are unknown parameters in input system, the package NOPH can automatically filter out the corresponding possible constraint conditions on these parameters; Thirdly, it may construct several different solutions simultaneously for the considered system. However, we should mention that the current version of the package NOPH only works in cases where f_i in Eq.(8.1) is either a polynomial or rational function. Theoretically speaking, f_i could represent more complicated functions, such as a piecewise function, a function involving absolute value function, etc. In addition, for n coupled self-excited oscillating systems, the n convergence-control parameters $\hbar_i$, $i = 1, \ldots, n$, can be introduced in the frame of HAM. In this way, the solution series may converge even faster. In future, these problems will be investigated and further improves to the package NOPH will be made so that the newer versions of NOPH are valid for more complicated periodic oscillations.

Acknowledgment

This work is partially supported by the National Natural Science Foundation of China (Approval No. 11071274).

Appendix A. The procedures contained in the NOPH

The package NOPH is composed of the six different procedures: **main**, **De_oper**, **Approx_nth**, **sol_analy**, **conti_higher** and **proc_pade**. We outline each of them as follows:

- **main**: This is the main procedure, its main task is to initialize relevant variables and to call other procedures.

- **De_oper**: The main task of this procedure is to change variables, and to define linear as well as nonlinear operators. Meanwhile, the input equation is decomposed to extract dependent and independent variables as well as unknown parameters.
- **Approx_nth**: In this procedure, follow the process and steps of HAM, we compute the first *h_order* components (including $u_i(\tau)$, ω_{i-1}, δ_{i-1}, $i = 1, \ldots, h\_order$) of the solution series.
- **sol_analy**: In this procedure, first normalize all solution expressions. If there are unknown parameters contained in *eqn* or *ini_con*, call the function **proc_pade** to compute the homotopy-Padé approximations of the frequency ω as well as the mean of motion δ. If there is no unknown parameter contained in *eqn* or *ini_con*, we compute the squared residual error of the input equation, the obtained result is a function denoted by $\Delta(\hbar, \tau)$. Further to eliminate τ by integrating Δ with respect to τ from 0 to 2π, the obtained result is denoted as $\Lambda(\hbar)$. By looking for the minimum value of the common logarithm of $\Lambda(\hbar)$ to determine the value of $\hbar$.
- **conti_higher**: The main task of this procedure is to continue to get more accurate approximations of the required solutions. It is to be stressed that this part of the calculation is free of the convergence-control parameter $\hbar$. Evidently, only when equations (8.1) and (8.3) do not contain any parameters, the function **conti_higher** will be called.
- **proc_pade**: This procedure is a complete implementation of the homotopy-Padé technique.

Appendix B. A simple user guide

To load the package `NOPH`, one proceeds as follows:

$> restart:$ initializing Maple.

$> currentdir(\text{“}D:/ect001a\text{”});$ setting current work directory.

Note that the source program and the sample file should be located in the current work directory.

$> read\text{“}oscillation.mpl\text{”}:$ reading the program file.

$> with(NOPH);$ loading the package NOPH.

In what follows three different types of examples are given to show how to use the package NOPH.

B.1. Example 1

Consider Duffing oscillation equation

$$\ddot{U} + U + \alpha U^3 = 0, \tag{B.1}$$

subject to initial conditions

$$U(0) = a, \ \dot{U}(0) = 0. \tag{B.2}$$

This is a single conservative oscillator with just odd nonlinearity. It is to be stressed that the package NOPH can solve single conservative oscillator completely automatically. For the system (B.1) and (B.2), two different cases are considered to show this point.

B.1.1. *All physical parameters are unknown*

In this case, there are two unknown parameters α and a in the system (B.1) and (B.2). For this system, one could run the main procedure as follows:

$> main([diff(U(t), t\$2) + U(t) + alpha * U(t)^3 = 0], [U(0) = a, D(U)(0) = 0], 5).$

As there are more than one unknown parameters in the input system, *sol_order* is omitted. *h_order* $= 5$, the governing equation should be put in a list, as should initial conditions. Then NOPH delivers the following results:

`The input equation and initial conditions are:`

$$\frac{\partial^2}{\partial t^2}U(t) + U(t) + \alpha U(t)^3 = 0,$$

$$U(0) = a, D(U)(0) = 0.$$

`The new EQ and initial conditions in` u `with the form:`

$$\omega^2 \frac{\partial^2}{\partial \tau^2}u(\tau) + u(\tau) + \alpha u(\tau)^3 = 0,$$

$$u(0) = a, D(u)(0) = 0,$$

`where:` $\tau = \omega t, \quad U(t) = u(\tau)$.

`It takes about 2.715 seconds to complete the first part computation;`

$$\omega_{P_{1,1}} = \frac{(279\,\alpha^2 a^4 + 768\,\alpha\,a^2 + 512)\sqrt{3\,\alpha\,a^2+4}}{2(285\,\alpha^2 a^4 + 768\,\alpha\,a^2 + 512)}$$

$$\omega_{P_{2,2}} = (574118847\,\alpha^7 a^{14} + 5452491348\,\alpha^6 a^{12} + 22121717760\,\alpha^5 a^{10}$$
$$+ 49711251456\,\alpha^4 a^8 + 66831974400\,\alpha^3 a^6 + 53758656512\,\alpha^2 a^4$$
$$+ 23957864448\,\alpha\,a^2 + 4563402752)/(2(195618321\,\alpha^6 a^1 2 + 1584935424\,\alpha^5 a^1 0$$
$$+ 5335944192\,\alpha^4 a^8 + 9556131840\,\alpha^3 a^6 + 9602662400\,\alpha^2 a^4 + 5133828096\,\alpha\,a^2$$
$$+ 1140850688)\sqrt{3\,\alpha\,a^2+4}).$$

`It takes about 0.281 seconds to complete the second part computation.`

`The total time is 2.996 seconds.`

B.1.2. *In the case of* $\alpha = 5$, $a = 1/2$

In this case, there is no unknown parameter in the system (B.1) and (B.2). To solve the systems (B.1) and (B.2), one could run the main procedure as follows:

$> main([diff(U(t), t\$2) + U(t) + 5 * U(t)^3 = 0], [U(0) = 1/2, D(U)(0) = 0], 4, 7).$

Here, $h\_order = 4, sol\_order = 7$. It means that the optimal value of $\hbar$ is determined by the first 4 components of solution series. Then continue to calculate from 5 to 7 order by the second part computation to get more accurate approximants.

For this example, the package `NOPH` outputs the following results:

`The input equation and initial conditions are:`

$$\frac{\partial^2}{\partial t^2}U(t) + U(t) + 5U(t)^3 = 0,$$

$$U(0) = 1/2, D(U)(0) = 0.$$

The new EQ and initial conditions in u with the form:

$$\omega^2 \frac{\partial^2}{\partial \tau^2} u(\tau) + u(\tau) + 5u(\tau)^3 = 0,$$

$$u(0) = 1/2, D(u)(0) = 0,$$

where:

$$\tau = \omega t, \quad U(t) = u(\tau).$$

It takes about 0.889 seconds to complete the first part computation.

The optimal value of $\hbar$ is: $\hbar = -1.0$.

The comparison graph between 5th-order and 7th-order approximations of $U(t)$:

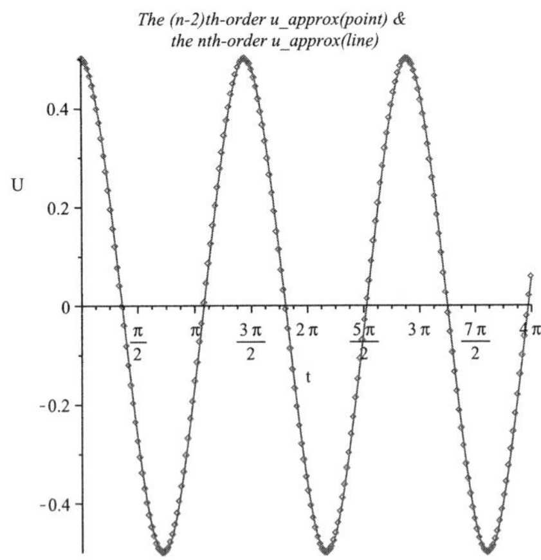

where: $n = 7$.

The HAM approximants of ω and the corresponding relative errors:

i	ω	errors
1	1.3919	$---$
2	1.3852	0.0049
3	1.3852	0.0000
4	1.3851	0.0000
5	1.3851	0.0000

The HAM-Padé approximants of ω and the corresponding relative errors:

i	ω	errors
1	1.3851	$---$
2	1.3851	0.0000

It takes about 1.809 seconds to complete the second part computation.

The total time is 5.912 seconds.

B.2. Example 2

Consider the Van der Pol equation

$$\ddot{U} + U = \varepsilon(1 - U(t)^2)\dot{U}(t), \tag{B.3}$$

subject to the initial condition

$$U(0) = c_1, \;\; \dot{U}(0) = 0. \tag{B.4}$$

This is a single self-excited oscillator. We solve it by considering two different cases.

B.2.1. *All physical parameters are unknown*

To solve the systems (B.3) and (B.4), one can run the main procedure as follows:

$> main([diff(U(t), t\$2) + U(t) = \epsilon * (1 - U(t)^2) * diff(U(t), t)],$
$[U(0) = c[1], D(U)(0) = 0], 7).$

As there are more than one unknown parameters in the input equation, *sol_order* is omitted. The program just perform the first part computation. Then the package NOPH delivers the following results:

`The input equation and initial conditions are:`

$$\frac{\partial^2}{\partial t^2}U(t) + U(t) - \epsilon(1 - U(t)^2)\frac{\partial}{\partial t}U(t) = 0,$$

$$U(0) = c_1, D(U)(0) = 0.$$

`The new EQ and initial conditions in` u `with the form:`

$$\omega^2\frac{\partial^2}{\partial \tau^2}u(\tau) + u(\tau) - \epsilon\omega\frac{\partial}{\partial \tau}u(\tau) + \epsilon\omega u(\tau)^2\left(\frac{\partial}{\partial \tau}u(\tau)\right) = 0,$$

$$u(0) = c_1, D(u)(0) = 0,$$

`where:`

$$\tau = \omega t, \quad U(t) = u(\tau).$$

`The possible solutions are:`

$$[\{\omega_0 = 1, c_{1,0} = 2\}, \quad \{\omega_0 = 1, c_{1,0} = -2\}].$$

`Please select one solution and input its order:`, such as input 1, then the NOPH further gives out:

`The 1st solution was selected.`

`It takes about 108.218 seconds to complete the first part computation.`

$$\omega_{P_{1,1}} = \frac{\varepsilon^2 + 32}{3\,\varepsilon^2 + 32}$$

$$\omega_{P_{2,2}} = \frac{49152 + 9\,\varepsilon^6 + 16640\,\varepsilon^2 + 960\,\varepsilon^4}{49152 + 45\,\varepsilon^6 + 19712\,\varepsilon^2 + 1920\,\varepsilon^4}$$

$$\begin{aligned}\omega_{P_{3,3}} = {} & (-3020544\,\varepsilon^8 - 1970749440\,\varepsilon^6 + 729\,\varepsilon^{12} - 109231931392\,\varepsilon^4 \\ & - 1175344644096 - 2382205288448\,\varepsilon^2 + 178848\,\varepsilon^{10}) \,/ \\ & (-73979136\,\varepsilon^8 - 4347887616\,\varepsilon^6 - 256206766080\,\varepsilon^4 + 5103\,\varepsilon^{12} \\ & - 1175344644096 - 2455664328704\varepsilon^2 + 598752\varepsilon^{10})\end{aligned}$$

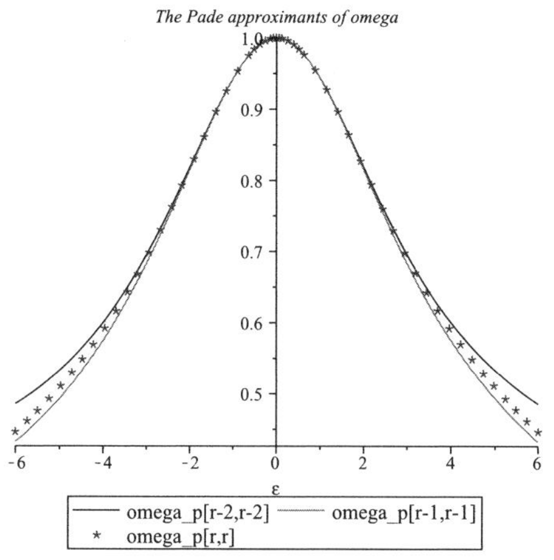

`where:` $r = 3$.

`It takes about 5.179 seconds to complete the second part computation.`

`The total time is 113.397 seconds.`

B.2.2. *In the case of* $\epsilon = 2$

In this case, no unknown parameter is involved in the system (B.3) and (B.4). One could run the main procedure as follows:

$> main([diff(U(t), t\$2) + U(t) = 2(1 - U(t)^2) * diff(U(t), t)],$
$[U(0) = c[1], D(U)(0) = 0], 12, 14).$

the `NOPH` outputs the following results:

`The input equation and initial conditions are:`

$$\frac{d^2}{dt^2}U(t) + U(t) = 2(1 - U(t)^2)\frac{d}{dt}U(t),$$

$$[U(0) = c_1,\ D(U)(0) = 0].$$

The new EQ and initial conditions in u with the form:

$$\left(\frac{d^2}{d\tau^2}u(\tau)\right)\omega^2 + u(\tau) - 2\left(\frac{d}{d\tau}u(\tau)\right)\omega + 2u(\tau)^2\left(\frac{d}{d\tau}u(\tau)\right)\omega = 0,$$

$$[u(0) = c_1, D(u)(0) = 0]$$

where: $[U(t) = u(\tau), \tau = \omega t]$.

The possible solutions are:

$[\{\omega_0 = 1, c_{1,0} = 0\}, \{\omega_0 = 1, c_{1,0} = -2\}, \{\omega_0 = 1, c_{1,0} = 2\}]$.

Please select one solution and input its order:, such as input 3, then the NOPH further delivers:

The 3rd solution was selected.

It takes about 9603.695 seconds to complete the first part computation.

The optimal value of $\hbar$ is: $\hbar = -0.6655$.

The comparison graph between 12th-order and 14th-order approximations of $U(t)$:

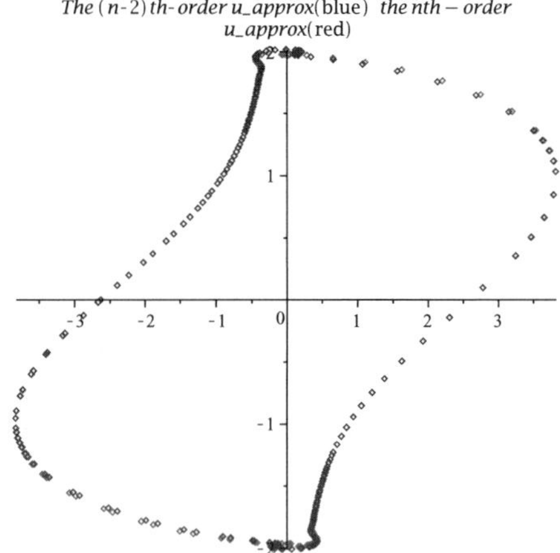

where: $n = 14$.

The HAM approximants of ω and c_1:

i	ω	c_1
1	1.0000	2.0000
2	0.8336	2.0000
3	0.8195	1.7694
4	0.8156	1.6918
5	0.8194	1.7405
6	0.8192	1.7827
7	0.8215	1.8016
8	0.8234	1.8056
9	0.8245	1.8042
10	0.8246	1.8029
11	0.8245	1.8010
12	0.8242	1.7988
13	0.8238	1.7970
14	0.8232	1.7964

The HAM-Padé approximants of ω and c_1:

i	ω	c_1
1	0.8182	2.0000
2	0.8147	1.7665
3	0.8171	1.7665
4	0.8257	1.8012
5	0.8238	1.8017
6	0.8232	1.7957

It takes about 860.540 seconds to complete the second part computation.

The total time is:10464.235 seconds.

B.3. Example 3

Consider a two directly coupled van der Pol oscillator

$$\begin{cases} \ddot{U}(t) - \varepsilon(1 - U^2(t))\dot{U}(t) + U(t) = \varepsilon\mu(V(t) - U(t)), \\ \ddot{V}(t) - \varepsilon(1 - V^2(t))\dot{V}(t) + p^2 V(t) = \varepsilon\mu(U(t) - V(t)), \end{cases} \tag{B.5}$$

with initial conditions

$$U(0) = a_1, \quad \dot{U}(0) = b_1, \quad V(0) = c_1, \quad \dot{V}(0) = 0, \tag{B.6}$$

where ε, μ, p are unknown physical parameters.

This system is a coupled self-excited system. We solve it by considering three different cases.

B.3.1. *All physical parameters are unknown*

To solve the systems (B.5) and (B.6), one can run the main procedure as follows:

$> main([diff(U(t), t\$2) - epsilon * (1 - U(t)^2) * diff(U(t), t) + U(t) = epsilon * mu * (V(t) - U(t)), diff(V(t), t\$2) - epsilon * (1 - V(t)^2) * diff(V(t), t) + p^2 * V(t) = epsilon * mu * (U(t) - V(t))], [U(0) = a[1], D(U)(0) = b[1], V(0) = c[1], D(V)(0) = 0], 5).$

For this example, the package `NOPH` outputs the following results:

`The input equation and initial conditions are:`

$$\frac{d^2}{dt^2}U(t) - \epsilon\,(1 - U(t)^2)\frac{d}{dt}U(t) + U(t) = \epsilon\,\mu\,(V(t) - U(t)),$$

$$\frac{d^2}{dt^2}V(t) - \epsilon\,(1 - V(t)^2)\frac{d}{dt}V(t) + p^2\,V(t) = \epsilon\,\mu(U(t) - V(t)),$$

$$U(0) = a_1, D(U)(0) = b_1, V(0) = c_1, D(V)(0) = 0.$$

`The new EQ and initial conditions in` u `with the form:`

$$\frac{d^2}{dt^2}u_1(t) - \epsilon\,(1 - u_1(t)^2)\frac{d}{dt}u_1(t) + u_1(t) + \epsilon\,\mu\,(u_1(t) - u_2(t)) = 0,$$

$$\frac{d^2}{dt^2}u_2(t) - \epsilon\,(1 - u_2(t)^2)\frac{d}{dt}u_2(t) + p^2\,u_2(t) + \epsilon\,\mu(u_2(t) - u_1(t)),$$

$$u_1(0) = a_1, D(u_1)(0) = b_1, u_2(0) = c_1, D(u_2)(0) = 0.$$

`where:`

$$\tau = \omega t,\ \ U(t) = u_1(\tau),\ \ V(t) = u_2(\tau).$$

`The nonlinear algebraic equations for` $a_{i,j}\&b_{i,j}$ `are:`

$$4(a_{1,0} - \epsilon\omega_0 b_{1,0} - {\omega_0}^2 a_{1,0} + \epsilon\mu a_{1,0} - \epsilon\mu c_{1,0}) + \epsilon\,\omega_0({a_{1,0}}^2 b_{1,0} + {b_{1,0}}^3) = 0,$$

$$-\epsilon\,\omega_0 {a_{1,0}}^3 + 4\,b_{1,0} - \epsilon\,\omega_0 a_{1,0}{b_{1,0}}^2 + 4\,\epsilon\,\omega_0 a_{1,0} + 4\,\epsilon\,\mu\,b_{1,0} - 4\,{\omega_0}^2 b_{1,0} = 0,$$

$$\epsilon\,\mu\,a_{1,0} + {\omega_0}^2 c_{1,0} - \epsilon\,\mu\,c_{1,0} - p^2 c_{1,0} = 0,$$

$$-\,\epsilon\,-\,4\,\omega_0 c_{1,0} + \omega_0 {c_{1,0}}^3 + 4\,b_{1,0}\mu = 0.$$

`The variables and parameters lists, respectively:`

$$[\omega_0, a_{1,0}, b_{1,0}, c_{1,0}],\ [\epsilon, \mu, p].$$

Then the package `NOPH` asks: “ `Please input additional conditions,` $\{\}$ `means no additional conditions`”.

Such as input $\{b_{1,0} = 0\}$ to simplify the above nonlinear algebraic equations, or else the `NOPH` outputs too many solutions and it is inconvenient for us to select a solution and input its order.

Then the package `NOPH` delivers:

`The reduced nonlinear algebraic equations read:`

$$4\,(-{\omega_0}^2 a_{1,0} + a_{1,0} + \epsilon\,\mu\,a_{1,0} - \epsilon\,\mu\,c_{1,0}) = 0,$$

$$\epsilon\,\omega_0 a_{1,0}({a_{1,0}}^2 - 4) = 0,$$

$$\epsilon\,\mu\,a_{1,0} + {\omega_0}^2 c_{1,0} - \epsilon\,\mu\,c_{1,0} - p^2 c_{1,0} = 0,$$

$$\epsilon\,\omega_0 c_{1,0}(-4 + {c_{1,0}}^2) = 0.$$

Next, the package `NOPH` asks:

`Please input a var and parameter list to be solved:`

If we further input $[p_1, p_2, p_3, 4, 2, 1]$, in which p_i means the ith element in the above parameter list, the integer i means the ith element in the variable list.

Then the `NOPH` outputs:

`The variables and parameters list to be solved:`

$[\epsilon,\, \mu,\, p,\, c_{1,0},\, a_{1,0},\, \omega_0]$.

`The possible solutions are:`

$$
\begin{aligned}
&\{p=1,\, \omega_0=1,\, a_{1,0}=2,\, b_{1,0}=0,\, c_{1,0}=2\},\\
&\{p=1,\, \omega_0=1,\, a_{1,0}=-2,\, b_{1,0}=0,\, c_{1,0}=-2\},\\
&\{p=1,\, \omega_0=\sqrt{1+2\,\varepsilon\,\mu},\, a_{1,0}=2,\, b_{1,0}=0,\, c_{1,0}=-2\},\\
&\{p=1,\, \omega_0=\sqrt{1+2\,\varepsilon\,\mu},\, a_{1,0}=-2,\, b_{1,0}=0,\, c_{1,0}=2\},\\
&\{p=-1,\, \omega_0=1,\, a_{1,0}=2,\, b_{1,0}=0,\, c_{1,0}=2\},\\
&\{p=-1,\, \omega_0=1,\, a_{1,0}=-2,\, b_{1,0}=0,\, c_{1,0}=-2\},\\
&\{p=-1,\, \omega_0=\sqrt{1+2\,\varepsilon\,\mu},\, a_{1,0}=2,\, b_{1,0}=0,\, c_{1,0}=-2\},\\
&\{p=-1,\, \omega_0=\sqrt{1+2\,\varepsilon\,\mu},\, a_{1,0}=-2,\, b_{1,0}=0,\, c_{1,0}=2\}.
\end{aligned}
\tag{B.7}
$$

`Select a solution and input its order:`, such as 4, the program further delivers:

`The 4th solution was selected.`

`The parameter constraint condition is:` $\{p=1\}$,
the [1,1] homotopy-Padé approximant

$$\omega_{P_{1,1}} = \frac{(\varepsilon^2+64\,\varepsilon\,\mu+32)\,\sqrt{2\,\varepsilon\,\mu+1}}{3\,\varepsilon^2+64\,\varepsilon\,\mu+32},$$

and the [2,2] homotopy-Padé approximant

$$
\begin{aligned}
\omega_{P_{2,2}} &= (16640\,\varepsilon^2+66560\,\varepsilon^3\,\mu+960\,\varepsilon^4+294912\,\varepsilon\,\mu+66560\,\varepsilon^4\,\mu^2\\
&+589824\,\varepsilon^2\,\mu^2+49152+393216\,\varepsilon^3\,\mu^3+1920\,\varepsilon^5\,\mu+9\,\varepsilon^6)\sqrt{2\,\varepsilon\,\mu+1}\\
&/\ (45\,\varepsilon^6+393216\,\varepsilon^3\,\mu^3+589824\,\varepsilon^2\,\mu^2+78848\,\varepsilon^4\,\mu^2+78848\,\varepsilon^3\,\mu\\
&+294912\,\varepsilon\,\mu+3840\,\varepsilon^5\,\mu+19712\,\varepsilon^2+49152+1920\,\varepsilon^4).
\end{aligned}
$$

B.3.2. *In the case of* $\varepsilon = 1/2$

From (B.7), it can be seen that there is parameter constraint $p = 1$ in each solution. Therefore, letting $\varepsilon = 1/2$, there will be only one unknown parameter μ contained in input system. In this case, one can run the main procedure as follows:

$> main([diff(U(t), t\$2) - 1/2*(1-U(t)^2)*diff(U(t), t) + U(t) = 1/2*mu*(V(t)-U(t)), diff(V(t), t\$2) - 1/2*(1-V(t)^2)*diff(V(t), t) + p^2*V(t) = 1/2*mu*(U(t)-V(t))], [U(0) = a[1], D(U)(0) = b[1], V(0) = c[1], D(V)(0) = 0], 5);$

Following the similar steps shown in the last subsection, the program automatically delivers (by selecting the 4th solution in (B.7) and setting $h\_order = 5$) the following results within 590 seconds.

`The parameter constraint condition is:` $\{p = 1\}$,

the [1,1] homotopy-Padé approximant

$$\omega_{P_{1,1}} = \frac{(128\,\mu + 129)\,\sqrt{\mu + 1}}{128\,\mu + 131},$$

and the [2,2] homotopy-Padé approximant

$$\omega_{P_{2,2}} = \frac{(9973504\,\mu + 9703424\,\mu^2 + 3415817 + 3145728\,\mu^3)\,\sqrt{\mu + 1}}{3468845 + 3145728\,\mu^3 + 9752576\,\mu^2 + 10075648\,\mu}.$$

`The comparison graph between` $\omega_{p[1,1]}$ `and` $\omega_{p[2,2]}$.

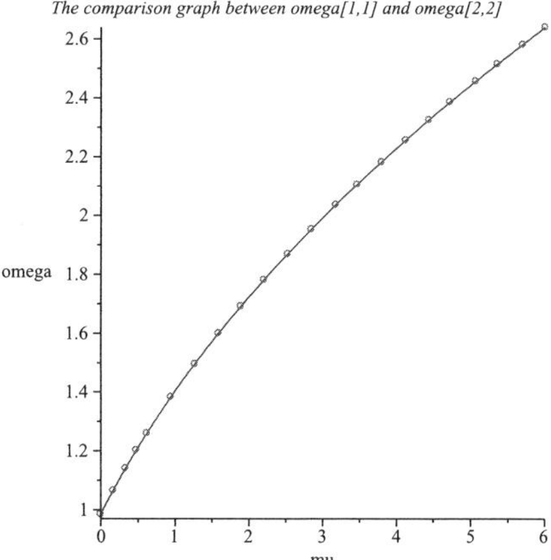

B.3.3. *In the case of* $\varepsilon = 1/2,\ \ \mu = 8$

In this case, there is no unknown parameter in input system, for there is parameter constraint $p = 1$ involved in each solution. One can run the main procedure as follows:

$> main([diff(U(t), t\$2) - 1/2 * (1 - U(t)^2) * diff(U(t), t) + U(t) = 4*(V(t) - U(t)), diff(V(t), t\$2) - 1/2*(1 - V(t)^2)*diff(V(t), t) + p^2*V(t) =$

$4 * (U(t) - V(t))], [U(0) = a[1], D(U)(0) = b[1], V(0) = c[1], D(V)(0) = 0], 7, 9)$.

Following the similar steps shown in the last subsection (by selecting the 1st solution, i.e.,$\{p = 1,\ \omega_0 = 1,\ a_{1,0} = 2,\ b_{1,0} = 0,\ c_{1,0} = 2\}$), the program automatically outputs $\hbar = -0.489$, approximants of ω, a_1, b_1, c_1, as well as comparison graphs for different order approximations of $U(t)$ and $V(t)$ within 755 seconds.

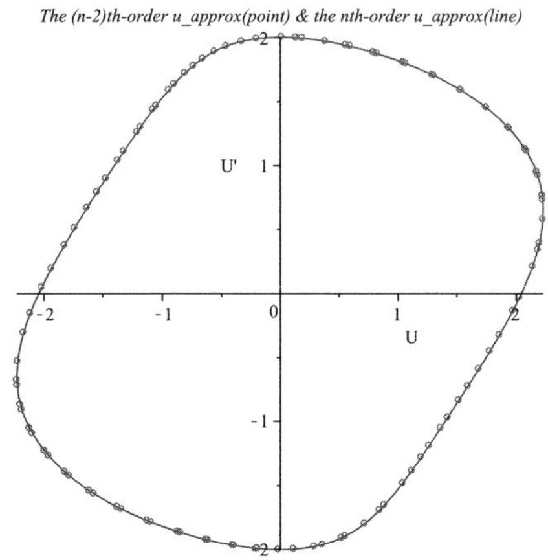

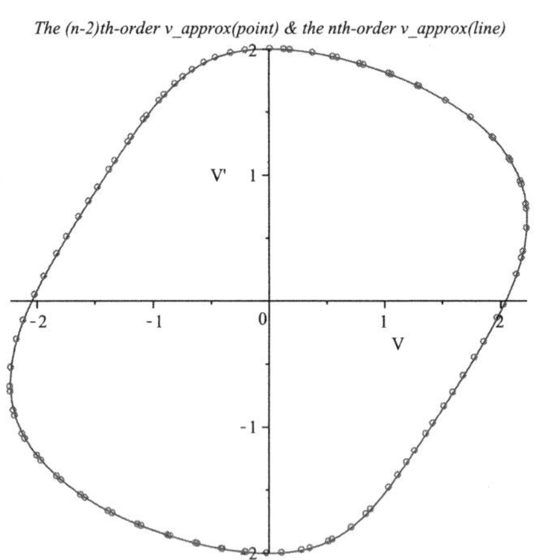

The approximants of ω, a_1, b_1 and c_1:

i	ω	a_1	b_1	c_1
1	1.0000	2.0000	0.0	2.0000
2	0.9847	2.0000	−0.3667	2.0000
3	0.9847	1.9689	−0.3692	1.9689
4	0.9847	1.9684	−0.3330	1.9684
5	0.9847	1.9698	−0.3322	1.9698
6	0.9847	1.9698	−0.3341	1.9698

The Padé approximants of ω, a_1, b_1 and c_1:

i	ω	a_1	b_1	c_1
1	0.9847	2.0000	−0.3692	2.0000
2	0.9847	1.9698	−0.3356	1.9698
3	0.9847	1.9698	−0.3341	1.9698

References

[1] M. V. Dyke, *Perturbation Methods in Fluid Mechanics.* The Parabolic Press, Stanford, CA (1975).

[2] A. H. Nayfeh, *Introduction to Perturbation Techniques.* Wiley, New York (1981).

[3] E. J. Hinch, *Perturbation methods.* Cambridge University Press, Cambridge (1991).

[4] A. W. Bush, *Perturbation Methods for Engineers and Scientists.* CRC Press, Boca Roton, FL. (1992).

[5] A. M. Lyapunov, The general problem of the stability of motion, *Int. J. Control.* **55**(3), 531–534 (1992).

[6] A. V. Karmishin, A. I. Zhukov, and V. G. Kolosov, *Methods of Dynamics Calculation and Testing for Thin-walled Structures.* Mashinostroyenie, Moscow (1990).

[7] G. Adomian, Nonlinear stochastic differential equations, *J. Math. Anal. Appl.* **55**(2), 441–452 (1976).

[8] S. Liao, *Beyond Perturbation: Introduction to the Homotopy Analysis Method.* Chapman & Hall/CRC Press, Boca Raton (2003).

[9] S. Liao, *Homotopy Analysis Method in Nonlinear Differential Equation.* Springer & Higher Education Press, Heidelberg & Beijing (2012).

[10] S. Liao, *The proposed homotopy analysis technique for the solution of nonlinear problems*, PhD thesis, Shanghai Jiao Tong University (1992).

[11] S. Liao, A uniformly valid analytic solution of two-dimensional viscous flow over a semi-infinite flat plate, *J. Fluid Mech.* **385**(1), 101–128 (1999).

[12] S. Liao and A. Campo, Analytic solutions of the temperature distribution in Blasius viscous flow problems, *J. Fluid Mech.* **453**, 411–425 (2002).

[13] S. Liao, On the homotopy analysis method for nonlinear problems, *Appl. Math. Comput.* **147**(2), 499–513 (2004).

[14] S. Liao, A new branch of solutions of boundary-layer flows over an impermeable stretched plate, *Int. J. Heat Mass Transfer.* **48**(12), 2529–2539 (2005).

[15] S. Liao and Y. Tan, A general approach to obtain series solutions of nonlinear differential equations, *Stud. Appl. Math.* **119**(4), 297–354 (2007).

[16] S. Liao, A new branch of solutions of boundary-layer flows over a permeable stretching plate, *Int. J. Nonlin. Mech.* **42**(6), 819–830 (2007).

[17] T. Hayat and M. Sajid, On analytic solution for thin film flow of a fourth grade fluid down a vertical cylinder, *Phys. Lett. A.* **361**(4), 316–322 (2007).

[18] M. Sajid, T. Hayat, and S. Asghar, Comparison between the HAM and HPM solutions of thin film flows of non-Newtonian fluids on a moving belt, *Nonlinear Dyn.* **50**(1-2), 27–35 (2007).

[19] S. Abbasbandy, The application of homotopy analysis method to nonlinear equations arising in heat transfer, *Phys. Lett. A.* **360**(1), 109–113 (2006).

[20] S. Abbasbandy, The application of homotopy analysis method to solve a generalized Hirota–Satsuma coupled KdV equation, *Phys. Lett. A.* **361**(6), 478–483 (2007).

[21] S. Abbasbandy, Homotopy analysis method for heat radiation equations, *Int. Commun. Heat Mass Transfer.* **34**(3), 380–387 (2007).

[22] Y. Liu and Z. Li, Homotopy analysis method for nonlinear differential equations with fractional orders, *Z. Naturforsch. A.* **63**(5-6), 241–247 (2008).

[23] Y. Liu, R. Yao, and Z. Li, An application of a homotopy analysis method to nonlinear composites, *J. Phys. A: Math. Theor.* **42**(12), 125205 (2009).

[24] Y. Liu, S. Liao, and Z. Li, Symbolic computation of strongly nonlinear periodic oscillations, *J Symb. Comput.* (2013).

[25] W. J. Wu, *Mechanical Theorem Proving in Geometries: Basic Principles.* Springer-Verlag, New York (1994).

[26] D. Wang, *An Implementation of the Characteristic Set Method in Maple.* Springer-Verlag, Wien New York (1995).

[27] D. Wang, *Elimination Practice: Software Tools and Applications.* Imperial College Press, London (2004).

[28] A. Beléndez, G. Bernabeu, J. Francés, D. Méndez, and S. Marini, An accurate closed-form approximate solution for the quintic Duffing oscillator equation, *Math. Comput. Model.* **52**(3), 637–641 (2010).

[29] R. H. Rand, Using Computer Algebra to Handle Elliptic Functions in the Method of Averaging. vol. 205, *Symbolic Computations and Their Impact on Mechanics*, ASME PVP (1992).

[30] H. Hu, Exact solution of a quadratic nonlinear oscillator, *J. Sound Vib.* **295** (1), 450–457 (2006).

[31] S. Chen and C. Chen, Application of the differential transformation method to the free vibrations of strongly non-linear oscillators, *Nonlinear Anal. R. World Appl.* **10**(2), 881–888 (2009).

Chapter 9

HAM-Based Mathematica Package BVPh 2.0 for Nonlinear Boundary Value Problems

Yinlong Zhao* and Shijun Liao†

School of Naval Architecture, Ocean and Civil Engineering
Shanghai Jiao Tong University, Shanghai 200240, China
**zhaoyl@sjtu.edu.cn, †sjliao@sjtu.edu.cn*

We issue the new version `BVPh 2.0` of the Mathematica package `BVPh`, a free software based on the homotopy analysis method (HAM) for nonlinear boundary-value and eigenvalue problems. The aim of the package `BVPh` is to develop a kind of analytic tool for as many nonlinear boundary value problems (BVPs) as possible such that multiple solutions of highly nonlinear BVPs can be conveniently found out, and that the infinite interval and singularities of governing equations and/or boundary conditions at multi-points can be easily resolved. Unlike its previous versions, `BVPh 2.0` works for systems of coupled nonlinear ordinary differential equations. It is user-friendly and free available online (`http://numericaltank.sjtu.edu.cn/BVPh.htm`). Different from numerical packages (such as `BVP4c`), it is based on the idea "computing numerically with functions instead of numbers". Especially, unlike other packages, the convergence of results given by the `BVPh 2.0` is guaranteed by means of the so-called convergence-control parameter in the frame of the homotopy analysis method. In this chapter, we briefly describe how to install and use the `BVPh 2.0` with a simple user's guide. Five typical examples (governed by up to four coupled ODEs) are used to illustrate the validity of the `BVPh 2.0`, and the corresponding input data of these examples for the `BVPh 2.0` are free available online (`http://numericaltank.sjtu.edu.cn/BVPh.htm`). The `BVPh 2.0` indeed provides us with an easy-to-use tool to efficiently solve various types of coupled linear/nonlinear ordinary differential equations.

Contents

9.1. Introduction . . . 362
9.2. Installation of the BVPh 2.0 . . . 366
9.3. Illustrative example . . . 367
9.4. Brief mathematical formulas . . . 372
9.4.1. Boundary value problems in a finite interval . . . 372
9.4.2. Eigenvalue-like problems in a finite interval . . . 375
9.4.3. Problems in a semi-infinite interval . . . 375
9.5. Approximation and iteration of solutions . . . 376
9.5.1. Polynomials . . . 377
9.5.2. Trigonometric functions . . . 377
9.5.3. Hybrid-base functions . . . 378
9.6. A simple user guide of the BVPh 2.0 . . . 380
9.6.1. Key modules . . . 380
9.6.2. Control parameters . . . 381
9.6.3. Input . . . 383
9.6.4. Output . . . 384
9.6.5. Global variables . . . 384
9.7. Examples . . . 385
9.7.1. Example 1: A system of ODEs in finite interval . . . 385
9.7.2. Example 2: A system of ODEs with algebraic property at infinity . . . 388
9.7.3. Example 3: A system of ODEs with an unknown parameter . . . 393
9.7.4. Example 4: A system of ODEs in different intervals . . . 397
9.7.5. Example 5: Iterative solutions of the Gelfand equation . . . 401
9.8. Conclusions . . . 405
Appendix A. Codes for examples . . . 405
A.1. Sample codes to run the illustrative example . . . 406
A.2. Input data of BVPh 2.0 for the illustrative example . . . 406
A.3. Input data of BVPh 2.0 for Example 1 . . . 408
A.4. Input data of BVPh 2.0 for Example 2 . . . 409
A.5. Input data of BVPh 2.0 for Example 3 . . . 411
A.6. Input data of BVPh 2.0 for Example 4 . . . 412
A.7. Input data of BVPh 2.0 for Example 5 . . . 414
References . . . 416

9.1. Introduction

Ordinary differential equations (ODEs) are widely used in mathematics, science and engineering. The so-called initial value problems (IVPs) specify some restrictions only at a single point. This kind of problems is often solved by means of numerical approach based on integration, such as the Runge–Kutta method. However, in many cases, a solution is described in a more complicated way. The so-called boundary value problems (BVPs) may specify some restrictions at more than one point. Unlike IVPs, BVPs might have multiple solutions, or may satisfy some restrictions at infinity. Hence, generally speaking, BVPs are more difficult to solve than IVPs.

Thanks to the development of mathematical software such as Maple, Mathematica and MATLAB, some packages are developed to solve nonlinear BVPs, such as the `BVP4c`, the `Chebfun`, the `NOPH`, and the `BVPh`. The famous MATLAB package `BVP4c` [1, 2] implements a collocation method, instead of a shooting method. The collocation polynomial provides a C^1-continuous solution that is fourth-order accurate uniformly in $[a, b]$. Mesh selection and error control are based on the residual of the continuous solution. However, it is not easy for the `BVP4c` to resolve the singularity in governing equations and/or boundary conditions. Besides, the `BVP4c` regards an infinite interval as a kind of singularity and replaces it by a finite one: this leads to the additional inaccuracy and uncertainty of solutions.

The `Chebfun` [3, 4] is a collection of algorithms on the "chebfun" objects written in MATLAB. It aims to combine the feel of symbolics with the speed of numerics. The basis of the `Chebfun` is Chebyshev expansions, fast Fouries transform, baryentric interpolation and so on. The idea "computing numerically with functions instead of numbers" [4] behind the `Chebfun` makes it have the potential to handle unbounded domains and singularities in an easy way. Although linear differential equations can be solved in a single step by `Chebfun` without iteration, only a few examples for nonlinear differential equations are given [3, 4]. Actually, `Chebfun` uses Newton's iteration to solve nonlinear problems. However, it is well known that the convergence of Newton's iteration is strongly dependent upon initial guesses and thus is not guaranteed. Besides, `Chebfun` searches for multiple solutions of nonlinear differential equations by using different guess approximations. However, it is not very clear how to choose these different guess approximations.

Based on the homotopy in topology, the so-called homotopy analysis method (HAM) was proposed by Liao [8–14] to gain analytic approximations of highly nonlinear problems. The HAM has some advantages over other traditional analytic approximation methods. First, unlike perturbation techniques, the HAM is independent of small/large physical parameters, and thus is valid in more general cases. Besides, different from all other analytic techniques, the HAM provides us a convenient way to guarantee the convergence of series solution. Furthermore, the HAM provides extremely large freedom to choose initial guess, equation-type and solution expression of linear sub-problems. It is found [13, 14] that lots of nonlinear BVPs in science, engineering and finance can be solved conveniently by means of the HAM, no matter whether the interval is finite or not.

Since the early HAM was proposed by Liao [9] in 1992, the HAM has been developed greatly in theory and widely applied to numerous nonlinear problems in lots of fields, with hundreds of publications. So, it is necessary to develop a HAM-based software to simplify the applications of the HAM. Based on the HAM, a Mathematica package `BVPh` 1.0— for nonlinear boundary value/eigenvalue problems with singularity and/or multipoint boundary conditions was issued by Liao [14] in May 2012, which is free available online (`http://numericaltank.sjtu.edu.cn/BVPh.htm`). Its aim is to develop a kind of analytic tool for as many nonlinear BVPs as possible such that multiple solutions of highly nonlinear BVPs can be conveniently found out, and that the infinite interval and singularities of governing equations and/or boundary conditions at multi-points can be easily resolved. As illustrated by Liao [14], the `BVPh 1.0` is valid for lots of nonlinear BVPs and thus is a useful tool in practice.

Currently, based on the HAM, the Maple package `NOPH` [5] for periodically oscillating systems of center and limit cycle types is developed, which delivers accurate approximations of frequency, mean of motion and amplitude of oscillation automatically. The `NOPH` combines Wu's elimination method and the homotopy analysis method (HAM). It is free available online (`http://numericaltank.sjtu.edu.cn/NOPH.htm`). Different from the `BVPh 1.0`, the `NOPH` is for periodic oscillations. This illustrates the general validity of the HAM for nonlinear problems once again.

It is a pity that the `BVPh 1.0` can only deal with BVPs of single ordinary differential equation (ODE), say, it can not solve systems of coupled ODEs. In this chapter we issue the new version `BVPh 2.0`, which works for many types of systems of coupled nonlinear ordinary differential equations (ODEs) in finite and/or semi-infinite intervals. Besides, new algorithms are used in some modules of `BVPh 2.0`. As a result, `BVPh 2.0` is much faster than `BVPh 1.0` in most cases. In this chapter, we illustrate how to use

`BVPh 2.0` to solve different kinds of systems of coupled nonlinear ODEs, including a system of two coupled ODEs in finite interval, a system of two coupled ODEs in semi-infinite interval, a system of two coupled ODEs with algebraic property at infinity, a system of three coupled ODEs with an unknown parameter to be determined, and a system of four coupled ODEs in different intervals.

This chapter is organized as follows. In § 9.2, we show how to install the package `BVPh 2.0`. In § 9.3, we take an example to illustrate how to use `BVPh 2.0` in detail. In § 9.4, we briefly describe some mathematical formulas based on which the `BVPh 2.0` is developed. In § 9.5, the iterative technique is illustrated for two typical kinds of base-functions in finite intervals. A simple user guide of `BVPh 2.0` is given in § 9.6. Some typical examples are given in § 9.7 to illustrate the potential of the `BVPh 2.0` and to show its validity. In § 9.8, some discussions are given. The reader is suggested to read § 9.3 at first for an illustrative example. If one is puzzled with some parameters, one is encouraged to search for them in § 9.6, and read the detailed description there.

In order to check the validity and correctness of the `BVPh 2.0`, we choose all of our typical examples from published articles, which were solved either analytically (by the HAM) or numerically before. As illustrated in this chapter, the package `BVPh 2.0` always gives results, which agree well with the published ones. The validity of the `BVPh 2.0` is checked for each example in the following way:

(1) the squared residual error usually decreases to as low as 10^{-10}, and decreases at least 6 orders of magnitude;
(2) the same physical quantities of interest are gained as the published ones;
(3) analytic solutions gained by `BVPh 2.0` agree well with the published ones.

Note that the package `BVPh 2.0` is developed with Mathematica 7.0. As some new features of Mathematica 7.0 are used, we strongly recommend you to use the `BVPh 2.0` in Mathematica 7.0 or higher version, since it might not work for some lower version of Mathematics.

9.2. Installation of the BVPh 2.0

The BVPh is a free/open-source software written in Mathematica for boundary value problems (BVPs). Its newest version BVPh 2.0 is available online (http://numericaltank.sjtu.edu.cn/BVPh.htm). The input data for this chapter's examples can also be found there. Since the commands of Mathematica are designed to be the same on different operating systems, the package written in Mathematica on Windows can be used in Mathematica on other operating systems.

The source file of the package BVPh 2.0 is BVPh2_0.m. The easiest way to load the package BVPh 2.0 to solve your problem is to put the file BVPh2_0.m and the input data for the problem, e.g., Example.m, in the same directory, then open a new notebook file and saved it as, e.g., runExample.nb, in the same directory and run the following codes.

```
(* Filename: runExample.nb *)
ClearAll["Global`*"];
SetDirectory[ToFileName[Extract["FileName" /.
NotebookInformation[EvaluationNotebook[]], {1},
FrontEnd`FileName]]];
<<BVPh2_0.m;
<<Example.m;
```

Note that the listings without an end-of-line semicolon is wrapped to fit the page width of this chapter. However, if you break the long command intentionally in Mathematica, it will not work as you expected. The above commands first clear all global variables, then set the current working directory to "the current directory". Here "the current directory" is where you put the source file BVPh2_0.m, the input data Example.m and the notebook file runExample.nb. The last two lines read in the package BVPh2_0.m and Example.m in the notebbook runExample.nb.

If you are familiar with Mathematica's file and directory operations, you can put the file BVPh2_0.m and the input data of the problem in different directories, then specifies the path where to get them. It is worth emphasizing that the pathname separator is "\\" under Windows, and "/" elsewhere. The sample codes to get the file BVPh2_0.m and the input data

`Example.m` in the notebook file `runExample.nb` are as follows on Windows.

```
(* Filename: runExample.nb *)
ClearAll["Global`*"];
<<"E:\\Package\\BVPh2_0.m";
<<"E:\\Project\\Example\\Example.m";
```

Here we assume the file `BVPh2_0.m` in the directory `E:\Package` and the input data `Example.m` in a different directory `E:\Project\Example`. The above sample codes may look like the following on Unix

```
(* Filename: runExample.nb *)
ClearAll["Global`*"];
<< "/home/user/Package/BVPh2_0.m";
<< "/home/user/Example/Example.m";
```

Here we assume `BVPh2_0.m` in the directory `/home/user/Package/` and the input data `Example.m` in a different directory `/home/user/Example`.

From now on, we will assume that the package `BVPh 2.0` has been successfully loaded so that the modules in the package are available. In the next section, we will use an illustrative example to show how to write the input data and how to get the approximations by `BVPh 2.0`.

9.3. Illustrative example

Consider a system of ODEs [6]

$$f''' - (f')^2 + ff'' + 2\lambda g + \beta[2ff'f'' - f^2 f'''] = 0, \tag{9.1}$$

$$g'' - f'g + fg' - 2\lambda f' + \beta[2ff'g' - f^2 g''] = 0, \tag{9.2}$$

subject to

$$f'(0) = 1, f(0) = 0, g(0) = 0, \tag{9.3}$$

$$f'(\infty) = 0, g(\infty) = 0, \tag{9.4}$$

where λ is rotation parameter, β is viscoelastic parameter, and the prime indicates the differentiation with respect to η. This system models two-dimensional flow of an upper convected Maxwell fluid in a rotating frame. Sajid [6] has solved it by the HAM.

To solve this problem by BVPh 2.0, we have to input the differential equations, boundary conditions, initial guesses and convergence-control parameters. The differential equations (9.1) and (9.2) can be coded as follows

```
TypeEQ = 1;
NumEQ = 2;
f[1,z_,{f_,g_},Lambda_]:=
    D[f,{z,3}]-D[f,z]^2+f*D[f,{z,2}]+2*la*g+
    beta*(2*f*D[f, z]*D[f,{z,2}]-f^2*D[f,{z,3}]);
f[2,z_,{f_,g_},Lambda_]:=
    D[g,{z,2}]-D[f,z]*g+f*D[g,z]-2*la*D[f,z]+
    beta*(2*f*D[f, z]*D[g, z]-f^2*D[g,{z, 2}]);
```

Here TypeEQ controls the type of governing equations: TypeEQ=1 corresponds to a system of ODEs without an unknown to be determined, TypeEQ=2 corresponds to a system of ODEs with an unknown, Lambda, to be determined. Since all the parameters in the problem will be given, we set TypeEQ to 1. Note that we use the delayed assignment SetDelayed(:=) in Mathematica to define these ODEs to avoid the evaluation when the assignment is made.

The boundary conditions (9.3) and (9.4) are defined in a semi-infinite interval, from 0 to $+\infty$. They are coded as

```
NumBC = 5;
BC[1,z_,{f_,g_}]:=(D[f, z]-1)/.z->0;
BC[2,z_,{f_,g_}]:=f/.z->0;
BC[3,z_,{f_,g_}]:=g/.z->0;
BC[4,z_,{f_,g_}]:=D[f,z]/.z->infinity;
BC[5,z_,{f_,g_}]:=g/.z->infinity;
```

Here NumBC is the number of boundary conditions of the problem. For this problem, we have 5 boundary conditions, so NumBC is set to 5. The symbol infinity is introduced in our package to denote ∞. When an expression contains infinity, the limit of the expression is computed as z approaches ∞. The delayed assignment (:=) is also used to avoid the evaluation when the assignment is made — the same reason as defining the differential equations.

For a multi-layer problem, the differential equations in the system are not necessarily in the same interval (see Example 4 in § 9.7). Hence, we have to give each equation its solution interval. To measure the accuracy of the approximate solutions, we have to compute the squared residual error over the corresponding solution interval. In practice, when the differential equation is defined in a semi-infinite interval, we simply truncate the infinite interval to a finite interval to compute the squared residual error, or it will take a lot of computation time. For this problem, the solution interval for each equation and the integral interval for the squared residual error are defined as

```
zL[1] = 0;
zR[1] = infinity;
zL[2] = 0;
zR[2] = infinity;
zRintegral[1] = 10;
zRintegral[2] = 10;
```

Here `zL[k]` (or `zR[k]`) is the left (or right) endpoint of the solution interval for the kth equation `f[k,z,{f,g},Lambda]`. And `zLintegral[k]` (or `zRintegral[k]`) is the left (or right) endpoint of the integral interval to compute the squared residual error for the kth equation. If the value of `zL[k]` (or `zR[k]`) is a finite number, `zLintegral[k]` (or `zRintegral[k]`) is set to the same value automatically. However, if any of them contains the symbol `infinity`, we have to set the corresponding endpoint of the integral interval to a finite value. That is why we write explicitly `zRintegral[1]=10` and `zRintegral[2]=10`. For this problem, the squared residual is integrated over the range [0, 10] for both equations.

The auxiliary linear operators for this problem are chosen as $\mathcal{L}_1 = \frac{\partial^3}{\partial \eta^3} - \frac{\partial}{\partial \eta}$, $\mathcal{L}_2 = \frac{\partial^2}{\partial \eta^2} - 1$, which are coded as

```
L[1,u_] := D[u,{z,3}]-D[u,z];
L[2,u_] := D[u,{z,2}]-u;
```

Here `L[k,u]` is the auxiliary linear operator corresponding to the kth equation. Note that i) η is the independent variable in the differential equations

(9.1) and (9.2), while z is the universal independent variable in the package BVPh 2.0; ii) the delayed assignment SetDelayed(:=) is used to define the operator; iii) u is a formal parameter.

For this problem, the initial guesses are $f_0 = 1 - e^{-z}$ and $g_0 = ze^{-z}$. They are coded as

```
U[1,0] = 1-Exp[-z];
U[2,0] = alpha*z*Exp[-z];
```

Here alpha is an introduced convergence-control parameter that will be determined later. U[k,0] is the initial guess of the kth equation. Note that U[k,0] and u[k,0] are usually the same in the package BVPh 2.0.

We want to solve this problem when the physical parameters $\beta = 1/5$ and $\lambda = 1/10$. These two parameters are coded as

```
beta = 1/5;
la = 1/10;
```

So far, we have defined all the input of this problem properly, except the convergence-control parameter c0[k] and alpha. Usually, the optimal values of the convergence-control parameters are obtained by minimizing the averaged squared residual error. For this problem, we get the approximate optimal values of c0[1], c0[2] and alpha by minimizing the squared residual error of the 3rd-order approximation as

```
GetOptiVar[3, {}, {c0[1],c0[2],alpha}];
```

The first parameter of GetOptiVar denotes which order approximation is used. Here 3 means the 3rd-order approximation is used. The second parameter denotes a list of constraints used in the optimization. When the second parameter of GetOptiVar is an empty list, it means the averaged squared residual is minimized without any constraint. Here we add no constraints to minimize the averaged squared residual error. The third parameter is a list of the variables to be optimized. Here we want to optimize c0[1], c0[2] and alpha. After some computation, it gives the optimized convergence-control parameters c0[1]=-1.26906, c0[2]=-1.19418 and alpha=-0.0657063.

Now we can use

```
BVPh[1,10]
```

to get the 10th-order approximation. If we are not satisfied with the accuracy of the 10th-order approximation, we can use BVPh[11,20], instead of BVPh[1,20], to get 20th-order approximation or higher order approximation.

The kth-order approximation of the ith differential equation is stored in U[i,k]. We can use

```
Plot[{U[1,20], U[2,20]}, {z,0,10},
AxesLabel->{"\[Eta]", ""},
PlotStyle->{{Thin, Red},{Dashed, Blue}}]
```

to plot the 20th-order approximate solution, which is shown in Fig. 9.1.

The accuracy of the kth-order approximation is measured by the averaged squared residual. We can use

```
ListLogPlot[Table[{2*i,ErrTotal[2*i]},{i,1,10}],
Joined->True,Mesh->All,
PlotRange->{{2,20},{10^(-15),10^(-5)}},
AxesLabel->{"m", "error"}];
```

to plot the curve of the total error versus the order of approximation, which is shown in Fig. 9.2. Note that ErrTotal[k] stores the total error of the system when the kth-order approximation is used, while Err[k] is a list that stores the error for each ODE in the system when the kth-order approximation is used.

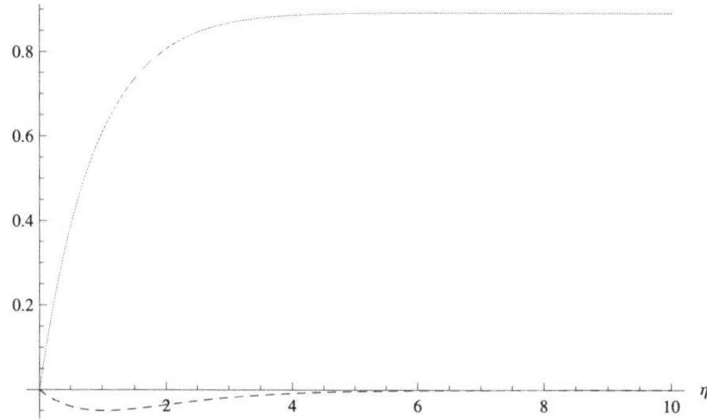

Fig. 9.1. The curve of $f(z)$ (solid) and $g(z)$ (dashed) for the illustrative example when $\beta = 1/5$, $\lambda = 1/10$.

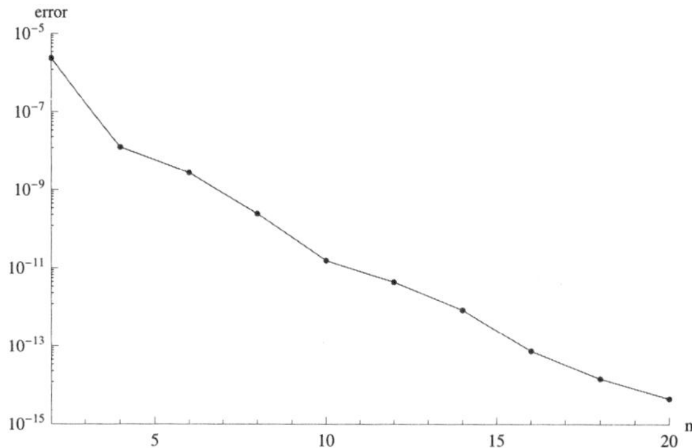

Fig. 9.2. Total error vs. order of approximation for the illustrative example when $\beta = 1/5$, $\lambda = 1/10$.

9.4. Brief mathematical formulas

The `BVPh 2.0` is based on the HAM. It is an extension of `BVPh 1.0` to systems of ODEs. Here the mathematical formulas are briefly described.

9.4.1. *Boundary value problems in a finite interval*

Consider a system of `NumEQ` ordinary differential equations (ODEs),

$$\mathcal{F}_i[z, u_1, u_2, \cdots] = 0,\ z \in [\mathtt{zL_i}, \mathtt{zR_i}],\ 1 \leqslant i \leqslant \mathtt{NumEQ}, \tag{9.5}$$

subject to the NumBC linear boundary conditions

$$\mathcal{B}_k[z, u_1, u_2, \cdots] = \gamma_k, \ 1 \leqslant k \leqslant \texttt{NumBC}, \tag{9.6}$$

where $\mathcal{F}_i$ is a nonlinear differential operator, $\mathcal{B}_k$ is a linear differential operator, z is the independent variable, $u_i(z)$ is a smooth function, NumEQ is a positive integer, γ_k, $\texttt{zL}_\texttt{i}$ and $\texttt{zR}_\texttt{i}$ are constants, respectively. The boundary conditions may be defined at multipoints including the two endpoints. Assume that at least one solution exists, and that all solutions are smooth.

Let $q \in [0, 1]$ denote an embedding parameter, $u_{i,0}(z)$ an initial guess of the solution $u_i(z)$, respectively. In the frame of the HAM, we construct such a continuous deformation $\phi_i(z; q)$ that, as q increases from 0 to 1, $\phi_i(z; q)$ varies continuously from the initial guess $u_{i,0}(z)$ to the true solution $u_i(z)$ of (9.5) and (9.6). Such kind of continuous deformations are governed by the so-called zeroth-order deformation equations

$$\begin{aligned}(1-q)\mathcal{L}_i[\phi_i(z;q) - u_{i,0}(z)] &= q c_{0,i} H_i(z) \mathcal{F}_i[\phi_1(z;q), \phi_2(z;q), \cdots],\\ z \in [\texttt{zL}_\texttt{i}, \texttt{zR}_\texttt{i}],\ q &\in [0,1],\ 1 \leqslant i \leqslant \texttt{NumEQ}\end{aligned} \tag{9.7}$$

where $\mathcal{L}_i$ is an auxiliary linear operator, $c_{0,i}$ is the so-called convergence-control parameter, $H_i(z)$ is an auxiliary function, corresponding to the ith governing equation in (9.5), respectively.

Note that the HAM provides us extremely large freedom to choose the auxiliary linear operator $\mathcal{L}_i$, the convergence-control parameter $c_{0,i}$ and the auxiliary function $H_i(z)$, $1 \leqslant i \leqslant$ NumEQ. Assume that all of them are properly chosen so that the homotopy-Maclaurin series

$$\phi_i(z;q) = u_{i,0}(z) + \sum_{k=1}^{+\infty} u_{i,k}(x) q^k \tag{9.8}$$

absolutely converges at $q = 1$, where

$$u_{i,m} = \mathcal{D}_m[\phi(z;q)] = \left. \frac{1}{m!} \frac{\partial^m \phi(z;q)}{\partial q^m} \right|_{q=0}. \tag{9.9}$$

Here, $\mathcal{D}_m$ is called the mth-order homotopy-derivative operator [8]. Then, we have the so-called homotopy-series solution

$$u_i(z) = u_{i,0} + \sum_{m=1}^{+\infty} u_{i,m}(z), \tag{9.10}$$

where $u_{i,m}$ is governed by the so-called mth-order deformation equation

$$\mathcal{L}_i[u_{i,m}(z) - \chi_m u_{i,m-1}(z)] = c_{0,i} H_i(z) \delta_{i,m-1}(z), \ z \in [\mathtt{zL_i}, \mathtt{zR_i}], \quad (9.11)$$

subject to the NumBC homogeneous linear boundary conditions

$$\mathcal{B}_k[z, u_1, u_2, \cdots] = 0, \ 1 \leqslant k \leqslant \mathtt{NumBC}, \quad (9.12)$$

where

$$\chi_m = \begin{cases} 0, & m \leq 1, \\ 1, & m > 1, \end{cases} \quad (9.13)$$

and

$$\begin{aligned} \delta_{i,m-1}(z) &= \mathcal{D}_{m-1}\{\mathcal{F}_i[\phi_1(z;q), \phi_2(z;q), \cdots]\} \\ &= \mathcal{D}_{m-1}\left\{\mathcal{F}_i\left[\sum_{i=0}^{m-1} u_{1,i}(z)q^i, \sum_{i=0}^{m-1} u_{2,i}(x)q^i, \cdots\right]\right\} \end{aligned} \quad (9.14)$$

can be easily obtained by means of the theorems proved in Ref. 8.

The mth-order approximation of $u_i(z)$ is given by

$$u_i(x) \approx U_{i,m}(x) = \sum_{k=0}^{m} u_{i,k}(z). \quad (9.15)$$

To measure the accuracy of the mth-order approximations $U_{i,m}$, $1 \leqslant i \leqslant$ NumEQ, the averaged squared residual error for the system (9.5) is defined as

$$E_m = \sum_{i=1}^{\mathtt{NumEQ}} \frac{\int_{\mathtt{zLintegral[i]}}^{\mathtt{zRintegral[i]}} |\mathcal{F}_i[U_{1,m}, U_{2,m}, \cdots]|^2 \, dz}{\mathtt{zRintegral[i]} - \mathtt{zLintegral[i]}}. \quad (9.16)$$

Here zLintegral[i] and zRintegral[i] are two endpoints of the integral interval over which the squared residual of the ith governing equation is integrated. Theoretically speaking, the smaller E_m, the more accurate the mth-order approximation $U_{i,m}(z)$. Given the initial guess $u_{i,0}(z)$, the auxiliary linear operator $\mathcal{L}_i$ and the auxiliary function $H_i(z)$, the squared residual error E_m is dependent on the convergence-control parameters $c_{0,i}$, $i = 1, 2, \cdots,$ NumEQ. Hence, the optimal values of $c_{0,i}$ can be determined by the minimum of E_m at some proper order m.

Note that i) for boundary value problems without an unknown to be determined, we should set the control parameter TypeEQ = 1 for the BVPh 2.0; ii) the intervals $[\mathtt{zL_i}, \mathtt{zR_i}]$ are not necessarily the same, so the package can solve multi-layer problems without any modification.

9.4.2. *Eigenvalue-like problems in a finite interval*

Consider a system of NumEQ ordinary differential equations (ODEs)

$$\mathcal{F}_i[z, u_1, u_2, \cdots, \lambda] = 0, \ z \in [\mathtt{zL_i}, \mathtt{zR_i}], \ 1 \leqslant i \leqslant \mathtt{NumEQ}, \tag{9.17}$$

subject to the NumBC linear boundary conditions,

$$\mathcal{B}_k[z, u_1, u_2, \cdots] = \gamma_k, \ 1 \leqslant k \leqslant \mathtt{NumBC}. \tag{9.18}$$

Note that there is an unknown parameter λ in the system (9.17), and the boundary conditions (9.18) include an additional boundary condition determining the unknown λ. This additional boundary condition is the zeroth boundary condition in BVPh 1.0. However, it is treated the same as the other boundary conditions in BVPh 2.0, that is, any one of the boundary conditions in (9.18) can be the additional boundary condition provided that NumEQ $-$ 1 is the order of the system (9.17).

All related formulas are the same as those given in § 9.4.1, except that a deformation $\Lambda(q)$ is also constructed such that $\Lambda(q)$ varies continuously from the initial guess λ_0 to λ as q increases from 0 to 1.

Note that, when the BVPh 2.0 is used to solve eigenvalue-like problems in a finite interval, we should set TypeEQ=2.

9.4.3. *Problems in a semi-infinite interval*

When the right endpoint $\mathtt{zR}_i$ of the solution interval in (9.5) is $+\infty$, the governing equation is defined in a semi-infinite interval $[\mathtt{zL}_i, +\infty]$. The BVPh can solve this kind of problem without truncating the domain. This feature is quite different from BVP4c, which regards a semi-infinite interval as a kind of singularity and replaces it by a finite one.

All related formulas are the same as those given in § 9.4.1, except that the finite intervals $z \in [\mathtt{zL_i}, \mathtt{zR_i}]$ are replaced by the semi-infinite ones $z \in [\mathtt{zL_i}, \infty]$. However, completely different initial guesses $u_{i,0}(z)$ and auxiliary linear operators $\mathcal{L}_i$ are used for this kind of problems, because their solutions are expressed by completely different base functions.

Note that i) when the BVPh 2.0 is used to solve problems in a semi-infinite interval, the symbol infinity is used to denote ∞ in input data; ii) the boundary conditions at infinity are considered as a limiting process;

iii) to get the error of the approximations, we have to provide finite intervals over which the squared residuals are integrated, as the numerical integration method is used in `BVPh 2.0`; iv) the integral interval for squared residual error can be set through `zLintegral[i]` and `zRintegral[i]`. For example, `zLintegral[i]=0` and `zRintegral[i]=10` means the squared residual of the ith governing equation is integrated over $[0, 10]$.

9.5. Approximation and iteration of solutions

Although the high-order deformation equations (9.11) are always linear, they are still not easy to solve in general, because the right-hand side term $\delta_{i,m-1}$ may be rather complicated. To solve this problem, we can approximate $\delta_{i,m-1}$ in the form

$$\delta_{i,m-1} \approx \sum_{k=0}^{N_t} b_k e_k(z)$$

where e_k denotes the base-function, N_t denotes the number of truncated terms, the coefficient b_k is uniquely determined by $\delta_{i,m-1}(z)$ and the base-functions $e_k(z)$.

To further accelerate the rate of convergence, we can employ the iteration approach by using the Mth-order approximation as a new initial guess $u_{i,0}^*(z)$, i.e.,

$$u_{i,0}(z) + \sum_{m=1}^{M} u_{i,m}(z) \to u_{i,0}^*(z).$$

The above expression provides us the so-called Mth-order iteration formula. In this way, the convergence of the homotopy-series can be greatly accelerated.

The iteration approach of the `BVPh 2.0` is currently possible only when the base functions are of polynomials, trigonometric functions and hybrid-base functions. We consider here the approximation of a smooth function $f(z)$ in these three different kinds of base functions. The approximation of a function in a semi-infinite interval is proposed in Ref. 20 in the frame of HAM.

9.5.1. *Polynomials*

It is well known that a smooth function $f(z)$ in the finite interval $z \in [-1, 1]$ can be well approximated by the Chebyshev polynomials

$$f(x) \approx \frac{a_0}{2} + \sum_{k=1}^{N_t} a_k T_k(z), \tag{9.19}$$

where $T_k(z)$ is the kth Chebyshev polynomial of the first kind, N_t denotes the number of Chebyshev polynomials, and

$$\begin{aligned} a_k &= \frac{2}{N_t+1} \sum_{i=1}^{N_t+1} f(x_i) T_k(x_i), \\ &= \frac{2}{N_t+1} \sum_{i=1}^{N_t+1} f\left[\cos\left\{\frac{\pi(i-\frac{1}{2})}{N_t+1}\right\}\right] \cos\left\{\frac{\pi k(i-\frac{1}{2})}{N_t+1}\right\}. \end{aligned}$$

When Chebyshev polynomial is used to approximate the boundary value/eigenvalue problems in a finite interval $z \in [a, b]$ by means of the `BVPh 2.0`, we should set `TypeL = 1`, `ApproxQ = 1`.

9.5.2. *Trigonometric functions*

It is well known that the Fourier series

$$\frac{a_0}{2} + \sum_{n=1}^{+\infty} \left(a_n \cos\frac{n\pi z}{a} + b_n \sin\frac{n\pi z}{a}\right)$$

of a continuous function $f(z)$ in a finite interval $z \in (-a, a)$ converges to $f(z)$ in the interval $z \in (-a, a)$, where

$$a_n = \frac{1}{a}\int_{-a}^{a} f(t)\cos\frac{n\pi t}{a}dt,\ b_n = \frac{1}{a}\int_{-a}^{a} f(t)\sin\frac{n\pi t}{a}dt.$$

For a continuous function $f(z)$ in $[0, a]$, we can define $f(z) = f(-z)$ in $z \in [-a, 0)$ and its Fourier series reads

$$f(z) = \frac{a_0}{2} + \sum_{n=1}^{+\infty} a_n \cos\frac{n\pi z}{a}. \tag{9.20}$$

Alternatively, we can define $f(z) = -f(-z)$ in $z \in [-a, 0)$ and its Fourier series reads

$$f(z) \sim \sum_{n=1}^{+\infty} b_n \sin\frac{n\pi z}{a}. \tag{9.21}$$

Note that in (9.21), we write "$\sim$" instead of "=", because it holds only if $f(0) = f(a) = 0$ is assumed.

In practice, we have the following approximations

$$f(z) \approx \frac{a_0}{2} + \sum_{n=1}^{N_t} a_n \cos\frac{n\pi z}{a} \tag{9.22}$$

or

$$f(z) \approx \sum_{n=1}^{N_t} b_n \sin\frac{n\pi z}{a}, \tag{9.23}$$

where N_t denotes the number of truncated terms.

When the above-mentioned trigonometric functions are used to solve boundary value/eigenvalue problems in a finite interval $z \in [0, a]$ by means of the `BVPh 2.0`, we should set `TypeL = 2`, `ApproxQ = 1` and `HYBRID = 0` with `TypeBase = 1` for the odd expression (9.23) and `TypeBase = 2` for the even expression (9.22), respectively.

9.5.3. *Hybrid-base functions*

Note that the first-order derivative of the even Fourier series (9.20) equals to zero at the two endpoints $z = 0$ and $z = a$, but the original function $f(z)$ may have arbitrary values of $f'(0)$ and $f'(a)$. So, in case of $f'(0) \neq 0$ and/or $f'(a) \neq 0$, one had to use many terms of the even Fourier series (9.20) so as to obtain an accurate approximation near the two endpoints. To overcome this disadvantage, we first express $f(z)$ by such a combination

$$f(z) \approx Y(z) + w(z), \tag{9.24}$$

where

$$Y(z) = \left(f'(0)z - \frac{f'(0) + f'(a)}{2a}z^2\right)\cos\frac{\pi z}{a} \tag{9.25}$$

and then approximate $w(z) = f(z) - Y(z)$ by the even Fourier series

$$w(z) \approx \frac{\bar{a}_0}{2} + \sum_{n=1}^{N_t} \bar{a}_n \cos\frac{n\pi z}{a} \tag{9.26}$$

with the Fourier coefficient

$$\bar{a}_n = \frac{2}{a}\int_0^a [f(t) - Y(t)]\cos\frac{n\pi t}{a}dt.$$

Here N_t is the number of truncated terms. Note that $Y(z)$ in (9.25) satisfies

$$Y'(0) = f'(0),\ Y'(a) = f'(a)$$

so that $w'(0) = w'(a) = 0$. Therefore, we often need a few terms of the even Fourier series to accurately approximate $w(z)$. Note also that both the trigonometric and polynomial functions are used in (9.24) to approximate $f(z)$. It is found that, by means of such kind of approximations based on hybrid-base functions, one often needs much less terms to approximate a given smooth function $f(z)$ in $[0, a]$ than the traditional Fourier series.

Alternatively, for a continuous function $f(z)$ in $[0, a]$, we can use

$$Y(z) = f(0) + \frac{f(a) - f(0)}{a} z \tag{9.27}$$

or

$$Y(z) = \frac{f(0) + f(a)}{2} + \frac{f(0) - f(a)}{2} \cos \frac{\pi z}{a}, \tag{9.28}$$

and approximate $w(z)$ by the odd Fourier series

$$w(z) \approx \sum_{n=1}^{N_t} \bar{b}_n \sin \frac{n\pi z}{a}, \tag{9.29}$$

where

$$\bar{b}_n = \frac{2}{a} \int_0^a [f(t) - Y(t)] \sin \frac{n\pi t}{a} dt.$$

Note that $Y(z)$ in (9.27) and (9.28) satisfies

$$Y(0) = f(0),\ Y(a) = f(a)$$

so that $w(0) = w(a) = 0$.

It is suggested to use the hybrid-base approximation (9.24) with (9.25) for an even function $f(z)$, and (9.24) with (9.27) or (9.28) for an odd function $f(z)$, respectively. If $f(z)$ is neither an odd nor even function, both of them work.

When the above-mentioned hybrid-base approximation is used, we have even larger freedom to choose the initial guess $u_0(z)$. For example, for a 2nd-order boundary value/eigenvalue problem in a finite interval $z \in [0, a]$, we may choose such an initial guess in the form

$$u_0(z) = B_0 + B_1 \cos \frac{\kappa\pi z}{a} + B_2 \sin \frac{\kappa\pi z}{a} \tag{9.30}$$

or

$$u_0(z) = (B_0 + B_1 z + B_2 z^2) \cos \frac{\kappa \pi z}{a}, \tag{9.31}$$

where B_0, B_1 and B_2 are determined by linear boundary conditions, κ is a positive integer.

When the above-mentioned trigonometric functions are used to solve boundary value/eigenvalue problems in a finite interval $z \in [0, a]$ by means of the BVPh 2.0, we should set `TypeL = 2`, `ApproxQ = 1` and `HYBRID = 1` with `TypeBase = 1` for the odd expression (9.29) and `TypeBase = 2` for the even expression (9.26), respectively.

9.6. A simple user guide of the BVPh 2.0

In this section, we will take a glance at the Mathematica package `BVPh 2.0`.

9.6.1. *Key modules*

`BVPh` The module `BVPh[k_,m_]` gives the kth to mth-order homotopy approximations of a system of ordinary differential equations (ODEs) subject to some boundary conditions. The system may have an unknown parameter (when `TypeEQ = 2`) or may not have an unknown parameter (when `TypeEQ = 1`). It is the basic module. For example, `BVPh[1,10]` gives the 1st-order to 10th-order homotopy-approximations. Thereafter, `BVPh[11,15]` further gives the 11th-order to 15th-order homotopy-approximations. For problems with an unknown parameter, the initial guess of the unknown parameter is determined by an algebraic equation. Thus, if there are more than one initial guesses of the unknown parameter, it is required to choose one among them by inputting an integer, such as 1 or 2, corresponding to the 1st or the 2nd initial guess of the unknown parameter, respectively.

`iter` The module `iter[k_,m_,r_]` provides us homotopy approximations of the kth to mth iteration by means of the rth-order iteration formula. It calls the basic module `BVPh`. For example, `iter[1,10,3]` gives homotopy-approximations of the 1st to 10th iteration by the 3rd-order iteration formula. Furthermore, `iter[11,20,3]` gives the homotopy-approximations of the 11th to 20th iterations. The iteration stops when the averaged squared residual error of the system is less than a critical value `ErrReq`, whose default is 10^{-20}.

`GetErr` The module `GetErr[k_]` gives the averaged squared residual error of the governing equation at the kth-order homotopy-approximation gained by the module `BVPh`. Note that, `errer[i,k]` provides the residual of the ith governing equation at the kth-order homotopy-approximation gained by `BVPh`, and `ErrTotal[k]` gives the total averaged squared residual error of the system at the kth-order homotopy-aprroximation gained by `BVPh`.

`hp` The module `hp[f_,m_,n_]` gives the `[m,n]` homotopy-padé approximation of a list of the homotopy-approximations `f`, where `f[[i+1]]` denotes the ith-order homotopy-approximation of the same governing equation.

`GetBC` The module `GetBC[i_,k_]` gives the ith boundary condition of the kth-order deformation equation.

9.6.2. *Control parameters*

`TypeEQ` A control parameter for the type of governing equations: `TypeEQ = 1` corresponding to a nonlinear problem without an unknown parameter, `TypeEQ = 2` corresponds to a nonlinear problem with an unknown parameter (called eigenvalue problem or eigenvalue-like problem), respectively.

`TypeL` A control parameter for the type of auxiliary linear operator: `TypeL = 1` corresponds to polynomial approximation, and `TypeL = 2` corresponds to a trigonometric approximation or a hybrid-base approximation, respectively.

`ApproxQ` A control parameter for approximation of solutions. When `ApproxQ = 1`, the right-hand side term of all higher-order deformation equations are approximated by Chebyshev polynomials (9.19), trigonometric functions in § 9.5.2, or by the hybrid-base functions in § 9.5.3. When `ApproxQ = 0`, there is no such kind of approximation. When `TypeL = 2`, `ApproxQ = 1` is valid only for problems in a finite interval $z \in [0, a]$, where $a > 0$ is a constant.

`HYBRID` A control parameter for the hybrid-base functions. When `HYBRID = 1`, hybrid-base functions are employed in approximation.

When `HYBRID = 0`, trigonometric functions without polynomials are employed in approximation. This parameter is usually used in conjunction with `TypeBase`, and is valid only when `TypeL = 2` and `ApproxQ = 1` for problems in a finite interval $z \in [0, a]$.

`TypeBase` A control parameter for the type of Fourier series approximation: `TypeBase = 1` corresponds to the odd Fourier approximation (9.23) or (9.29), `TypeBase = 2` corresponds to the even Fourier approximation (9.22) or (9.26), respectively. This parameter is usually used in conjunction with `HYBRID`, and is valid only when `TypeL = 2` and `ApproxQ = 1` for problems in a finite interval $z \in [0, a]$.

`Ntruncated` A control parameter to determine the number of truncated terms used to approximate the right-hand side of higher-order deformation equations. The larger `Ntruncated`, the better the approximations, but the more CPU time. It is valid only when `ApproxQ = 1`. The default is 10.

`NtermMax` A positive integer used in the module `truncated`, which ignores all polynomial terms whose order is higher than `NtermMax`. The default is 90.

`ErrReq` A critical value of the averaged squared residual error of governing equations to stop the computation. The default is 10^{-20}.

`NgetErr` A positive integer used in the module `BVPh`. The averaged squared residual error of governing equations is calculated when the order of approximation is divisible by `NgetErr`. The default is 2.

`Nintegral` Number of discrete points with equal space, which are used to numerically calculate the integral of a function. It is used in the module `int`. The default is 50.

`ComplexQ` A control parameter for complex variables. `ComplexQ = 1` corresponds to the existence of complex variables in governing equations and/or boundary conditions. `ComplexQ = 0` corresponds to the nonexistence of such kind of complex variables. The default is 0.

`FLOAT` A control parameter for floating-point computation. When `FLOAT = 1`, floating-point numbers are employed in computation. When `FLOAT = 0`, rational numbers are employed in computation. The default is 1.

9.6.3. *Input*

`NumEQ` The number of governing equations.

`f[i_,z_,{u_,···},lambda_]` The ith governing equation with or without an unknown parameter, corresponding to $\mathcal{F}_i[z, u, \cdots]$ or $\mathcal{F}_i[z, u, \cdots, \lambda]$ in either a finite interval $z \in [a, b]$ or a semi-infinite interval $z \in [b, +\infty)$, where a and b are bounded constants. Note that the formal parameter `lambda` denotes the unknown parameter λ to be determined, or the eigenvalue λ for eigenvalue problems, but has no meaning at all for problems without an unknown parameter λ, or non-eigenvalue problems.

`NumBC` The number of boundary conditions.

`BC[k_,z_,{u_,···}]` The kth boundary condition corresponding to $\mathcal{B}_k[z, u, \cdots]$, where $1 \leq k \leq$ `NumBC`. Note that the symbol `infinity` denotes ∞ in boundary conditions.

`U[i,0]` The initial guess $U_{i,0}(z)$, i.e., $u_{i,0}(z)$.

`c0[i]` The convergence-control parameter $c_{0,i}$, corresponding to the ith governing equation.

`H[i_,z_]` The auxiliary function corresponding to the ith governing equation. The default is `H[i_,z_]:=1`.

`L[i_,f_]` The auxiliary linear operator corresponding to the ith governing equation.

`zL[i]` The left endpoint of the interval of the solution corresponding to the ith governing equation. Note that intervals of the solutions are not necessarily the same, especially for multi-layer flow problem.

`zR[i]` The right endpoint of the interval of the solution corresponding to the ith governing equation.

`zLintegral[i]` The left endpoint of the integral interval to compute the averaged squared residual error of the ith governing equation. When the left endpoint of the solution interval for the ith governing equation is a finite number, `zLintegral[i]` is automatically set to `zL[i]`. Otherwise, the user has to specify the value of `zLintegral[i]`.

`zRintegral[i]` The right endpoint of the integral interval to compute the averaged squared residual error of the ith governing equation. When the right endpoint of the solution interval for the ith governing equation is a finite number, `zRintegral[i]` is automatically set to `zR[i]`. Otherwise, the user has to specify the value of `zRintegral[i]`.

9.6.4. *Output*

`U[i,k]` The kth-order homotopy-approximation of the solution to the ith governing equation given by the basic module `BVPh`.

`V[i,k]` The kth-iteration homotopy-approximation of the solution to the ith governing equation given by the iteration module `iter`.

`Lambda[k]` The kth-order homotopy-approximation of the eigenvalue λ or the unknown parameter λ given by the basic module `BVPh`.

`LAMBDA[k]` The kth-iteration homotopy-approximation of the eigenvalue λ or the unknown parameter λ given by the iteration module `iter`.

`error[i,k]` The residual of the ith governing equation given by the kth-order homotopy-approximation (obtained by the basic module `BVPh`).

`Err[k]` A list of the averaged squared residual error of each governing equation given by the kth-order homotopy-approximation (obtained by the basic module `BVPh`).

`ErrTotal[k]` The total of the averaged squared residual error for each governing equation given by the kth-order homotopy-approximation (obtained by the basic module `BVPh`).

`ERR[k]` A list of the averaged squared residual error of each governing equation given by the kth-iteration homotopy-approximation (obtained by the iteration module `iter`).

`ERRTotal[k]` The total of the averaged squared residual error for each governing equation given by the kth-iteration homotopy-approximation (obtained by the iteration module `iter`).

9.6.5. *Global variables*

All control parameters and output variables mentioned above are global. Besides these, the following variables and parameters are also global.

`z` The independent variable z.

`u[i,k]` The solution to the kth-order deformation equation of the ith governing equation.

`lambda[k]` A constant variable, corresponding to λ_k.

`delta[i,k]` A function dependent upon z, corresponding to the right-hand side term $\delta_{i,k}(z)$ in the high-order deformation equation.

`L[i,w]` The ith auxiliary linear operator $\mathcal{L}_i$ applied to w.

`Linv[i,f]` The inverse operator of $\mathcal{L}_i$, corresponding to $\mathcal{L}_i^{-1}$, applied to f.

`sNum` A positive integer to determine which initial guess λ_0 is chosen when there are multiple solutions of λ_0.

9.7. Examples

More examples are given in this section to show the usage the package `BVPh 2.0`.

9.7.1. *Example 1: A system of ODEs in finite interval*

Consider a system of coupled ODEs [7]

$$(1+K)f'''' - ReMf'' + 2Reff''' - Kg'' = 0, \tag{9.32}$$

$$\left(1+\frac{K}{2}\right)g'' - ReK[2g - f''] + Re[2fg' - f'g] = 0, \tag{9.33}$$

subject to

$$f(0) = 0,\ f(1) = 0,\ f'(1) = 1,\ f''(0) = 0, \tag{9.34}$$

$$g(1) = 0,\ g(0) = 0, \tag{9.35}$$

where K is the ratio of viscosities, Re is the Reynolds number and M is the Hartman number. Hayat [7] has solved this problem by the HAM.

Here we solve this problem by `BVPh 2.0`. Since there are two ODEs in system (9.32)–(9.33) without an unknown to be determined, we have `NumEQ = 2` and `TypeEQ=1`. The system is input as

```
TypeEQ = 1;
NumEQ = 2;
f[1,z_,{f_,g_},Lambda_]:=(1+K)*D[f,{z,4}]
    -Rey*M*D[f,{z,2}]+2*Rey*f*D[f,{z,3}]-K*D[g,{z,2}];
f[2,z_,{f_,g_},Lambda_]:=(1+K/2)*D[g,{z,2}]
    -Rey*K*(2*g-D[f,{z,2}])+Rey*(2*f*D[g,z]-D[f,z]*g);
```

The six boundary conditions are defined as

```
NumBC = 6;
BC[1,z_,{f_,g_}]:=f/.z->0;
BC[2,z_,{f_,g_}]:=f/.z->1;
BC[3,z_,{f_,g_}]:=(D[f, z]-1)/.z->1;
BC[4,z_,{f_,g_}]:=D[f,{z,2}]/.z->0;
BC[5,z_,{f_,g_}]:=g/.z->1;
BC[6,z_,{f_,g_}]:=g/.z->0;
```

Now let us input the solution intervals

```
zL[1]=0;    zR[1]=1;
zL[2]=0;    zR[2]=1;
```

Since all the solution intervals are finite, we do not have to specify the integral interval to compute the averaged squared residual error.

The initial guesses are chosen as $f_0 = (z^3 - z)/2$ and $g_0 = 0$. They are input as

```
U[1,0] = (z^3-z)/2;
U[2,0] = 0;
```

The auxiliary linear operators are chosen as $\mathcal{L}_1 = \frac{\partial^4}{\partial z^4}$ and $\mathcal{L}_2 = \frac{\partial^2}{\partial z^2}$. They are defined as

```
L[1,u_]:=D[u,{z,4}];
L[2,u_]:=D[u,{z,2}];
```

Note that we use the delayed assignment `SetDelayed(:=)` to define these linear operators.

Without loss of generality, let us consider the case when $Re = M = 2$ and $K = 1/2$. These physical parameters are input as

```
Rey = M = 2;
K = 1/2;
```

At this time, we have input all the data for this problem, except the convergence-control parameters `c0[k]`. Hayat [7] chose the convergence-control parameters c0[1]=c0[2]=-0.7 through $\hbar$-curve. Here we minimize the averaged squared residual error of the 4th-order approximations to get optimal values for c0[k]

```
GetOptiVar[4,{},{c0[1],c0[2]}];
```

The convergence-control parameters c0[1] and c0[2] are found to be about -0.5825 and -0.721452, respectively.

Then we call the main module BVPh to get the 20th-order approximations

```
BVPh[1, 20];
```

The 20th-order approximations are stored in U[i,20],i=1,2, while the corresponding averaged squared residual error of the system is ErrTotal[20]. We can use

```
Plot[{U[1,20],U[2,20]},{z,0,1},AxesLabel->{"z",""},
PlotStyle->{{Thin, Red}, {Dashed, Blue}},
PlotRange->{{0, 1}, {-0.2, 0.2}}]
```

to plot the 20th-order approximations, which is shown in Fig. 9.3. This figure agrees with Hayat's Figs. 9 and 12 when $M = 2$, $Re = 2$ and $K = 0.5$. The 20th-order approximations give the values of $f''(1) = 3.61076396287$ and $g'(1) = -0.738463496789$, which are the same with Hayat's result [7]. The total error ErrTotal[k] of the system for every two order of approximations is plotted by the command

```
ListLogPlot[Table[{2 i, ErrTotal[2*i]}, {i, 1, 10}],
Joined -> True, Mesh -> All,
PlotRange -> {{2, 20}, {10^(-34), 1}},
AxesLabel -> {"m", "error"}]
```

in Fig. 9.4. We can see from it that the error decreases beautifully.

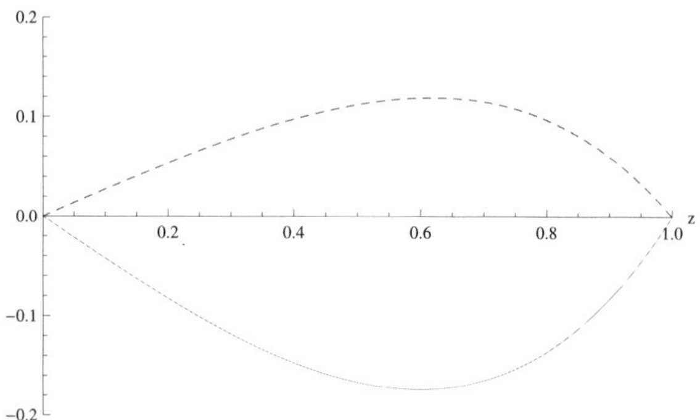

Fig. 9.3. The curve of $f(z)$ (solid), $g(z)$ (dashed) for Example 1.

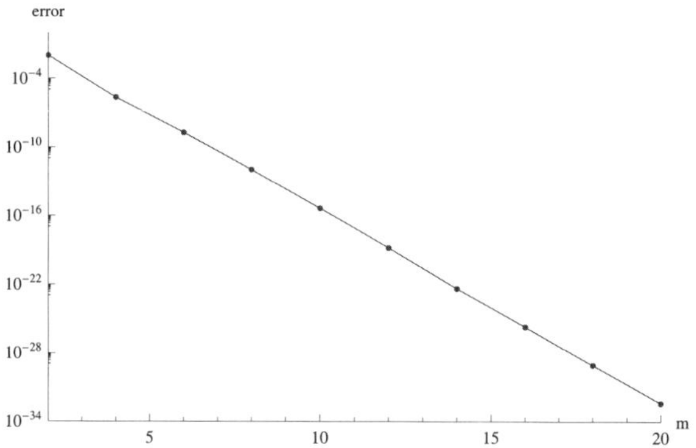

Fig. 9.4. Total error vs. order of approximation for Example 1.

9.7.2. *Example 2: A system of ODEs with algebraic property at infinity*

Consider a set of two coupled nonlinear differential equations [13]

$$f'''(\eta) + \theta(\eta) - f'^2 = 0, \tag{9.36}$$

$$\theta''(\eta) = 3\sigma f'(\eta)\theta(\eta), \tag{9.37}$$

subject to

$$f(0) = f'(0) = 0,\ \theta(0) = 1,\ f'(+\infty) = \theta(+\infty) = 0, \tag{9.38}$$

where the prime denotes differentiation with respect to the similarity variable η, σ is the Prandtl number, $f(\eta)$ and $\theta(\eta)$ relate to the velocity profile and temperature distribution of the boundary layer, respectively. Liao [13] employed the HAM to solve this system analytically. Now we use the `BVPh 2.0` to solve it.

Under the transformation

$$\xi = 1 + \lambda\eta, \ F(\xi) = f'(\eta), \ S(\xi) = \theta(\eta), \tag{9.39}$$

Eqs. (9.36) and (9.37) become

$$\lambda^2 F''(\xi) + S(\xi) - F^2(\xi) = 0, \tag{9.40}$$

$$\lambda^2 S''(\xi) = 3\sigma F(\xi) S(\xi), \tag{9.41}$$

subject to

$$F(1) = 0, \ S(1) = 1, \ F(+\infty) = S(+\infty) = 0. \tag{9.42}$$

Since there are two ODEs in system (9.40)–(9.41) without an unknown to be determined, we have `NumEQ = 2` and `TypeEQ=1`. This new system is defined as

```
TypeEQ = 1;
NumEQ = 2;
f[1,z_,{F_,S_},Lambda_]:=la^2*D[F,{z,2}]+S-F^2;
f[2,z_,{F_,S_},Lambda_]:=la^2*D[S,{z, 2}]-3*sigma*F*S;
```

The four boundary conditions (9.42) are defined as

```
NumBC = 4;
BC[1,z_,{F_, S_}] := F /. z -> 1;
BC[2,z_,{F_, S_}] := (G - 1) /. z -> 1;
BC[3,z_,{F_, S_}] := F /. z -> infinity;
BC[4,z_,{F_, S_}] := G /. z -> infinity;
```

Now let us input the solution intervals and integral intervals to compute averaged squared residual error

```
zL[1] = 1;    zR[1] = infinity;
zL[2] = 1;    zR[2] = infinity;
zRintegral[1] = 10;
zRintegral[2] = 10;
```

The initial guesses are chosen as $F_0 = \gamma(\xi^{-2} - \xi^{-3})$, $S_0 = \xi^{-4}$, and they are input as

```
U[1, 0] = gamma*(z^(-2) - z^(-3));
U[2, 0] = z^(-4);
```

The auxiliary linear operators are $\mathcal{L}_F = \frac{\xi}{3}\frac{\partial^2}{\partial \xi^2} + \frac{\partial}{\partial \xi}$ and $\mathcal{L}_S = \frac{\xi}{5}\frac{\partial^2}{\partial \xi^2} + \frac{\partial}{\partial \xi}$, which are defined as

```
L[1, u_] := D[u, {z, 2}]*z/3 + D[u, z];
L[2, u_] := D[u, {z, 2}]*z/5 + D[u, z];
```

Without loss of generality, let us consider the case when $\sigma = 1$, $\gamma = 3$ and $\lambda = 1/3$. We use the same convergence-control parameters `c0[1]` $=$ `c0[2]`$= -1/2$ as Liao [13]. These physical parameters and the control parameters `c0[k]` are defined as

```
sigma = 1;    gamma = 3;
la = 1/3;    c0[1] = -1/2;    c0[2] = -1/2;
```

Then we call the main module `BVPh`

```
BVPh[1, 20];
```

to get the 20th-order approximation. If we are not satisfied with the accuracy of the 20th-order approximation, we can use `BVPh[21,40]`, instead of `BVPh[1,40]`, to get 40th-order approximation or higher order approximation.

Note that `U[1,40]` and `U[2,40]` are the 40th-order approximations of the transformed system (9.40), (9.41) and (9.42). To plot the curve of the 40th-order approximations for the original problem, we first replace z with $1+\lambda\eta$ to obtain the 40th-order approximations for $f'(\eta)$ and $g(\eta)$, then plot the curve we want. This is done in Mathematica by the following command

```
trans = {z -> 1 + la*\[Eta]};
Plot[Evaluate[{U[1, 40], U[2, 40]} /. trans],
{\[Eta], 0, 10},PlotRange -> {{0, 10}, {0, 1}},
AxesLabel -> {"\[Eta]", ""},
PlotStyle -> {{Thin, Red}, {Dashed, Blue}}]
```

and the curve is shown in Fig. 9.5. Here `trans={z->1+la*\[Eta]}` is the corresponding transformation, and `\[Eta]` is the symbol η in Mathematica.

The total error `ErrTotal[k]` of the transformed system for every two order approximations is plotted in Fig. 9.6 by the following command

```
ListLogPlot[Table[{2 i, ErrTotal[2*i]}, {i, 1, 20}],
Joined -> True, Mesh -> All,
PlotRange -> {{2, 40}, {10^(-10), 0.01}},
AxesLabel -> {"m", "error"}]
```

Note that `ErrTotal[k]` not only measures the accuracy of the kth-order approximations for the transformed problem, but also measures the corresponding approximations for the original problem.

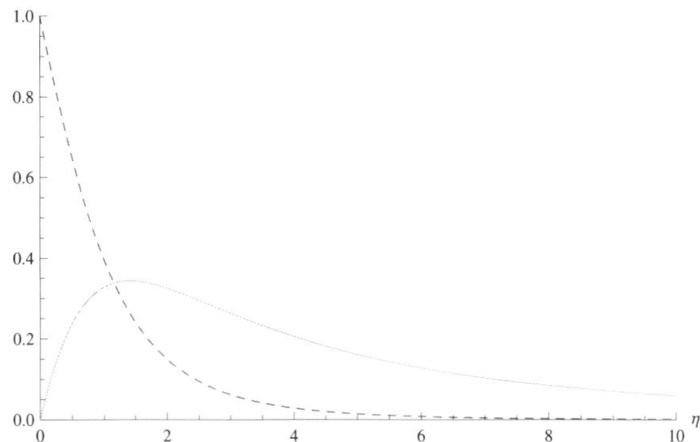

Fig. 9.5. The curve of $f'(\eta)$ (solid) and $\theta(\eta)$ (dashed) for Example 2.

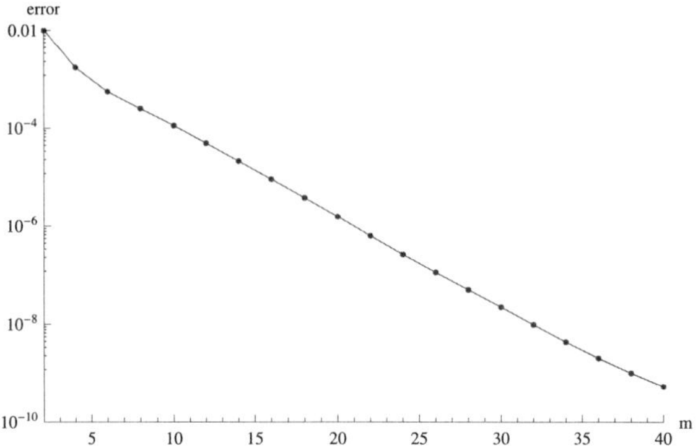

Fig. 9.6. Total error vs. order of approximation for Example 2.

The 40th-order approximations of $f''(0)$ and $g''(0)$ are 0.693268 and -0.769879, respectively. Kuiken's numerical result is $f''(0) \approx 0.693212$ and $g'(0) \approx -0.769861$. To get more accurate result, we have two choices. One is to call the module `BVPh` to get higher order approximation as before, the other is to apply the Padé approximation to the current approximations. The latter is done by calling the module `hp` as follows

```
hp[Table[D[(U[1,i]/.trans),\[Eta]]/.\[Eta]->0,
{i,0,40}],20,20]
hp[Table[D[(U[2,i]/.trans),\[Eta]]/.\[Eta]->0,
{i,0,40}],20,20]
```

which give 0.693212 and -0.769861, the $[20, 20]$ homotopy-Padé approximations of $f''(0)$ and $g'(0)$, respectively.

Note that we can compare the curve of $2n$th-order approximation and the $[n, n]$ homotopy-Padé approximation in a simple and efficient way. Here we compare `U[1,40]` and the $[20, 20]$ homotopy-Padé approximation of `U[1,i]`, $i = 0 \cdots 40$, in the Mathematica by the following command.

```
Plot[{U[1, 40]/.trans, hp[Table[U[1,i]/.trans,
{i, 0, 40}],20, 20]},{\[Eta],0,10},PlotRange->Full,
AxesLabel->{"\[Eta]", ""},
PlotStyle->{{Thin,Red},{Dashed,Blue}}
```

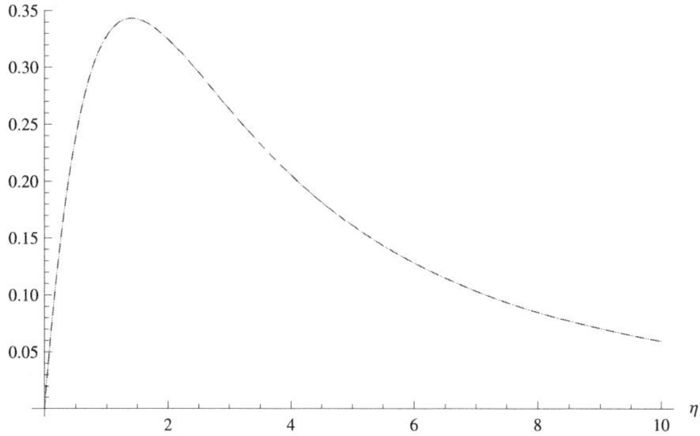

Fig. 9.7. The curve of 40th-order approximation of $f'(\eta)$ (solid) and $[20, 20]$ homotopy-Padé approximations of $f'(\eta)$ (dashed) for Example 2.

The comparison is shown in Fig. 9.7. From it we can see that the two are almost the same. This validate the convergence of the approximations to some extent. The above command is very efficient, because the `Plot` command in Mathematica first substitute the sample points into the expression and then applies the `hp` to a list of numerical values, rather than applies the `hp` to a list of expressions and then substitute the sample points into the resulting expression.

9.7.3. *Example 3: A system of ODEs with an unknown parameter*

Consider a system of ODEs [15]

$$U'' + (GrPr)\theta - N_r\phi + \sigma = 0, \tag{9.43}$$

$$\theta'' + N_b\theta'\phi' + N_t(\theta')^2 + N_b + N_t - U = 0, \tag{9.44}$$

$$\phi'' + \frac{N_t}{N_b}\theta'' - L_eU = 0, \tag{9.45}$$

subject to

$$U(-1) = U(1) = 0,\ \theta(-1) = \theta(1) = 0,\ \phi(-1) = \phi(1) = 0, \tag{9.46}$$

with an additional condition

$$\int_0^1 U dY = RePr, \tag{9.47}$$

where Gr is the Grashof number, Pr the Prandtl number, Nr the buoyancy ratio, σ the pressure parameter, Nb the Brownian motion parameter, Nt the thermophoresis parameter, Le the Lewis number, and Re the Reynolds number. All of the above parameters will be given for a special case except σ, which is to be determined from the system. Xu [15] solved this problem by the HAM.

Here we solve this problem by `BVPh 2.0`. Since there are three ODEs in system (9.43)–(9.45) with an unknown σ to be determined, we have `NumEQ = 3` and `TypeEQ=2`. The system is input as

```
TypeEQ = 2;     NumEQ = 3;
f[1,z_,{f_,g_,s_},sigma_] :=
    D[f,{z,2}]+Gr*Pr*g-Nr*s+sigma;
f[2,z_,{f_,g_,s_},sigma_] :=
    D[g,{z,2}]+Nb*D[g,z]*D[s,z]+Nt*(D[g,z])^2-f;
f[3,z_,{f_,g_,s_},sigma_] :=
    D[s,{z,2}]+Nt/Nb*D[f,{z,2}]-Le*f;
```

The seven boundary conditions, including the additional condition (9.47), are defined as

```
NumBC = 7;
BC[1,z_,{f_,g_,s_}] :=f/.z->-1;
BC[2,z_,{f_,g_,s_}] :=f/.z->1;
BC[3,z_,{f_,g_,s_}] :=g/.z->-1;
BC[4,z_,{f_,g_,s_}] :=g/.z->1;
BC[5,z_,{f_,g_,s_}] :=s/.z->-1;
BC[6,z_,{f_,g_,s_}] :=s/.z->1;
BC[7,z_,{f_,g_,s_}] :=Integrate[f,{z,0,1}]-Ra*Pr;
```

Now let us input the solution intervals

```
zL[1] = -1;     zR[1] = 1;
zL[2] = -1;     zR[2] = 1;
zL[3] = -1;     zR[3] = 1;
```

Since all the solution intervals are finite, we do not have to specify the integral interval to compute the averaged squared residual error.

The initial guesses are chosen as $U_0 = \epsilon_1 - 3(-25 + \epsilon_1)z^2/2 + 5(-15 + 2\epsilon_1)z^4/2$, $\theta_0 = \epsilon_2(1 - z^2)$ and $\phi_0 = \epsilon_3(1 - z^2)$, where ϵ_1, ϵ_2 and ϵ_3 are constants to be optimized. They are input as

```
U[1,0]=eps1-3/2*(-25+4eps1)z^2+5/2*(-15+2eps1)*z^4;
U[2,0]=eps2*(1-z^2);
U[3,0]=eps3*(1-z^2);
```

The auxiliary linear operators are chosen as $\mathcal{L}_1 = \mathcal{L}_2 = \mathcal{L}_3 = \frac{\partial^2}{\partial Y^2}$. They are defined as

```
L[1, u_] := D[u, {z, 2}];
L[2, u_] := D[u, {z, 2}];
L[3, u_] := D[u, {z, 2}];
```

Note that we use the delayed assignment `SetDelayed(:=)` to define these linear operators.

Without loss of generality, let us consider the case when $Nr = 3/20$, $Nt = Nb = 1/20$, $Le = 10$, $Gr = 5$, $Pr = 1$, and $Re = 5$. These physical parameters are input as

```
Nr = 3/20;     Nt = 1/20;
Nb = 1/20;     Le = 10;
Gr = 5;    Pr = 1;    Ra = 5;
```

At this time, we have input all the data for this problem, except the convergence-control parameters `c0[k]`, `eps1`, `eps2` and `eps3`. We minimize the averaged squared residual error of the 3th-order approximations to get optimal values for these parameters by the module `GetOptiVar` as follow

```
c0[1] = c0[2] = c0[3] = h;
GetOptiVar[3, {}, {eps1, eps2, eps3, h}];
```

Note that we put constraints `c0[1]=c0[2]=c0[3]` on `c0[1]`, `c0[2]` and `c0[3]` to simplify the computation. There is no constraint on `eps1`, `eps2` and `eps3`.

After some computation, we get optimal values for all the convergence-control parameters `c0[1]` $=$ `c0[2]` $=$ `c0[3]` ≈ -0.769452, `eps1` ≈ 7.56408, `eps2` ≈ -2.58887 and `eps3` ≈ -30.0044. Now we can use

```
BVPh[1,10]
```

to get the 10th-order approximation. If we are not satisfied with the accuracy of the 10th-order approximation, we can use `BVPh[11,20]` to get 20th-order approximation or higher order approximation.

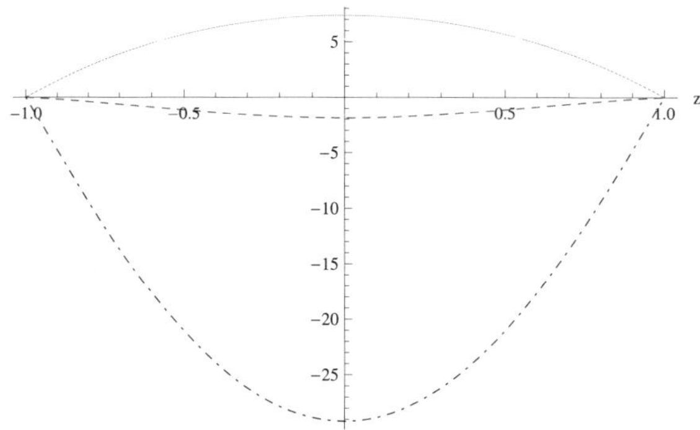

Fig. 9.8. The curve of U (solid), θ (dashed) and $\phi(z)$ (dot dashed) for Example 3.

The 20th-order approximations of U, θ and ϕ are stored in `U[1,20]`, `U[2,20]` and `U[3,20]`, the 20th-order approximation of σ is stored in `Lambda[19]`, while the corresponding averaged squared residual error of the system is stored in `ErrTotal[20]`. `Lambda[19]` is about 18.272555944, which is the same with Xu's result [15]. The 20th-order approximations are plotted in Fig. 9.8. The total error of the system for every two order of approximations is plotted in Fig. 9.9.

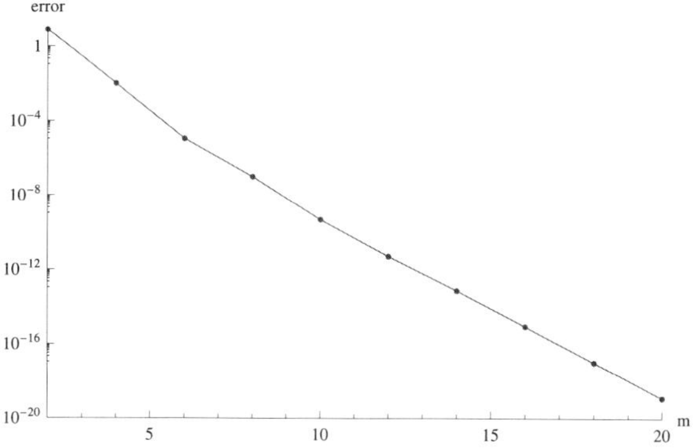

Fig. 9.9. Total error vs. order of approximation for Example 3.

9.7.4. *Example 4: A system of ODEs in different intervals*

Consider a two-phase flow [16]:

(i) Region 1

$$\frac{d^2u_1}{dy^2} + \frac{Gr}{Re}\sin(\phi)\theta_1 = P, \tag{9.48}$$

$$\frac{d^2\theta_1}{dy^2} + PrEc\left(\frac{du_1}{dy}\right)^2 = 0, \tag{9.49}$$

(ii) Region 2

$$\frac{d^2u_2}{dy^2} + \frac{Gr}{Re}\frac{nbh^2}{\lambda}\sin(\phi)\theta_2 - \frac{M^2h^2}{\lambda}u_2 = \frac{h^2}{\lambda}P, \tag{9.50}$$

$$\frac{d^2\theta_2}{dy^2} + EcPr\frac{\lambda}{\lambda_T}\left(\frac{du_2}{dy}\right)^2 + EcPr\frac{h^2}{\lambda_T}M^2u_2^2 = 0, \tag{9.51}$$

subject to

$$u_1(1) = 1,\ \theta_1(1) = 1, \tag{9.52}$$

$$u_1(0) = u_2(0),\ \theta_1(0) = \theta_2(0), \tag{9.53}$$

$$u_1'(0) = \frac{\lambda}{h}u_2'(0),\ \theta_1'(0) = \frac{\lambda_T}{h}\theta_2'(0), \tag{9.54}$$

$$u_2(-1) = 0,\ \theta_2(-1) = 0, \tag{9.55}$$

where Gr is the Grashof number, Ec is the Eckert number, Pr is the Prandtl number, Re is the Reynolds number, M is the Hartmann number and P

is the dimensionless pressure gradient. This model describes a two-fluid magnetohydrodynamic Poiseuille–Couette flow and heat transfer in an inclined channel. Umavathi [16] investigate this model analytically by regular perturbation method and numerically by finite difference technique. The `BVPh 2.0` can solve this problem (9.48)–(9.55) directly without difficulty. Since all the parameters in the system will be given, we have `NumEQ = 4` and `TypeEQ=1`. The system is input as

```
TypeEQ = 1;
NumEQ = 4;
f[1,z_,{u1_,s1_,u2_,s2_},Lambda_]:=
    D[u1, {z, 2}]+ Gr/Ra*Sin[phi]*s1 - P ;
f[2,z_,{u1_,s1_,u2_,s2_},Lambda_]:=
    D[s1, {z, 2}] + Pr*Ec*(D[s1, z])^2;
f[3,z_,{u1_,s1_,u2_,s2_},Lambda_]:=D[u2,{z, 2}]-h^2/lamb*P
    +Gr/Ra*Sin[phi]*n*b*h^2/lamb*s2-M^2*h^2/lamb*u2;
f[4,z_,{u1_,s1_,u2_,s2_},lambda_]:=D[s2,{z,2}]
    +Pr*Ec*lamb/lambT*D[u2,z]^2+Pr*Ec*h^2/lambT*M^2*u2^2;
```

The eight boundary conditions (9.52)–(9.55) are defined as

```
NumBC=8;
BC[1,z_,{u1_,s1_,u2_,s2_}]:=(u1-1)/.z->1;
BC[2,z_,{u1_,s1_,u2_,s2_}]:=(u1-u2)/.z->0;
BC[3,z_,{u1_,s1_,u2_,s2_}]:=u2/.z->-1;
BC[4,z_,{u1_,s1_,u2_,s2_}]:=(D[u1,z]-D[u2,z]*lamb/h)/.z->0;
BC[5,z_,{u1_,s1_,u2_,s2_}]:=(s1-1)/.z->1;
BC[6,z_,{u1_,s1_,u2_,s2_}]:=(s1-s2)/.z->0;
BC[7,z_,{u1_,s1_,u2_,s2_}]:=s2/.z->-1;
BC[8,z_,{u1_,s1_,u2_,s2_}]:=(D[s1,z]-D[s2,z]*lambT/h)/.z->0;
```

Now let us input the solution intervals

```
    zL[1] = 0;  zR[1] = 1; (* u1 *)
    zL[2] = 0;  zR[2] = 1; (* s1 *)
    zL[3] = -1; zR[3] = 0; (* u2 *)
    zL[4] = -1; zR[4] = 0; (* s2 *)
```

Note that the solution intervals are not the same. Since all the solution intervals are finite, we do not have to specify the integral interval to compute the averaged squared residual error.

The initial guesses are chosen as $u_{1,0} = \frac{\lambda}{h}(z - z^2) + 1$, $\theta_{1,0} = \frac{z\lambda_T}{h} + (1 - \frac{\lambda_T}{h})z^2$, $u_{2,0} = 1 + z$ and $\theta_{2,0} = z + z^2$. They are input as

```
U[1, 0] = (z - z^2)*lamb/h+1;               (* u1 *)
U[2, 0] = z*lambT/h +(1-lambT/h)*z^2;       (* s1 *)
U[3, 0] = 1 + z;                            (* u2 *)
U[4, 0] = z^2 + z;                          (* s2 *)
```

The auxiliary linear operators are chosen as $\mathcal{L}_1 = \mathcal{L}_2 = \mathcal{L}_3 = \mathcal{L}_4 = \frac{\partial^2}{\partial y^2}$. They are defined as

```
L[1,u_]:=D[u,{z,2}];
L[2,u_]:=D[u,{z,2}];
L[3,u_]:=D[u,{z,2}];
L[4,u_]:=D[u,{z,2}];
```

Note that we use the delayed assignment `SetDelayed(:=)` to define these linear operators, and z is the independent variable in the package.

Without loss of generality, let us consider the case when $Pr = 7/10$, $Ec = 1/100$, $P = -5$, $b = 1$, $n = 1$, $Re = 1$, $M = 2$, $Gr = 5$, $h = 1$, $\lambda = 1$, $\lambda_T = 1$, and $\phi = \pi/6$. These physical parameters are input as

```
P = -5;   b = 1;
n = 1;    Ra = 1;
M = 2;    Gr = 5;
lamb = 1;  lambT = 1;
h = 1;  phi = Pi/6;
Pr = 7/10; Ec = 1/100;
```

At this time, we have input all the data for this problem, except the convergence-control parameters `c0[k]`. We minimize the averaged squared residual error of the 4th-order approximations to obtain optimal values for `c0[k]` by the command

```
GetOptiVar[4,{},{c0[1],c0[2],c0[3],c0[4]}];
```

Note that the second parameter of GetOptiVar is an empty list, which means that we give no constraint on the convergence-control parameters c0[k].

After some time, we obtain the optimal values for c0[k], which reads c0[1] ≈ -0.898166, c0[2] ≈ -0.946828, c0[3] ≈ -0.780946 and c0[4] ≈ -1.12363. Then we call the main module BVPh to get the 30th-order approximations

```
BVPh[1, 30];
```

The 30th-order approximations for u_1, θ_1, u_2, θ_2 are stored in U[1,30], U[2,30], U[3,30] and U[4,30], respectively, while the corresponding averaged squared residual error of the system is ErrTotal[30]. The 30th-order approximations are plotted in Fig. 9.10. The value of $\theta(y)$ agrees with Umavathi's result [16] (black dots), as shown in Fig. 9.10. The 30th-order approximation of $\theta(y)$ gives the heat transfer rate $Nu_+ = \theta_1'(1) = 0.8860625$ and $Nu_- = \theta_2'(1) = 1.122312$, which agrees with $Nu_+ = 0.88606$ and $Nu_- = 1.12230$ in Umavathi's [16] Table 3. The total error ErrTotal[k] of the system for every two order of approximations is plotted in Fig. 9.11.

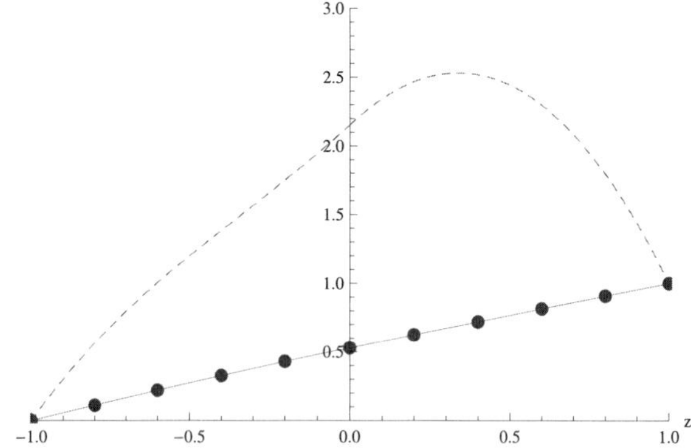

Fig. 9.10. The curve of $u(y)$ (solid) and $\theta(y)$ (dashed) for Example 4. The black dots are the values for $\theta(y)$ obtained by Umavathi [16].

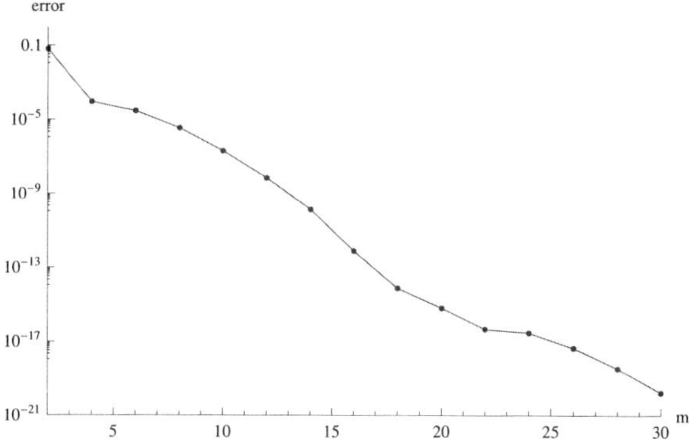

Fig. 9.11. Total error vs. order of approximation for Example 4.

9.7.5. *Example 5: Iterative solutions of the Gelfand equation*

When the problem is defined in a finite interval, the `BVPh 1.0` can solve it using an iterative approach. The `BVPh 2.0` has inherited this feature. However, there are some minor differences in the input.

Consider the Gelfand equation [17–19]

$$u'' + (K-1)\frac{u'}{z} + \lambda e^u = 0, \; u'(0) = u(1) = 0, \tag{9.56}$$

where the prime denotes the differentiation with respect to z, $K \geq 1$ is a constant, $u(z)$ and λ denote eigenfunction and eigenvalue, respectively. Following Liao [14] , an additional boundary condition

$$u(0) = A \tag{9.57}$$

is added to distinguish different eigenfunctions.

To solve this problem by `BVPh 2.0`, we have to input the differential equations, boundary conditions, and initial guesses. Since the problem is a single ODE with an unknown λ to be determined, we set `NumEQ = 1` and `TypeEQ=2`. The differential equation can be coded as follows

```
TypeEQ = 2;
NumEQ = 1;
f[1,z_,{u_},lambda_] :=
D[u,{z,2}] +(K-1)*D[u,z]/z+lambda*Exp[u];
```

The three boundary conditions, including the additional condition (9.57), are defined as

```
NumBC = 3;
BC[1, z_, {u_}] := (u-A)/. z -> 0;
BC[2, z_, {u_}] := D[u,z]/. z -> 0;
BC[3, z_, {u_}] := u /. z -> 1;
```

Now let us input the solution intervals

```
zL[1] = 0;    zR[1] = 1;
```

Since the solution interval is finite, we do not have to specify the integral interval to compute the averaged squared residual error.

The initial guess is chosen as $U_0 = \frac{A}{2}[1 + \cos(\pi z)]$, which is input into Mathematica as

```
U[1,0] = A/2*(1 + Cos[Pi*z]);
```

The auxiliary linear operator is chosen as $\mathcal{L} = \frac{\partial^2}{\partial z^2} + \left(\frac{\pi}{a}\right)^2$, which is defined in Mathematica as

```
L[1,f_] :=  D[f,{z,2}]+Pi^2*f;
```

Note that we use the delayed assignment `SetDelayed(:=)` to define the linear operator.

Without loss of generality, let us consider the case when $A = 1$ and $K = 2$. The physical parameters are input as

```
K = 2;    A = 1;
```

Because we want to approximate the right-hand sides using the hybrid-base function and use an iterative approach to get the approximations, the control parameters in `BVPh 2.0` are modified to

```
TypeL       =  2;
HYBRID      =  1;  (* hybrid-base functions *)
TypeBase    =  2;  (* even Fourier series   *)
ApproxQ     =  1;
Ntruncated  =  30;
```

Here `TypeL = 2`, `HYBRID = 1` and `ApproxQ = 1` together mean that the right-hand sides of all high-order deformation equations is approximated by the hybrid-base approximations. `TypeBase = 2` means the even expression (9.26) is used (`TypeBase = 1` also applies to this problem). `Ntruncated = 20` means $N_t = 30$.

At this time, we have input all the data for this problem, except the convergence-control parameter `c0[1]`. To get optimal `c0[1]`, we minimize the averaged squared residual error of the 6th-order approximations. This is done in `BVPh 2.0` by calling the module `GetOptiVar`

```
GetOptiVar[6, {}, {c0[1]}];
```

After some computation, we get the optimal value for the convergence-control parameter `c0[1]`$= -0.522418\cdots$. Now we can use the 3rd-order iteration HAM approach

```
iter[1,6,3]
```

to get the desired approximation. Here 6 means the iteration times. After about 40 seconds, the 6th iteration gives the eigenvalue 1.90921, which is the same with Liao's result [14]. The kth iteration approximations of u and λ are stored in `V[1,k]`, and `LAMBDA[k]`, while the corresponding averaged squared residual error is stored in `ERRTotal[k]`. The 6th iteration approximation is plotted in Fig. 9.12 by

```
Plot[V[1,6],{z,0,1},AxesLabel->{"z", "u(z)"}]
```

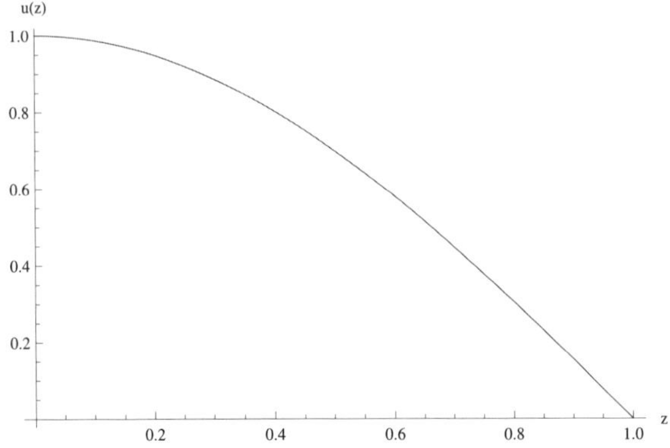

Fig. 9.12. The curve of the eigenfunction $u(z)$ corresponding to the eigenvalue $\lambda = 1.90921$ when $A = 1$ and $K = 2$ for Example 5.

The total error `ERRTotal[k]` of the problem for each iteration is plotted in Fig. 9.13 by the command

```
ListLogPlot[Table[{i, ERRTotal[i]}, {i, 1, 6}],
PlotRange->{{1,6},{10^-10,0.01}},Joined->True, Mesh->All,
AxesLabel->{"m","error"}]
```

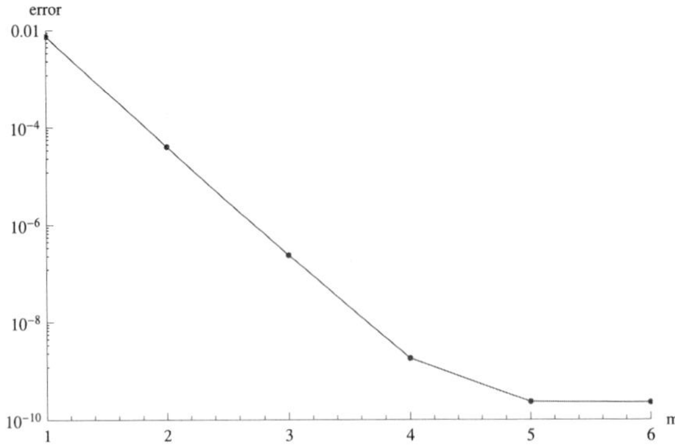

Fig. 9.13. Total error for each iteration vs. iteration times m for Example 5.

9.8. Conclusions

The homotopy analysis method (HAM) has been successfully applied to solve lots of nonlinear problems in science and engineering. Based on the HAM, the Mathematica package `BVPh 1.0` was issued by Liao in May, 2012. The aim of `BVPh` is to provide an analytic tool for as many nonlinear BVPs as possible in the frame work of the HAM.

However, `the BVPh 1.0` can only deal with problems of a single ODE. Unlike `BVPh 1.0`, the new version `BVPh 2.0` works for many types of systems of coupled nonlinear ODEs. In this chapter, we briefly describe how to install and use the `BVPh 2.0`. Five typical examples are employed to demonstrate the validity of `BVPh 2.0`, including a system of two coupled ODEs in finite interval, a system of two coupled ODEs in semi-infinite interval, a system of two coupled ODEs with algebraic property at infinity, a system of three coupled ODEs with an unknown parameter to be determined, and a system of four coupled ODEs in different intervals. Besides, new algorithms are used in some modules of `BVPh 2.0`. Hence, `BVPh 2.0` is much faster than `BVPh 1.0` in most cases. The package `BVPh 2.0` and all input data for these examples are free available online at `http://numericaltank.sjtu.edu.cn/BVPh.htm`.

It is well known that the iterative method can gain accurate approximations more efficiently by means of the HAM. The `BVPh 2.0` has also inherited the feature of `BVPh 1.0` to solve the problems in finite interval using the iterative method. For problems in semi-infinite interval, an iterative method for two typical kinds of based functions is proposed [20]. This iterative approach will be employed in the future version of `BVPh`.

Appendix A. Codes for examples

Here are the input data for all the examples in this chapter. Note that the listings without an end-of-line semicolon is wrapped to fit the page width. However, if you break the long command intentionally in Mathematica, it will run as multi-line commands and may not work as you expected.

A.1. Sample codes to run the illustrative example

```
(*             Filename: runIllustrative.nb                *)

(* 1. Clear all global variables                           *)
ClearAll["Global`*"];

(* 2. Read in the package BVPh 2.0                         *)
<< "E:\\Package\\BVPh2_0.m"

(* 3. Set the current working directory to                 *)
(*    "the current directory"                              *)
SetDirectory[ToFileName[Extract["FileName" /.
NotebookInformation[EvaluationNotebook[]], {1},
FrontEnd`FileName]]];

(* 4. Read in your input data in current directory  *)
(* Note that the two files runIllustrative.nb and   *)
(* Illustrative.m are in the current directory      *)
<< Illustrative.m
```

A.2. Input data of BVPh 2.0 for the illustrative example

```
(*             Filename: Illustrative.m                  *)
Print["The input file ",$InputFileName," is loaded !"];

(* Modify  control parameters in BVPh if necessary  *)

(*             Define the governing equation             *)
TypeEQ = 1;
NumEQ = 2;
f[1,z_,{f_,g_},Lambda_]:=
    D[f,{z,3}]-D[f,z]^2+f*D[f,{z,2}]+2*la*g+
    beta*(2*f*D[f, z]*D[f,{z,2}]-f^2*D[f,{z,3}]);
f[2,z_,{f_,g_},Lambda_]:=
```

```
    D[g,{z,2}]-D[f,z]*g+f*D[g,z]-2*la*D[f,z]+
    beta*(2*f*D[f, z]*D[g, z]-f^2*D[g,{z, 2}]);

(*              Define Boundary conditions              *)
NumBC = 5;
BC[1,z_,{f_,g_}]:=(D[f, z]-1)/.z->0;
BC[2,z_,{f_,g_}]:=f/.z->0;
BC[3,z_,{f_,g_}]:=g/.z->0;
BC[4,z_,{f_,g_}]:=D[f,z]/.z->infinity;
BC[5,z_,{f_,g_}]:=g/.z->infinity;

(* solution interval and integral interval for error *)
zL[1] = 0;
zR[1] = infinity;
zL[2] = 0;
zR[2] = infinity;
zRintegral[1] = 10;
zRintegral[2] = 10;

(*              Define initial guess                    *)
U[1,0] = 1-Exp[-z];
U[2,0] = ahpha*z*Exp[-z];

(*        Define the auxiliary linear operator          *)
L[1,u_] := D[u,{z,3}]-D[u,z];
L[2,u_] := D[u,{z,2}]-u;

(*              Define physical parameters              *)
beta = 1/5;
la = 1/10;

(*                 Print input data                     *)
PrintInput[{f[z],g[z]}];

(*                 Get optimal c0                       *)
```

```
GetOptiVar[3, {}, {c0[1],c0[2],alpha}];

(*          Gain 10th-order HAM approximation        *)
BVPh[1, 10];
```

A.3. Input data of BVPh 2.0 for Example 1

```
(*              Filename: Example1.m                 *)
Print["The input file ",$InputFileName," is loaded !"];

(* Modify control parameters in BVPh if necessary  *)
ErrReq=10^-30;

(*           Define the governing equation           *)
TypeEQ = 1;
NumEQ = 2;
f[1,z_,{f_,g_},Lambda_]:=(1+K)*D[f,{z,4}]
   -Rey*M*D[f,{z,2}]+2*Rey*f*D[f,{z,3}]-K*D[g,{z,2}];
f[2,z_,{f_,g_},Lambda_]:=(1+K/2)*D[g,{z,2}]
   -Rey*K*(2*g-D[f,{z,2}])+Rey*(2*f*D[g,z]-D[f,z]*g);

(*              Define Boundary conditions           *)
NumBC = 6;
BC[1,z_,{f_,g_}]:=f/.z->0;
BC[2,z_,{f_,g_}]:=f/.z->1;
BC[3,z_,{f_,g_}]:=(D[f, z]-1)/.z->1;
BC[4,z_,{f_,g_}]:=D[f,{z,2}]/.z->0;
BC[5,z_,{f_,g_}]:=g/.z->1;
BC[6,z_,{f_,g_}]:=g/.z->0;

(* solution interval and integral interval for error *)
zL[1]=0;
zR[1]=1;
zL[2]=0;
zR[2]=1;
```

```
(*                  Define initial guess                *)
U[1,0] = (z^3-z)/2;
U[2,0] = 0;

(*         Define the auxiliary linear operator         *)
L[1,u_]:=D[u,{z,4}];
L[2,u_]:=D[u,{z,2}];

(*                Define physical parameters            *)
Rey = M = 2;
K = 1/2;

(*                    Print input data                  *)
PrintInput[{f[z],g[z]}];

(*                    Get optimal c0                    *)
GetOptiVar[4,{},{c0[1],c0[2]}];

(*          Gain 20th-order HAM approximation           *)
BVPh[1, 20];
```

A.4. Input data of BVPh 2.0 for Example 2

```
(*                Filename: Example2.m                  *)
Print["The input file ",$InputFileName," is loaded !"];

(*  Modify control parameters in BVPh if necessary      *)

(*          Define the governing equation               *)
TypeEQ = 1;
NumEQ = 2;
f[1,z_,{F_,S_},Lambda_]:=la^2*D[F,{z,2}]+S-F^2;
f[2,z_,{F_,S_},Lambda_]:=la^2*D[S,{z, 2}]-3*sigma*F*S;
```

```
(*             Define Boundary conditions                  *)
NumBC = 4;
BC[1,z_,{F_, S_}] := F /. z -> 1;
BC[2,z_,{F_, S_}] := (G - 1) /. z -> 1;
BC[3,z_,{F_, S_}] := F /. z -> infinity;
BC[4,z_,{F_, S_}] := G /. z -> infinity;

(* solution interval and integral interval for error *)
zL[1] = 1;
zR[1] = infinity;
zL[2] = 1;
zR[2] = infinity;
zRintegral[1] = 10;
zRintegral[2] = 10;

(*              Define initial guess                       *)
U[1, 0] = gamma*(z^(-2) - z^(-3));
U[2, 0] = z^(-4);

(*       Defines the auxiliary linear operator             *)
L[1, u_] := D[u, {z, 2}]*z/3 + D[u, z];
L[2, u_] := D[u, {z, 2}]*z/5 + D[u, z];

(*       Define physical and control parameters            *)
sigma = 1;
gamma = 3;
la = 1/3;
c0[1] = -1/2;
c0[2] = -1/2;

(*                  Print input data                       *)
PrintInput[{f[z], g[z]}];

(*        Gain 20th-order HAM approximation                *)
BVPh[1, 20];
```

A.5. Input data of BVPh 2.0 for Example 3

```
(*            Filename: Example3.m                       *)
Print["The input file ",$InputFileName," is loaded !"];

(* Modify  control parameters in BVPh if necessary *)

(*            Define the governing equation              *)
TypeEQ = 2;
NumEQ = 3;
f[1,z_,{f_,g_,s_},sigma_] :=
  D[f,{z,2}]+Gr*Pr*g-Nr*s+sigma;
f[2,z_,{f_,g_,s_},sigma_] :=
  D[g,{z,2}]+Nb*D[g,z]*D[s,z]+Nt*(D[g,z])^2-f;
f[3,z_,{f_,g_,s_},sigma_] :=
  D[s,{z,2}]+Nt/Nb*D[f,{z,2}]-Le*f;

(*             Define boundary conditions                *)
NumBC = 7;
BC[1,z_,{f_,g_,s_}] :=f/.z->-1;
BC[2,z_,{f_,g_,s_}] :=f/.z->1;
BC[3,z_,{f_,g_,s_}] :=g/.z->-1;
BC[4,z_,{f_,g_,s_}] :=g/.z->1;
BC[5,z_,{f_,g_,s_}] :=s/.z->-1;
BC[6,z_,{f_,g_,s_}] :=s/.z->1;
BC[7,z_,{f_,g_,s_}] :=Integrate[f,{z,0,1}]-Ra*Pr;

(*              Define solution interval                 *)
zL[1] = -1;
zR[1] = 1;
zL[2] = -1;
zR[2] = 1;
zL[3] = -1;
zR[3] = 1;
```

```
(*       Defines the auxiliary linear operator      *)
L[1, u_] := D[u, {z, 2}];
L[2, u_] := D[u, {z, 2}];
L[3, u_] := D[u, {z, 2}];

(*             Define physical parameters            *)
Nr = 1/5;
Nt = 1/20;
Nb = 1/20;
Le = 10;
Gr = 5;
Pr = 1;
Ra = 5;

(*              Define initial guess                 *)
U[1,0]=eps1-3/2*(-25+4eps1)z^2+5/2*(-15+2eps1)*z^4;
U[2,0]=eps2*(1 - z^2);
U[3,0]=eps3*(1 - z^2);

(*                Print input data                   *)
PrintInput[{f[z], g[z], s[z]}];

(*    Get optimal convergence-control parameters     *)
c0[1] = c0[2] = c0[3] = h;
GetOptiVar[3, {}, {eps1, eps2, eps3, h}];

(*         Gain 10th-order HAM approximation         *)
BVPh[1, 20];
```

A.6. Input data of BVPh 2.0 for Example 4

```
(*            Filename: Example4.m                   *)
Print["The input file ",$InputFileName," is loaded !"];

(*  Modify control parameters in BVPh if necessary  *)
```

```
(*        Define the governing equation                   *)
TypeEQ = 1;
NumEQ = 4;
f[1,z_,{u1_,s1_,u2_,s2_},Lambda_]:=
D[u1, {z, 2}]+ Gr/Ra*Sin[phi]*s1 - P ;
f[2,z_,{u1_,s1_,u2_,s2_},Lambda_]:=
D[s1, {z, 2}] + Pr*Ec*(D[s1, z])^2;
f[3,z_,{u1_,s1_,u2_,s2_},Lambda_]:=D[u2,{z, 2}]
-h^2/lamb*P+Gr/Ra*Sin[phi]*n*b*h^2/lamb*s2
-M^2*h^2/lamb*u2;
f[4,z_,{u1_,s1_,u2_,s2_},lambda_]:=D[s2,{z,2}]
+Pr*Ec*lamb/lambT*D[u2,z]^2+Pr*Ec*h^2/lambT*M^2*u2^2;

(*           Define Boundary conditions                   *)
    NumBC=8;
    BC[1,z_,{u1_,s1_,u2_,s2_}]:=(u1-1)/.z->1;
    BC[2,z_,{u1_,s1_,u2_,s2_}]:=(u1-u2)/.z->0;
    BC[3,z_,{u1_,s1_,u2_,s2_}]:=u2/.z->-1;
    BC[4,z_,{u1_,s1_,u2_,s2_}]:=
        (D[u1,z]-D[u2,z]/m/h)/.z->0;
    BC[5,z_,{u1_,s1_,u2_,s2_}]:=(s1-1)/.z->1;
    BC[6,z_,{u1_,s1_,u2_,s2_}]:=(s1-s2)/.z->0;
    BC[7,z_,{u1_,s1_,u2_,s2_}]:=s2/.z->-1;
    BC[8,z_,{u1_,s1_,u2_,s2_}]:=
        (D[s1,z]-D[s2,z]/K/h)/.z->0;

(*              Define solution interval                  *)
    zL[1]=0; (* u1 *)
    zR[1]=1;
    zL[2]=0; (* s1 *)
    zR[2]=1;
    zL[3]=-1;(* u2 *)
    zR[3]=0;
    zL[4]=-1;(* s2 *)
```

```
   zR[4]=0;

(*                 Define initial guess                    *)
   U[1, 0] = (z - z^2)*lamb/h+1;              (* u1 *)
   U[2, 0] = z*lambT/h +(1-lambT/h)*z^2;      (* s1 *)
   U[3, 0] = 1 + z;                           (* u2 *)
   U[4, 0] = z^2 + z;                         (* s2 *)

(*       Define the auxiliary linear operator              *)
   L[1,u_]:=D[u,{z,2}];
   L[2,u_]:=D[u,{z,2}];
   L[3,u_]:=D[u,{z,2}];
   L[4,u_]:=D[u,{z,2}];

(*           Define physical parameters                    *)
   P = -5;  b = 1;
   n = 1;   Ra = 1;
   M = 2;   Gr = 5;
   lamb = 1;  lambT = 1;
   h = 1;  phi = Pi/6;
   Pr = 7/10; Ec = 1/100;

(*                  Print input data                       *)
PrintInput[{u1[z], s1[z], u2[z], s2[z]}];

(*                  Get optimal c0                         *)
GetOptiVar[4,{},{c0[1],c0[2],c0[3],c0[4]}]

(*       Gain 10th-order HAM approximation                 *)
BVPh[1,30];
```

A.7. Input data of BVPh 2.0 for Example 5

```
(*          Filename: Example5.m                           *)
Print["The input file ",$InputFileName," is loaded !"];
```

```
(* Modify  control parameters in BVPh if necessary  *)
TypeL       =  2;
HYBRID      =  1; (* hybrid-base functions *)
TypeBase    =  2; (* even Fourier series   *)
ApproxQ     =  1;
Ntruncated  =  20;

(*       Define the governing equation               *)
TypeEQ = 2;
NumEQ = 1;
f[1,z_,{u_},lambda_] :=
  D[u,{z,2}] +(K-1)*D[u,z]/z+lambda*Exp[u];

(*          Define Boundary conditions               *)
NumBC = 3;
BC[1, z_, {u_}] := (u-A)/. z -> 0;
BC[2, z_, {u_}] := D[u,z]/. z -> 0;
BC[3, z_, {u_}] := u /. z -> 1;

(*             Define solution interval              *)
zL[1] = 0;
zR[1] = 1;

(*                Define initial guess               *)
U[1,0] =  A/2*(1 + Cos[Pi*z]);

(*       Define the auxiliary linear operator        *)
L[1,f_] :=  D[f,{z,2}]+Pi^2*f;

(*          Define physical parameters               *)
K  =  2;
A=1;

(*               Print input data                    *)
```

```
PrintInput[{u[z]}];

(*                    Get optimal c0                    *)
GetOptiVar[6,{},c0[1]];

(*                    Print input data                  *)
PrintInput[{u[z]}];

(*           Use 3rd-order iteration approach           *)
iter[1,6,3]
```

Acknowledgment

This work is partly supported by National Natural Science Foundation of China under Grant No. 11272209 and State Key Laboratory of Ocean Engineering under Grant No. GKZD010056.

References

[1] L. F. Shampine, I. Gladwell and S. Thompson, *Solving ODEs with MATLAB*. Cambridge University Press, Cambridge (2003).

[2] J. Kierzenka, L. F. Shampine, A BVP solver based on residual control and the Matlab PSE, *ACM TOMS.* **27**(3), 299–316 (2001).

[3] Z. Battles and L. N. Trefethen, An extension of Matlab to continuous functions and operators, *SIAM J. Sci. Comput.* **25**(5), 1743–1770 (2004).

[4] L. N. Trefethen, Computing numerically with functions instead of numbers, *Math. Comput. Sci.* **1**, 9–19 (2007).

[5] Y. P. Liu, S. J. Liao and Z. B. Li, Symbolic computation of strongly nonlinear periodic oscillations, *J. Symb. Comput.* **55**, 72–95 (2013).

[6] M. Sajid, Z. Iqbal, T. Hayat and S. Obaidat, Series solution for rotating flow of an upper convected Maxwell fluid over a strtching sheet, *Commun. Theor. Phys.* **56**(4), 740–744 (2011).

[7] T. Hayat, M. Nawa and A. A. Hendi, Heat transfer analysis on axisymmetric MHD flow of a micropolar fluid between the radially stretching sheets, *J. Mech.* **27**(4), 607–617 (2011).

[8] S. J. Liao, Notes on the homotopy analysis method: Some definitions and theorems, *Commun. Nonlinear Sci. Numer. Simulat.* **14**(4), 983–997 (2009).

[9] S. J. Liao, *Proposed homotopy analysis techniques for the solution of nonlinear problem*. Ph.D. thesis, Shanghai Jiao Tong University (1992).

[10] S. J. Liao, A uniformly valid analytic solution of two-dimensional viscous flow over a semi-infinite flat plate, *J. Fluid Mech.* **385**, 101–128 (1999).

[11] S. J. Liao, On the analytic solution of magnetohydrodynamic flows of non-newtonian fluids over a stretching sheet, *J. Fluid Mech.* **488**, 189–212 (2003).

[12] S. J. Liao, Series solutions of unsteady boundary-layer flows over a stretching flat plate, *Stud. Appl. Math.* **117**(3), 239–263 (2006).

[13] S. J. Liao, *Beyond Perturbation—Introductioin to the Homotopy Analysis Method.* Chapman & Hall/CRC Press, Boca Raton (2003).

[14] S. J. Liao, *Homotopy Analysis Method in Nonlinear Differential Equations.* Springer-Verlag Press, New York (2012).

[15] H. Xu, T. Fan and I. Pop, Analysis of mixed convection flow of a nanofluid in a vertical channel with the Buongiorno mathematical model, *Int. Commun. Heat Mass.* **44**, 15–22 (2013).

[16] J. C. Umavathi, I. C. Liu and J. Prathap Kumar. Magnetohydrodynamic Poiseuille-Couette flow and heat transfer in an inclined channel, *J. Mech.* **26**(4), 525–532 (2010).

[17] J. P. Boyd, An analytical and numerical study of the two-dimensional Bratu equation, *J. Sci. Comput.* **1**(2), 183–206 (1986).

[18] J. Jacobsen and K. Schmitt, The Liouville-Bratu-Gelfand problem for radial operators, *J. Differ. Equations.* **184**, 283–298 (2002).

[19] J. S. McGough, Numerical continuation and the Gelfand problem, *Appl. Math. Comput.* **89**(1-3), 225–239 (1998).

[20] Y. L. Zhao, Z. L. Lin and S. J. Liao, An iterative analytical approach for nonlinear boundary value problems in a semi-infinite domain, *Comput. Phys. Commun.* **184**(9): 2136–2144 (2013).